CHEMICAL VAPOR DEPOSITION - RECENT ADVANCES AND APPLICATIONS IN OPTICAL, SOLAR CELLS AND SOLID STATE DEVICES

Edited by **Sudheer Neralla**

Chemical Vapor Deposition - Recent Advances and Applications in Optical, Solar Cells and Solid State Devices
http://dx.doi.org/10.5772/61559
Edited by Sudheer Neralla

Contributors

Hitoshi Habuka, Chang-Seop Lee, Yura Hyun, Aide Torres-Huerta, Pawel Madejczyk, Frank Mendoza, Tej Limbu, Brad Weiner, Gerardo Morell, Milana Vasudev, Haiping Zhou, Shuyan Xu, Shaoqing Xiao, J. Alarcón-Salazar, R. López-Estopier, E. Quiroga-González, A. Morales-Sánchez, J. Pedraza-Chávez, Mariano Aceves-Mijares, Ignacio Enrique Zaldivar Huerta, Nong Moon Hwang, Vahid Mohammadi, Stoyan Nihtianov

Published by InTech

Janeza Trdine 9, 51000 Rijeka, Croatia

Publishing Process Manager Edi Lipovic

Technical Editor SPi Global

Cover InTech Design team

First published August, 2016

Printed in Croatia

Additional hard copies can be obtained from orders@intechopen.com

Chemical Vapor Deposition - Recent Advances and Applications in Optical, Solar Cells and Solid State Devices, Edited by Sudheer Neralla

p. cm.

Print ISBN 978-953-51-2572-3

Online ISBN 978-953-51-2573-0

For EU product safety concerns:
IN TECH d.o.o., Prolaz Marije Krucifikse Kozulić 3, 51000 Rijeka, Croatia,
info@intechopen.com or visit our website at intechopen.com.

Meet the editor

Dr. Sudheer Neralla is a Materials Scientist at the Jet-Hot High Performance Coatings, a nanotechnology-based coating industry in Burlington, NC, USA. He is also involved as an Adjunct Research Scientist with the NSF Engineering Research Center at North Carolina Agricultural and Technical State University, Greensboro, NC. Dr. Neralla received his doctoral degree in mechanical engineering from the North Carolina Agricultural and Technical State University, Greensboro, NC. His research interests include synthesis of nanomaterials, thin films, nanoindentation, friction and wear, and corrosion study of coatings and biodegradable Mg-based alloys.

Contents

Preface

Over the past few years, chemical vapor deposition (CVD) methods have undergone significant changes and have embraced the technological updates to enable growth of novel materials, including nanostructures, thin films, and multiphase materials with focus on electronic, physical, and optical properties. These updates have helped overcome some of the limitations, such as synthesis temperatures, materials, and field of applications, which would have limited CVD methods to high-temperature applications otherwise. One of the major advantages of CVD technique is its ability to coat any shape, and with the advent of metal-organic chemical vapor deposition (MOCVD) and plasma-enhanced chemical vapor deposition (PECVD) techniques, deposition temperatures have drastically reduced.

CVD has now evolved into the most widely used technique for growth of thin films in electronics industry. Several books on CVD methods have emerged in the past, and thus the scope of this book goes beyond providing fundamentals of the CVD process.

The book is divided into two sections. Section 1 covers authors' works on the synthesis of various nanomaterials and thin films using CVD methods. This section includes Chapters 1 through 5.

Chapter 1 presents preparation and characterization of carbon nanofibers and composites by chemical vapor deposition method. Effect of synthesis temperature and of metal catalyst concentration on the electrochemical characteristics is studied.

Chapter 2 summarizes the nonclassical crystallization in the growth of thin films and nanostructures by CVD. Several variables such as surface conductivity, flow rate, substrate position, and nanoparticle size are studied in deposition of diamond, ZnO, and silicon.

Chapter 3 discusses the growth of HgCdTe heterostructures using MOCVD technique. Electrical and chemical characterization of HgCdTe structures is described and infrared photodiodes were constructed using these heterostructures.

In Chapter 4, a parametric study of synthesis of bilayer graphene on copper using hot filament chemical vapor deposition method is presented. Synthesis process and the parameters' effect are discussed.

In Chapter 5, a method for in situ observation of CVD is introduced. A langasite crystal microbalance is used to evaluate the surface chemical reactions in a CVD reactor.

Section 2 is devoted to recent advances in materials synthesis using CVD and their applications such as photodetectors, optical sources, solar cell, and solid-state devices.

In Chapter 6, a new method, PureB, is introduced for deposition of boron at low temperatures using CVD method for application as photodetectors. Different models behind PureB growth are discussed.

Chapter 7 presents work on the synthesis of silicon-rich oxide using low-pressure CVD method. Parameters affecting the stoichiometry of the silicon oxide are analyzed.

Chapter 8 reports high-density plasma CVD of Si-based materials for solar cell applications. High-frequency plasma-enhanced CVDmethod has been employed to overcome the limitations of traditional PECVD methods.

In Chapter 9, CVD methods for synthesis of thin films with application in solid-state devices are discussed. Growth of Pt-YSZ and Pt-ZrO_2 ceramic-metallic composites were developed and evaluated in solid-state devices.

Chapter 10 provides an overview of plasma-enhanced chemical vapor deposition and recent advanced applications.

Through this book, an effort is made to bring together most recent works in the area of CVD. I would like to express my sincere thanks to all the participating authors of this book for their valuable contributions.

Dr. Sudheer Neralla
Jet-Hot High Performance Coatings
NSF-Engineering Research Center
North Carolina A&T State University
Greensboro, NC-USA

Section 1

CVD: Overview and Synthesis of Micro/Nano Structures

Chapter 1

Preparation and Characterization of Carbon Nanofibers and its Composites by Chemical Vapor Deposition

Chang-Seop Lee and Yura Hyun

Additional information is available at the end of the chapter

http://dx.doi.org/10.5772/63755

Abstract

Hydrocarbon gas or carbon monoxide was pyrolyzed by chemical vapor deposition (CVD), and carbon nanofiber (CNF) synthesis was performed using transition metals such as Ni, Fe, and Co as catalysts. When synthesizing carbon nanofibers using the CVD method, experimental variables are temperature, catalysts, source gas, etc. Especially, the particle size of the catalyst is the most important factor in determining the diameter of carbon nanofibers. Hydrocarbon gases, such as CH_4, C_2H_4, benzene, and toluene are used as the carbon source, and in addition to these reaction gases, nonreactive gases such as H_2, Ar, and N_2 gases are used for transportation. Synthesis occurs at a synthesis temperature of 600–900°C, and catalyst metals such as Ni, Co, and Fe are definitely required when synthesizing CNFs. Therefore, it is possible to synthesize CNFs in selective areas through selective deposition of such catalyst metals. In this study, CNFs were synthesized by CVD. Ethylene gas was employed as the carbon source for synthesis of CNFs with H_2 as the promoting gas and N_2 as the balancing gas. Synthesized CNFs can be used in various applications, such as composite materials, electromagnetic wave shielding materials, ultrathin display devices, carbon semiconductors, and anode materials of Li secondary batteries. In particular, there is an increasing demand for light-weight, small-scale, and high-capacity batteries for portable electronic devices, such as notebook computers or smartphones along with the recent issue of fossil energy depletion. Accordingly, CNFs and their silicon-series composites are receiving attention for use as anode materials for lithium secondary batteries that are eco-friendly, light weight, and high capacity.

Keywords: carbon nanofibers, transition metal catalyst, chemical vapor deposition, composite, Li ion batteries

1. Introduction

Chemical vapor deposition (CVD) is widely used as a surface treatment technology for materials. CVD forms a solid-state thin film mostly on the surface and is used not only to produce high-purity bulk materials and powder but also to manufacture composite materials through infiltration techniques.

CVD is used to deposit a wide variety of materials. Most of the elements in the periodic table deposited in the pure element are formed by CVD technology. However, they are deposited mostly in the compound form rather than the pure element form. CVD can make precursor gases flow to one or more heated objects in a chamber to coat the desired compound. A chemical reaction occurs on the hot surface, and this leads to the deposition of a thin film on the surface. This reaction also produces the unreacted precursor gas and the chemical byproduct discharged from the chamber at the same time.

CVD can deposit many kinds of materials and can be applied to broad areas, so the synthesis condition is also diverse. CVD synthesis can occur in a high- or low-temperature reactor, the pressure ranges from sub-Torr pressures to above-atmospheric pressures, regardless of the kind of catalyst, and the reaction temperature can range from 200 to 1600°C to diversify the synthesis condition.

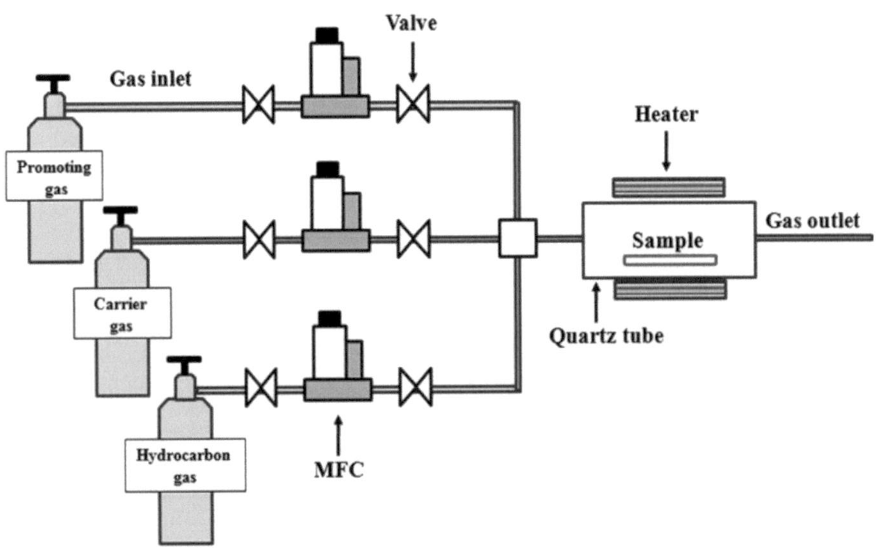

Figure 1. The schematic diagram of a tube-furnace CVD system.

Microfabrication processes widely use CVD to deposit materials in various forms, including monocrystalline, polycrystalline, amorphous, and epitaxial. These materials include silicon (SiO_2, germanium, carbide, nitride, and oxynitride), carbon (fiber, nanofibers, nanotubes, diamond, and graphene), fluorocarbons, filaments, tungsten, and titanium nitride [1, 2].

CVD is a technology used to deposit a solid-state thin film on the substrates from vapor species through chemical reaction. One of the most unique characteristics of CVD synthesis is that chemical reaction plays an important role, so it is comparable to other thin film deposition technologies. **Figure 1** shows the schematic drawing of a typical tube-furnace CVD system. Gas flows are regulated by mass flow controllers (MFCs) and fed into the reactor through a gas-distribution unit. Chemical deposition takes place in the reactor, which is heated by outside heaters.

During the CVD process, the reaction gas is supplied to the reactor through mass flow controllers (MFCs) controlling the flow rate of the gas being passed. In addition, the mixture gas device mixes the gases evenly before they flow into the reactor. The chemical reaction occurs in the reactor and solid-state materials are deposited on the substrates. A heater is placed around the reactor to provide reaction at high temperatures. CVD is used not only to create a solid-state thin film on the surface and produce high-purity bulk materials and powder but also to manufacture composite materials through infiltration techniques. CVD is used to deposit various materials on solid surfaces.

A characteristic feature of CVD technique is its excellent throwing power, enabling the production of coatings of uniform thickness and properties with low porosity even on substrates with complicated shapes. Another characteristic feature is the possibility of localized or selective deposition on patterned substrates [1–4].

In this chapter, we describe the preparation process for carbon nanofibers (CNFs) and their silicon/silicon oxide composites using the chemical vapor deposition method and investigate the physicochemical and electrochemical characteristics of the prepared materials for the application of anode materials in Li secondary batteries.

2. Synthesis and characterization of CNFs on transition metal catalysts by CVD

2.1. Preparation of transition metal catalysts

In this study, transition metal catalysts were prepared through the coprecipitation method and then used in the synthesis of CNFs. In order to prepare the metal catalysts with different compositions, the mass of the precursor was first calculated according to the ratio of the metal content required.

Solution A was composed of aluminum nitrate, which helps to generate alumina (Al_2O_3) to serve as a supporter for the transition metal catalysts in the transition metal nitrate, dissolved in distilled water. With the foregoing supporter working to capture the nanometal catalyst, the coagulation phenomenon occurs when the temperature is increased up to the temperature for the synthesis of carbon nanofibers without a supporter because of the unstable nanometal particles. The usage of a supporter helps carbon nanofibers grow without a clustered catalyst and thereby serves as a matrix that prevents catalyst coagulation.

Meanwhile, it is preferred to mix passive metals to control the interparticle coagulation of transition metals, such as Fe, Co, and Ni, which all have catalytic activity against the reaction gas during high-temperature reaction. This study employed a mixture of the foregoing Solution A and another Solution B, which was composed of ammonium molybdate and distilled water.

Solution C was made of ammonium carbonate, which served as a precipitator to precipitate the transition metals and the aluminum included in the foregoing Solution A. Precipitation was induced by gradual blending of the mixture composed of Solutions A and B and the mixture of Solution C. This step was followed by agitation for stability of the precipitation.

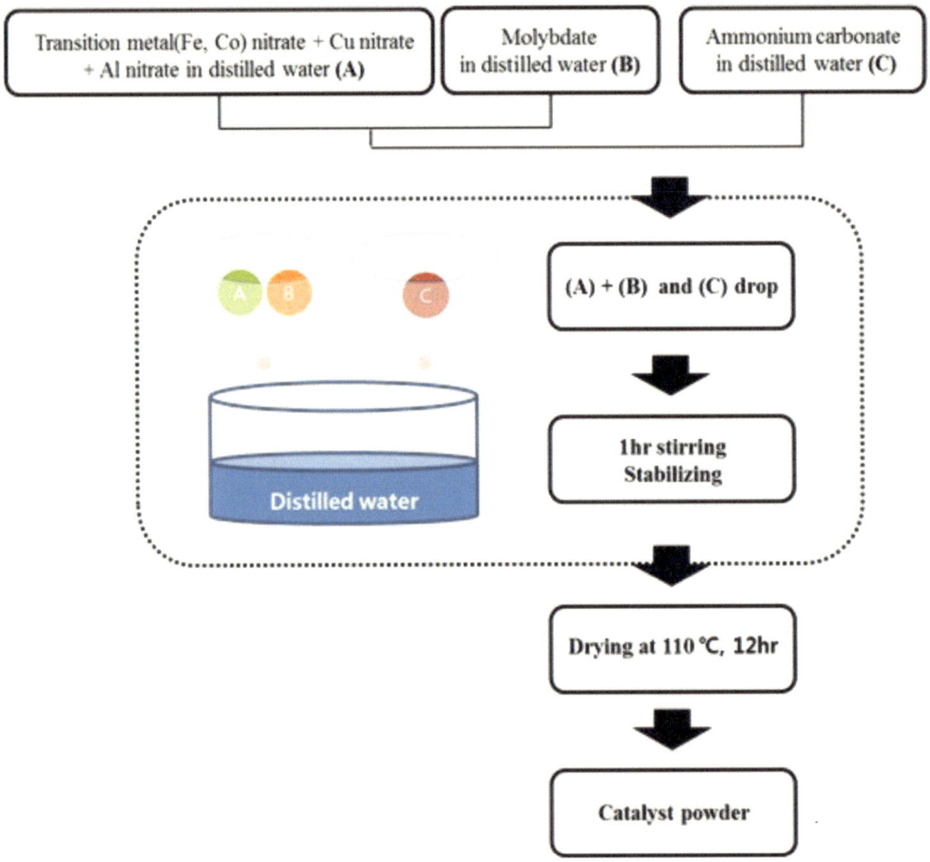

Figure 2. Preparation process transition metal catalysts.

These solutions were sufficiently stirred to stabilize the precipitates; moisture was removed by filtering; and they were dried for more than 12 h in a 110°C oven. Fully dried precipitates were made into powder, and this powder of a metal catalyst was used as the catalyst in the synthesis of carbon nanofibers. The preparation process for the catalysts is shown in **Figure 2** [5–13].

2.2. Synthesis of CNFs

Chemical vapor deposition (CVD) method was employed to synthesize carbon nanofibers in horizontal tube furnace. The schematic diagram of the reaction apparatus, manufactured as metal heating element and horizontal quartz reaction tube in 80 mm (diameter) × 1400 mm (length), was demonstrated in **Figure 3**.

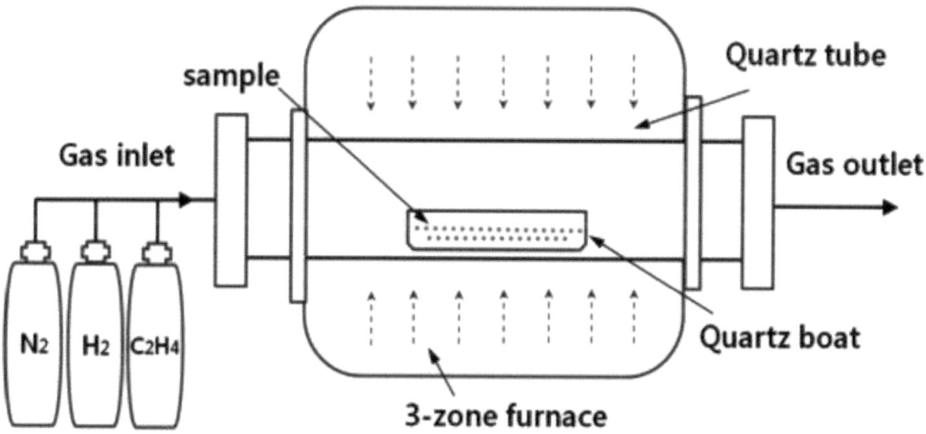

Figure 3. Schematic diagram of CVD apparatus for preparation of CNFs.

Flux of reaction gas was regulated by electronic mass flow controller (MFC); ethylene gas (C_2H_4) was used to grow carbon nanofibers; and $2H_2$ gas was used as promoting gas for gas phase reaction, whereas N_2 gas was used for stabilization of reaction. Following are conditions of synthesis reaction.

A prepared metal catalyst was evenly spread on a quartz boat, placed into reactor under N_2 atmosphere, and temperature was increased to 10°C/min. When the temperature was reached 700°C, it was maintained for 30 min; 20% H_2 gas balanced with N_2 gas were flown into all together; and then H_2 gas balanced with N_2 gas and 20% ethylene balanced with N_2 gas were flown into the reactor for 1 h. Ethylene and H_2 gas were shut off after the reaction was completed; N_2 gas was passed with the reactor atmosphere inactive until room temperature was reached. Then, carbon nanofibers were synthesized [11, 14, 15].

2.3. Synthesis of CNFs on iron and copper catalysts

2.3.1. *Scanning Electron Microscope (SEM)*

Carbon nanofibers are synthesized when pyrolyzed hydrocarbons contact metal particles at high temperature. The microstructure of the synthesized carbon nanofibers was observed by SEM and is shown in **Figure 4**. As shown in **Figure 4(a)–(d)**, carbon nanofibers grew both in the case of synthesis by Fe catalyst only as well as with an Fe:Cu weight ratio of 7:3, 5:5, and 3:7. In addition, it is known that the fiber diameters average 25–35 nm. Since physical properties may vary depending on diameter size, the diameters of carbon nanofibers can be adjusted according to the weight ratio of catalysts to meet specific purposes.

It was found that the carbon nanofibers grew slightly in the SEM image in (e) but not in (f). Here, it is believed that Fe played the role of a positive catalyst, whereas Cu played the role of a negative catalyst [9, 11].

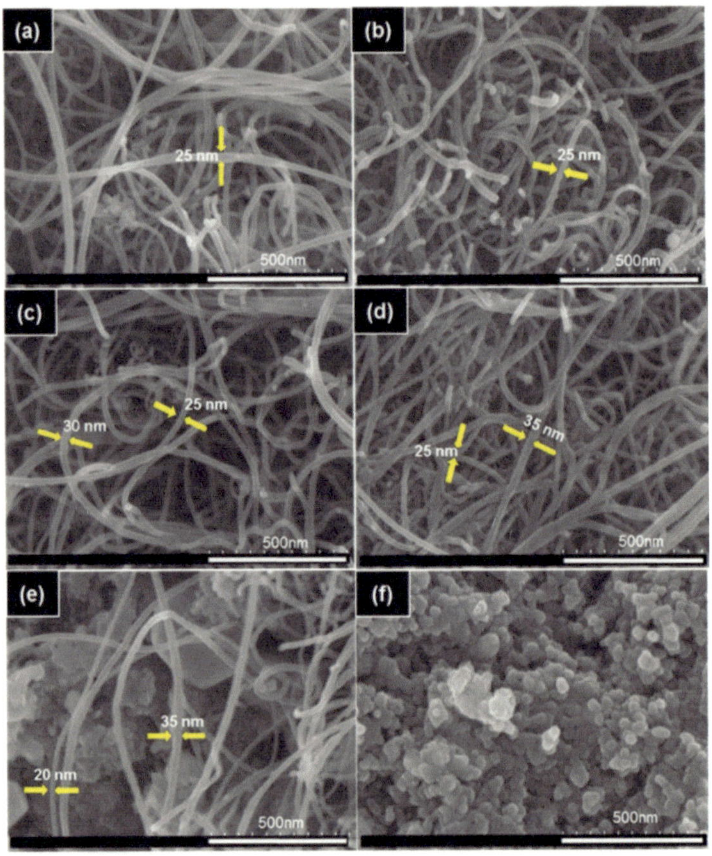

Figure 4. SEM images of CNFs synthesized from ethylene at 700°C under different concentrations of Fe and Cu catalysts. (a) Fe:Cu = 10:0, (b) Fe:Cu = 7:3, (c) Fe:Cu = 5:5, (d) Fe:Cu = 3:7, (e) Fe:Cu = 1:9, and (f) Fe:Cu = 0:10.

2.3.2. Brunauer–Emmett–Teller (BET)

A comparison was performed by measuring the surface area (m^2/g) of respective carbon nanofibers using a measuring instrument for specific surface area. When the weight ratio of Fe and Cu was 3:7, the highest BET surface area was found to be 305 m^2/g. With weight ratios of 289, 264, 250, and 77 m^2/g, the respective BET surface areas were 10:0, 5:5, 7:3, and 1:9.

Synthesized carbon nanofibers usually have wide specific surface area, which makes them good at storing energy, and thus, they can be used as electrodes materials for capacitor or lead storage batteries, or lithium ion secondary batteries [8, 16].

2.3.3. X-ray Diffraction (XRD)

Figure 5 shows the XRD results for the change of crystal quality according to the Fe and Cu weight ratio. It was confirmed that both carbon nanofibers synthesized with Fe catalyst only and those synthesized with Fe:Cu catalysts at the weight ratios of 7:3 and 5:5 showed carbon peaks with the highest strength.

Most carbon peaks had high strength except for the nanofibers synthesized with Fe and Cu at the weight ratio of 1:9. Therefore, the ratio of pure carbon nanofibers with excellent crystal quality was confirmed to be high [9, 11].

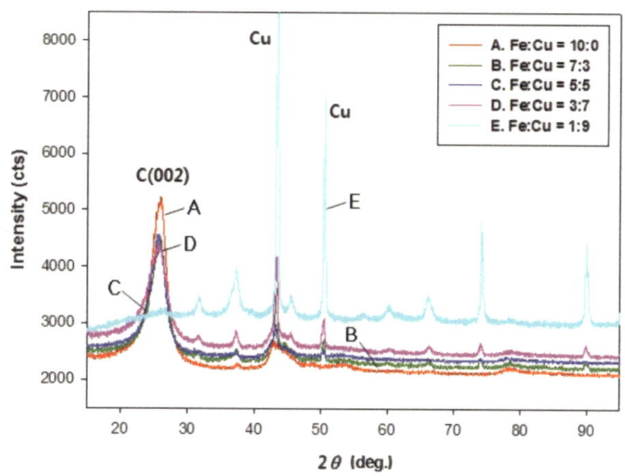

Figure 5. Change in carbon nanofiber crystal quality according to weight ratio of Fe and Cu.

3. Synthesis and characterization of SiO_2/CNF composites by CVD

3.1. Synthesis and electrochemical performance of mesoporous SiO_2–CNF composite on Ni foam

3.1.1. Synthesis of catalysts and mesoporous SiO_2

Binders, electronic conducting additives, and current collectors constituting the electrode are very important factors in the manufacturing process for batteries because the overall performance of the battery depends on the performance of these materials.

When the volume changes repeatedly during the adsorption and desorption of lithium, bonds of active materials become weaker or the conductive additives and contact resistance increase in the electrode. In particular, because active materials such as silicon or tin are used in high-capacity electrodes, the volume changes are even bigger and thus the bond strength between the collector and the anode active materials become weaker.

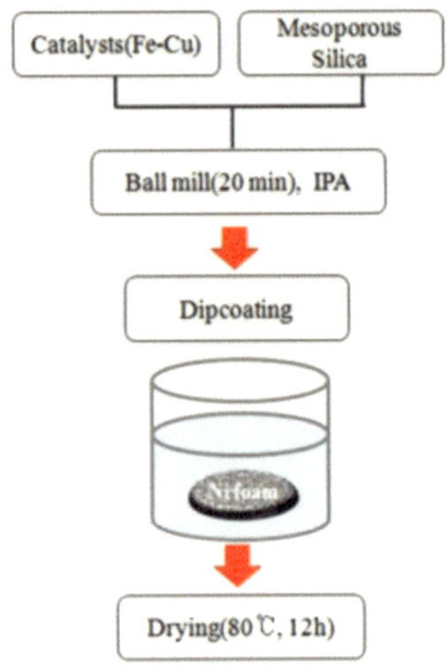

Figure 6. Deposition of catalysts and mesoporous SiO_2 on Ni foam.

Therefore, in this study, we tried to increase the bond strength between the collector and the anode active materials as well as to improve the problem regarding the volume expansion of the electrode by synthesizing CNFs and mesoporous SiO_2–CNF composites directly on the collector, Ni foam, using the CVD method without a binder [17, 18] (**Figure 6**).

3.1.2. Synthesis of CNFs and mesoporous SiO_2–CNF composites

CNFs and mesoporous SiO_2–CNF composites were synthesized in the quartz reactor using chemical vapor deposition. The CVD apparatus used in this experiment is shown in **Figure 7**. C_2H_4/N_2(20/80 vol%) gas was used as the carbon source for the synthesis of carbon nanofibers. H_2/N_2(20/80 vol%) and N_2(99%) were used as the promotion gas for the gas phase reaction and the carrier gas, respectively.

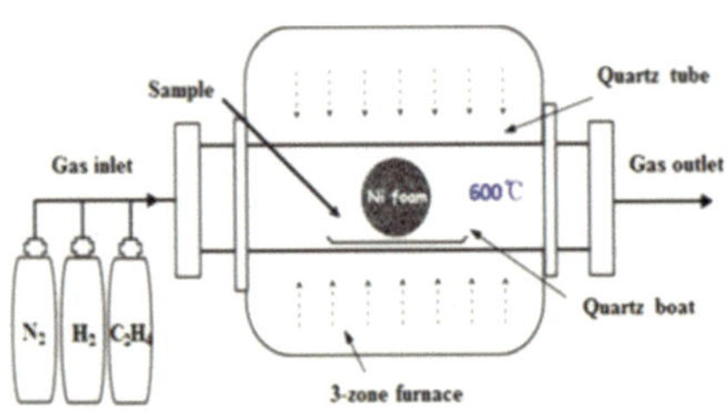

Figure 7. Schematic diagram of CVD apparatus for the preparation of CNFs and mesoporous SiO_2–CNF composites.

After the Fe–Cu catalyst or Fe–Cu/mesoporous SiO_2 deposited on Ni foam was placed in the reaction furnace, the temperature was raised by 10°C/min while the nitrogen atmosphere was maintained. At 600°C, nitrogen and hydrogen gases were flowed while the temperature was maintained for 30 min. Hydrogen and ethylene gases were flowed for 10 min. After the reaction was completed, CNFs and mesoporous SiO_2–CNF composites were synthesized by cooling to room temperature with nitrogen gas [17, 18].

3.1.3. Fabrication process of anode materials for lithium secondary batteries

A three-electrode cell was prepared by applying CNFs and mesoporous SiO_2–CNF composites as anode active materials of lithium secondary batteries. Three-electrode cell was assembled in the glove box filled with Ar gas and was assembled as a half cell. The scheme for cell assembly is shown in **Figure 8**. Prepared active materials were used for the working electrode while lithium was used for the counter and reference electrode. A glass fiber separator was used as the separator membrane. 1 M $LiClO_4$ was employed as the electrolyte and dissolved in a mixture of EC (ethylene carbonate):PC (propylene carbonate) in a 1:1 volume ratio [17, 19–21].

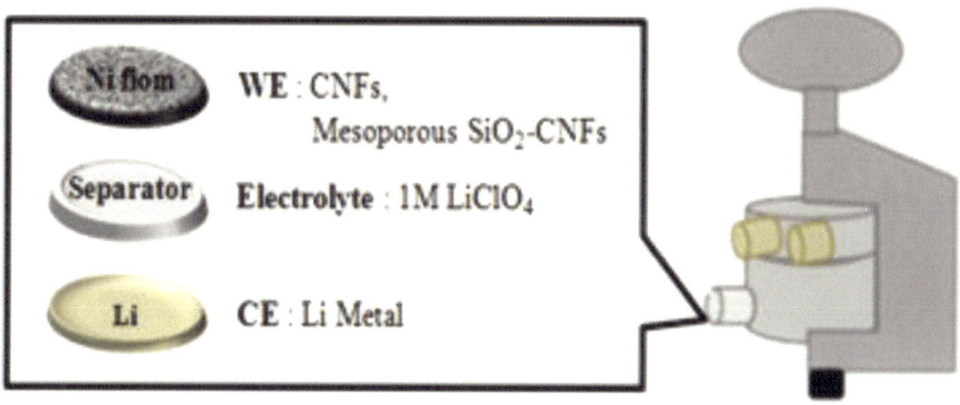

Figure 8. Fabrication scheme of lithium secondary batteries.

3.1.4. Scanning Electron Microscope (SEM)

SEM images of the CNFs and mesoporous SiO_2–CNF composites synthesized on Fe–Cu, Fe–Cu/mesoporous SiO_2, and mesoporous SiO_2-deposited Ni foam using the CVD method were obtained. Analysis of the SEM images showed that CNFs and mesoporous SiO_2–CNF composites grew on Fe–Cu, Fe–Cu/mesoporous SiO_2, and mesoporous SiO_2-deposited Ni foam. The average diameter of the grown CNFs was 25–100 nm [18] (**Figure 9**).

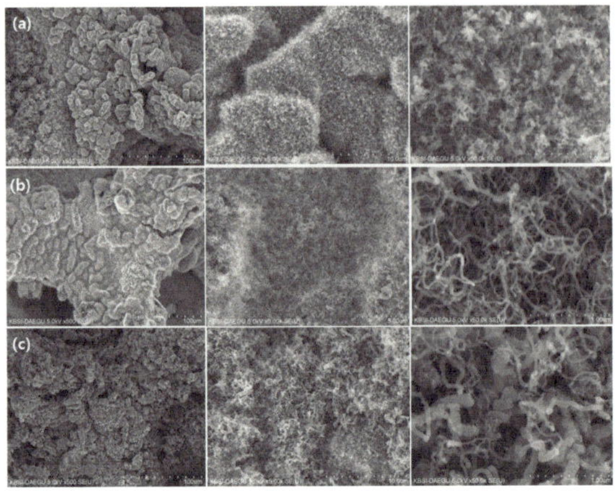

Figure 9. SEM images of CNFs and mesoporous SiO_2–CNF composites. (a) CNF–BF/Fe–Cu/Ni foam, (b) CNF–BF/Fe–Cu/mesoporous SiO_2/Ni foam, and (c) CNF–BF/mesoporous SiO_2/Ni foam.

3.1.5. Transmission electron microscopy (TEM)

TEM was measured to determine the development of porosity in the synthesized mesoporous SiO_2 materials as well as the structure of the synthesized CNFs and mesoporous SiO_2–CNF composites. Panel **Figure 10(a)** shows that mesoporous SiO_2 with uniform porosity was synthesized. As Shown in **Figure 10(b)**, CNFs were synthesized with a hollow tube-like structure in various diameters. As shown in **Figure 10(c)**, the CNFs were surrounded by mesoporous SiO_2 in the mesoporous SiO_2–CNF composites. Panel **Figure 10(d)** shows the elemental mapping from analyzing Si and O atoms. The overall distributions of mesoporous silica were examined [18] (**Figure 10**).

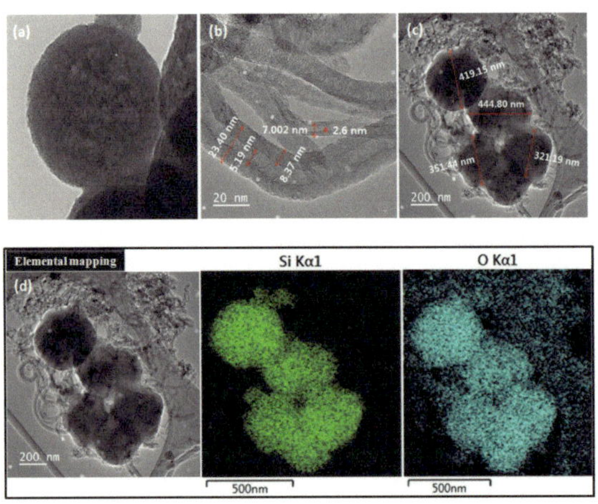

Figure 10. TEM images of CNFs and mesoporous SiO_2–CNF composites. (a) mesoporous SiO_2, (b) CNF–BF/Fe–Cu/Ni foam, (c) CNF–BF/mesoporous SiO_2/Ni foam, and (d) elemental mapping of CNF–BF/mesoporous SiO_2/Ni foam.

http://dx.doi.org/10.5772/63755

3.1.6. Cycle performances

Charging–discharging characteristics were examined by employing a current of 100 mA/g in order to investigate electrochemical characteristics such as the capacity and cycle ability of the three-electrode cell synthesized by applying CNFs and the mesoporous SiO_2–CNF composites synthesized in this study as anode active materials. The electrochemical characteristics of the three-electrode cell were examined with and without the binder.

As shown in **Figure 11(a)**, when CNFs synthesized following the deposition of Fe–Cu catalyst on Ni foam were used as anode active materials, the initial capacity (256 mAh/g) was reduced to 231 mAh/g after 30 cycles, resulting in a retention rate of 90.2%. As shown in **Figure 11(b)**, when mesoporous SiO_2–CNF composites synthesized after the deposition of Fe–Cu catalyst and mesoporous SiO_2 on Ni foam were used as anode active materials, the initial capacity (289 mAh/g) was reduced to 169 mAh/g after 30 cycles, for a retention rate of 58.5%. As shown in **Figure 11(c)**, when mesoporous SiO_2–CNF composites synthesized after the deposition of mesoporous SiO_2 on Ni foam were used as anode active materials, the initial capacity (2420 mAh/g) was reduced to 2092 mAh/g after 30 cycles. The retention rate was 86.4%.

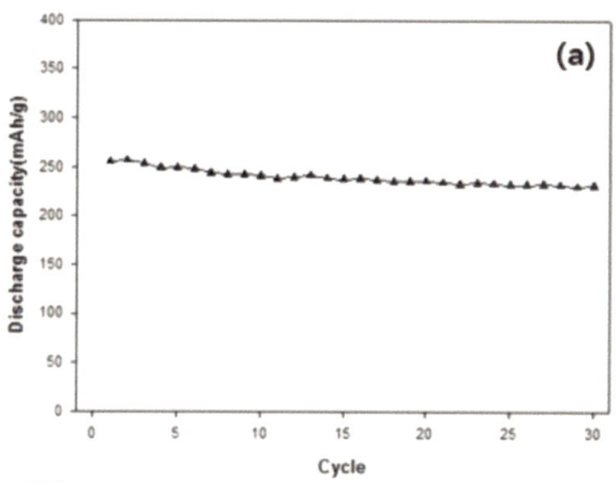

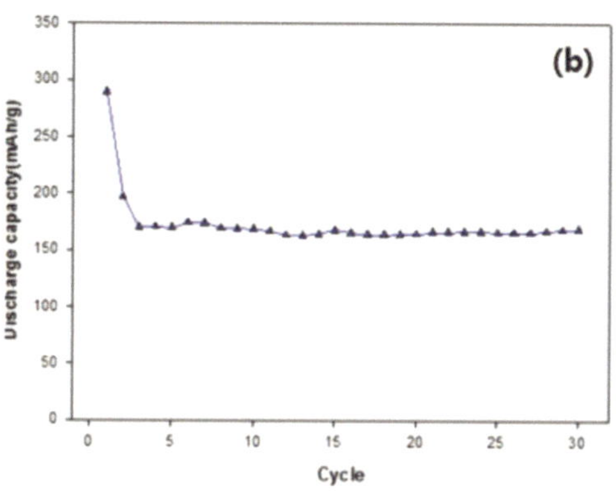

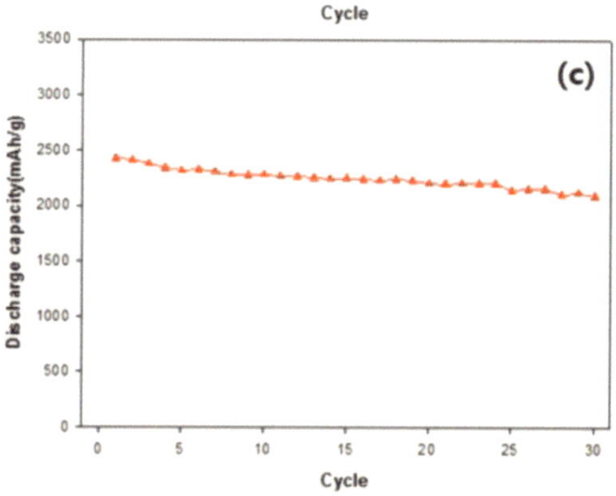

Samples	Max. discharge capacity (mAh/g)	After 30 cycles	
		Discharge capacity	Retention rate (%)
(a)	256	231	90.2
(b)	289	169	58.5
(c)	2,420	2,092	86.4
Retention rate (%) = discharge capacity at 30 cycle/discharge capacity at maximum x 100			

Figure 11. Cycle performance of CNFs and mesoporous SiO_2–CNF composites without binder up to 30 cycles. (a) CNF–BF/Fe–Cu/Ni foam, (b) CNF–BF/Fe–Cu/mesoporous SiO_2/Ni foam, and (c) CNF–BF/mesoporous SiO_2/Ni foam.

The discharging capacity of mesoporous SiO_2–CNF composites was higher than that of CNFs due to the high theoretical capacity of Si. As shown in (b) and (c), the capacity varied depending on the preparation methods used for mesoporous SiO_2–CNF composites. The mesoporous SiO_2–CNF composites synthesized after Fe–Cu catalyst and mesoporous SiO_2 deposited on Ni foam did not show relatively good performance compared to those synthesized without a binder. The reason could be that SiO_2 could not play a role because more CNFs grew in the presence of the Fe–Cu catalyst.

On the other hand, as shown in **Figure 11(c)**, mesoporous SiO_2–CNF composites were synthesized after mesoporous $SiO_2$was deposited on Ni foam without a catalyst. The CNFs grew because the Ni foam served as the catalyst. Thus, panel (c) had a higher capacity due to the mesoporous SiO_2 than panel **Figure 11(b)**, in which many CNFs were grown. The CNFs also grew appropriately. Thus, the retention rate was relatively high [18] (**Figure 11**).

3.2. Synthesis and electrochemical performance of SiO_2/CNF composite on Ni-Cu/C-fiber textiles

3.2.1. Deposition of catalysts

The electrophoretic deposition method was used to deposit Ni and Cu catalysts onto C-fiber textiles, and a schematic diagram of the experimental apparatus used in electrophoretic deposition is displayed in **Figure 12**. The C-fiber textiles were used as the cathode and a carbon electrode was employed as an anode, with a distance of 85 mm between each electrode. Three experimental conditions were employed in depositing the catalyst onto the C-fiber textiles. Ni was deposited onto the C-fiber textiles with a nickel(II) acetate tetrahydrate aqueous solution (Ni) while Ni and Cu were deposited onto the C-fiber textiles with a mixed solution of nickel(II) acetate tetrahydrate and copper(II) acetate monohydrate (Ni–Cu). For the third condition, Cu

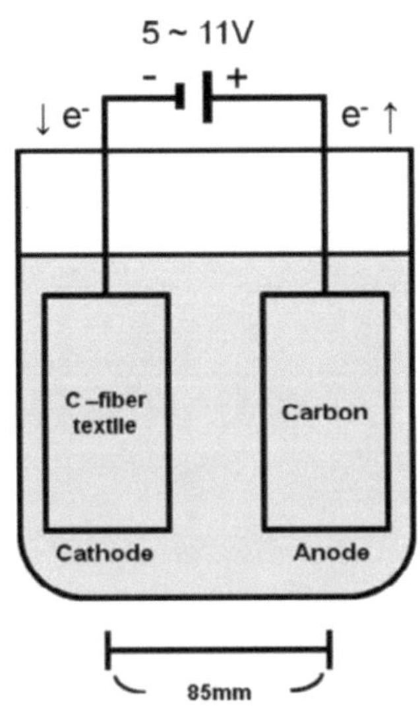

Figure 12. Electrophoretic deposition apparatus used in the deposition of catalysts.

was predeposited onto the C-fiber textiles and Ni was subsequently deposited onto the same C-fiber textile in a nickel(II) acetate tetrahydrate aqueous solution (Ni/Cu) [11, 14, 16, 22, 23].

3.2.2. Reduction

A reduction step was applied. This was done to convert the metal oxides on the surface of the C-fiber textiles into elemental nickel and copper using a tube furnace. H_2 gas mixed with N_2 gas was used for the reduction process, and the flux of the reaction gas was regulated by MFC. The reactor temperature was increased at the rate of 12°C/min, until it reached 700°C. Once the temperature reached 700°C, N_2 gas mixed with 20% H_2 gas was flowed into the reactor. This reduction process was performed for 2 h [23, 24].

3.2.3. Growth of CNFs

CNFs were synthesized onto C-fiber textiles using the CVD method in a horizontal tube furnace after the reduction process was completed. The prepared metal catalyst was evenly spread on a quartz boat, which was then placed into the reactor under an N_2 atmosphere, and the reactor temperature was increased to 12°C/min. Once the temperature reached 700°C, this temperature was maintained for 30 min; 20% H_2 gas balanced with N_2 gas was flowed into the reactor. Then, for 3 h, the H_2 gas balanced with N_2 gas and 20% ethylene balanced with N_2 gas were flowed together into the reactor. The flow of ethylene and H_2 gases was cut off after the reaction was completed. N_2 was then passed through the reactor under an inactive reactor atmosphere to cool it down to room temperature [23, 24].

3.2.4. Oxidation and SiO_2 coating on CNFs

For SiO_2 coating on the surface of CNFs, the hydroxyl group was introduced as an anchor group. This was performed by oxidizing the hydroxyl group for half an hour in 80°C nitric acid and rinsing with distilled water. Then, for the synthesis of a composite of SiO_2-coated CNFs, TEOS was dissolved in ethyl alcohol followed by the dispersion of the CNFs grown on C-fiber textiles in the solution and addition of ammonia water for a 24 h reaction at 50°C [23, 24].

3.2.5. Fabrication of anode materials for lithium secondary batteries

The as-prepared CNFs were grown on C-fiber textiles without any binders and the conducting compounds were used as working electrodes for the fabrication of a conventional three-electrode cell. Lithium was used as the counter and reference electrode. A glass fiber separator was used as the separator membrane. 1 M $LiClO_4$ was employed as the electrolyte and dissolved in a mixture of EC (ethylene carbonate):PC (propylene carbonate) in a 1:1 volume ratio [23].

3.2.6 Scanning Electron Microscope (SEM)

SEM images of CNFs grown with the Ni (a), Ni–Cu (b), and Ni/Cu (c) catalysts deposited onto C-fiber textiles are shown in **Figure 13**. As shown in **Figure 13(a)**, Y-shaped CNFs were grown with an average diameter of 40 nm using the Ni catalyst only, representing the growth of CNF branches that stem from a single origin. Meanwhile, in **Figure 13(b)**, another type of Y-shaped CNFs stemming from a single catalyst in various directions is shown. This figure is relevant to the size of the catalysts created because of the differences in the average diameters. Furthermore, in **Figure 13(c)**, helically grown CNFs with a uniform diameter of 33 nm are shown. With Ni deposited onto the predeposited C-fiber textiles, no Y-shaped carbon nanofiber can be observed in **Figure 13(c)** due to the tendency of the catalyst deposit and the introduction of Cu to affect the growth mechanism of CNFs [23].

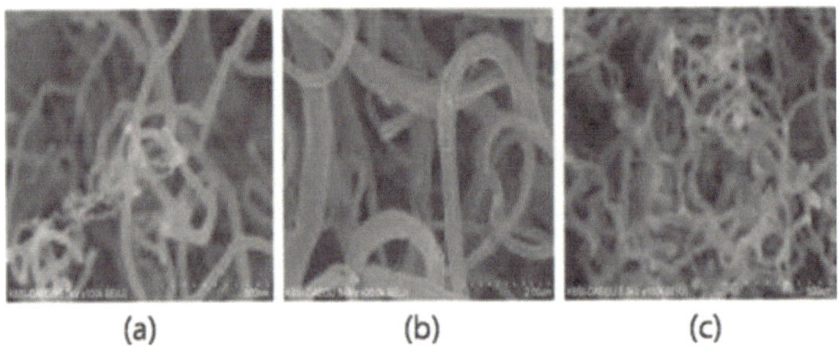

Figure 13. SEM images of CNFs grown on the catalysts Ni (a), Ni–Cu (b), and Ni/Cu (c) on C-fiber textiles.

3.2.7. Transmission electron microscopy (TEM)

TEM images were analyzed in order to investigate the structure of the SiO_2-coated layer in the SiO_2/CNF composite after the growth of CNFs onto C-fiber textiles. These images are shown

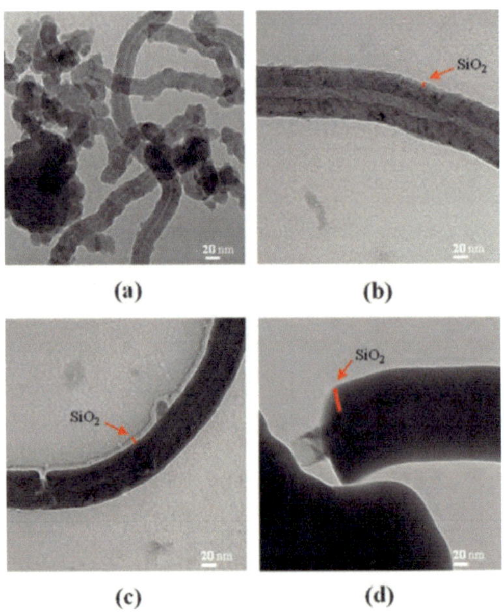

Figure 14. TEM images of CNFs (a) and SiO_2/CNF composite (b)–(d).

in **Figure 14**. As shown in **Figure 14(a)**, the TEM image of CNFs illustrates the multilayer graphite forming wires with a central microhollow. As for the TEM images from the SiO_2/CNF composite **Figure 14(b)–(d)**, they represent the SiO_2 from the output of TEOS hydrolysis, which was uniformly coated onto the CNFs to obtain a layered structure [23, 24].

3.2.8. Cycle performances

The SiO_2/CNF composite was subjected to a repeated cycling test at a current density of 100 mA g^{-1} within a voltage window of 0.1–2.6 V. For comparison, the CNF electrode was tested at the same condition. The cycling performances of the CNFs and the SiO_2/CNF composite electrodes for Li secondary batteries are shown in **Figure 15**. The early-stage discharge capacity of the CNF electrode was 300 mAh/g and a near-stable discharge capacity was maintained for 30 cycles. In the case of the SiO_2/CNF composite, a comparatively high discharge capacity of 2053 mAh/g was observed in the second cycle, and the discharge capacity of the 29th cycle was significantly reduced to 1295 mAh/g, with 63% capacity retention as compared to that of the second cycle. This indicates that the discharge capacity of the CNF electrode nearly reached its theoretical capacity (372 mAh/g) and showed no decline. The SiO_2/CNF composite had a high discharge capacity of 2053 mAh/g, but the cycle performance was not as good as that of the CNFs [23].

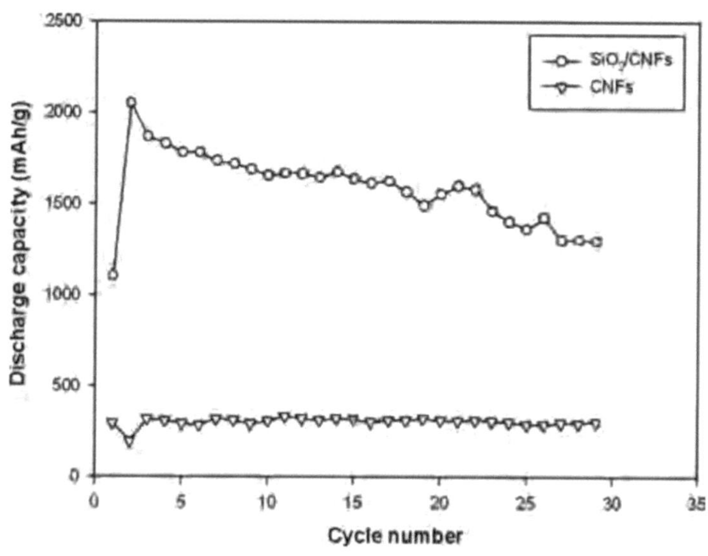

Figure 15. Discharge capacity of CNFs and SiO_2/CNF composite.

4. Conclusions

CNFs were synthesized by using CVD and the effects of synthesis conditions on the growth of CNFs were investigated by controlling the synthesis temperature and the concentration ratio of transition metal catalysts. Physiochemical and electrochemical characteristics of the grown CNFs were investigated using various spectroscopic and electrochemical techniques. Based on these CNFs, SiO_2–CNF composites were synthesized, and the physiochemical characteristics of the SiO_2/CNF composites as well as their electrochemical characteristics as anode materials of lithium secondary batteries were investigated.

(1) CNFs were synthesized by ethylene decomposition using CVD based on Fe and Cu catalysts. According to the SEM measurements, the CNFs had 15–35 nm diameter. In addition, according to the measured specific surface areas (m^2/g) of carbon nanofibers using BET, the synthesized CNFs had the largest specific surface area of 77–305 m^2/g [9, 11].

(2) CNFs were synthesized by ethylene decomposition using CVD with Co and Cu catalysts. According to the SEM measurements, the CNFs had 20–35 nm diameter. In addition, according to the measured specific surface areas (m^2/g) of the carbon nanofibers using BET, the synthesized CNFs had the largest specific surface area of 178–306 m^2/g [11, 14].

(3) CNFs and mesoporous SiO_2–CNF composites were synthesized using Fe–Cu binary catalysts with the CVD method. According to the results of SEM measurements, the average diameter of grown CNFs along with mesoporous SiO_2 was 25–100 nm. According to the results of galvanostatic charging and discharging, the discharging capacity of mesoporous SiO_2–CNF composites was higher than that of CNFs due to the high theoretical capacity of Si. In particular, mesoporous SiO_2–CNF composites synthesized without binders after mesoporous SiO_2 was deposited on Ni foam showed the highest charging and discharging capacity and retention rate. The initial capacity (2420 mAh/g) was reduced to 2092 mAh/g after 30 cycles for a retention rate of 86.4% [18].

(4) CNFs were grown with the CVD method onto C-fiber textiles, based upon Ni, Ni–Cu, and Ni/Cu catalysts, followed by TEOS hydrolysis to coat SiO_2 onto the CNFs. The conclusion of the results is as follows. CNFs grown on Ni/C-fiber textiles were synthesized with a diameter of 40 nm and showed a consistent Y-shaped branch morphology. CNFs grown on Ni–Cu/C-fiber textiles were synthesized with a diameter of 300 nm and had a multidirectional Y-shaped branch morphology. CNFs grown on Ni/Cu/C-fiber textiles appeared to be the most uniform CNFs and had a diameter of 33 nm. Based on galvanostatic charge–discharge, the SiO_2/CNF composites featured a much more excellent discharge capacity of 1295 mAh/g compared to the CNF, which remained at 304 mAh/g, after 29 cycles. Further, a fairly decent capacity retention, 63% compared to the first two cycles, was observed after 20 cycles [23].

Acknowledgements

This research was financially supported by the Ministry of Education, Science Technology (MEST) and National Research Foundation of Korea (NRF) through the Human Resource Training Project for Regional Innovation (No. 2015035858).

Author details

Chang-Seop Lee* and Yura Hyun

*Address all correspondence to: surfkm@kmu.ac.kr

Department of Chemistry, Keimyung University, Daegu, South Korea

References

[1] Peter M. Martin, editor. Handbook of Deposition Technologies for Films and Coatings: Science, Applications and Technology. 3rd ed.UK: Elsevier; 2009. pp. 400–459.

[2] Jong-Hee Park, T.S. Sudarshan, editor. Chemical Vapor Deposition. 1st ed. USA: ASM International; 2001. pp. 1–22.

[3] Jinbo Bai, Aïssa Allaoui, Effect of the length and the aggregate size of MWNTs on the improvement efficiency of the mechanical and electrical properties of nanocomposites —experimental investigation. 2003;34(8):689. DOI:10.1016/S1359-835X(03)00140–4

[4] Orna Breuer, Uttandaraman Sundararaj. Big Returns from Small Fibers: A Review of Polymer/Carbon Nanotube Composites. Polymer Composites. 2004;25(6):630–645. DOI: 10.1002/pc.20058

[5] Jae-Seok Lim, Seong-Young Lee, Sei-Min Park, Myung-Soo Kim. Preparation of Carbon Nanofibers by Catalytic CVD and Their Purification. Carbon Letters. 2005;6(1):31–40.

[6] Krijn P. de Jong, John W. Geus. Carbon Nanofibers: Catalytic Synthesis and Applications. Catalysis Reviews: Science and Engineering. 2000;42(4):481–510. DOI: 10.1081/CR-100101954

[7] Taeyun Kim, Karina Mees, Ho-Seon Park, Monika Willert-Porada, Chang-Seop Lee. Growth of Carbon Nanofibers Using Resol-Type Phenolic Resin and Cobalt(II) Catalyst. Journal of Nanoscience and Nanotechnology. 2013;13(11):7337–348. DOI: http://dx.doi.org/10.1166/jnn.2013.7852

[8] Jia-Qi Huang, Wancheng Zhu, Ling Hu, Qiang Zhang, Fei Wei. Robust Growth of Herringbone Carbon Nanofibers on Layered Double Hydroxide Derived Catalysts and Their Applications as Anodes for Li-ion Batteries. Carbon. 2013;62:393–404. DOI: 10.1016/j.carbon.2013.06.023

[9] Yura Hyun, Haesik Kim, Chang-Seop Lee. Synthesis of Carbon Nanofibers on Iron and Copper Catalysts by Chemical Vapour Deposition. Advanced Materials Research. 2013;750–752:265–275. DOI: 10.4028/www.scientific.net/AMR.750-752.265

[10] Taeyun Kim, Hosun Park, Monica Willert Porada, Chang-Seop Lee. Growth of Carbon Nanofibers Using Resol-Type Phenolic Resin and Cobalt(II) Catalyst. Journal of

Nanoscience and Nanotechnology. 2013;13(11):7337–7348. DOI: http://dx.doi.org/10.1166/jnn.2013.7852

[11] Yura Hyun, Eun-Sil Park, Karina Mees, Ho-Seon Park, Monika Willert-Porada, Chang-Seop Lee. Synthesis and Characterization of Carbon Nanofibers on Transition Metal Catalysts by Chemical Vapour Deposition. Journal of Nanoscience and Nanotechnology. 2015;15(9):7293–7304. DOI: http://dx.doi.org/10.1166/jnn.2015.10588

[12] Eun-Sil Park, Jong-Ha Choi, Chang-Seop Lee. Synthesis and Characterization of Vapor-grown Si/CNF and Si/PC/CNF Composites Based on Co–Cu Catalysts. Bulletin of the Korean Chemical Society. 2015;36(5):1366–1372. DOI: 10.1002/bkcs.10262

[13] Eunyi-Jang, Heai-Ku Park, Jong-Ha Choi, Chang-Seop Lee. Synthesis and Characterization of Carbon Nanofibers Grown on Ni and Mo Catalysts by Chemical Vapor Deposition. Bulletin of the Korean Chemical Society. 2015;36(5):1452–1459. DOI: 10.1002/bkcs.10285

[14] Eun-Sil Park, Jong-Won Kim, Chang-Seop Lee, Synthesis and Characterization of Carbon Nanofibers on Co and Cu Catalysts by Chemical Vapor Deposition. Bulletin of the Korean Chemical Society. 2014;35(6):1687–1695. DOI : 10.5012/bkcs.2014.35.6.1687

[15] Sang-Won Lee, Chang-Seop Lee. Growth and Characterization of Carbon Nanofibers on Fe/C-fiber Textiles Coated by Deposition-Precipitation and Dip-Coating. Journal of Nanoscience and Nanotechnology. 2015;15(9):7317–7326. DOI: http://dx.doi.org/10.1166/jnn.2015.10585

[16] Sang-Won Lee, Chang-Seop Lee. Electrophoretic Deposition of Iron Catalyst on C-Fiber Textiles for the Growth of Carbon Nanofibers. Journal of Nanoscience and Nanotechnology. 2014;14(11): 8619–8625. DOI: http://dx.doi.org/10.1166/jnn.2014.9960

[17] Yura Hyun, Heai-Ku Park, Ho-Seon Park, Chang-Seop Lee. Characteristics and Electrochemical Performance of Si-Carbon Nanofibers Composite as Anode Material for Binder-Free Lithium Secondary Batteries. Journal of Nanoscience and Nanotechnology. 2015;15(11):8951–8960. DOI: http://dx.doi.org/10.1166/jnn.2015.11553

[18] Yura Hyun, Jin-Yeong Choi, Heai-Ku Park, Jae Young Bae, Chang-Seop Lee. Synthesis and Electrochemical Performance of Mesoporous SiO_2-Carbon Nanofiber Composite as Anode Materials for Lithium Secondary Batteries. Materials Research Bulletin. Forthcoming.

[19] Eun-Sil Park, Heai-Ku Park, Ho-Seon Park, Chang-Seop Lee. Synthesis and Electrochemical Properties of CNF–Si Composites as an Anode Material for Li Secondary Batteries. Journal of Nanoscience and Nanotechnology. 2015;15(11):8961–8970. DOI: http://dx.doi.org/10.1166/jnn.2015.11554

[20] Eunyi-Jang, Heai-Ku Park, Chang-Seop Lee. Synthesis and Application of Si/Carbon Nanofiber Composites based on Ni and Mo Catalysts for Anode Material of Lithium

Secondary Batteries. Journal of Nanoscience and Nanotechnology. Forthcoming. DOI: 10.1166/jnn.2016.12229

[21] Yura Hyun, Jin-Young Choi, Heai-Ku Park, Chang-Seop Lee. Synthesis and Electrochemical Performance of Ruthenium Oxide-coated Carbon Nanofibers as Anode Materials for Lithium Secondary Batteries. Applied Surface Science. Forthcoming. DOI: http://dx.doi.org/10.1016/j.apsusc.2016.01.095

[22] Sang-Won Lee, Karina Mees, Ho-Seon Park, Monika Willert-Porada, Chang-Seop Lee. Synthesis of Carbon Nanofibers on C-fiber Textiles by Thermal CVD Using Fe Catalyst. Advanced Materials Research. 2013;750–752:280–292. DOI: 10.4028/www.scientific.net/AMR.750–752.280

[23] Ki-Mok Nam, Heai-Ku Park, Chang-Seop Lee. Synthesis and Electrochemical Properties of Carbon Nanofibers and SiO_2/Carbon Nanofiber Composite on Ni-Cu/C-Fiber Textiles. Journal of Nanoscience and Nanotechnology. 2015;15(11):8989–8995. DOI: http://dx.doi.org/10.1166/jnn.2015.11555

[24] Ki-Mok Nam, Karina Mees, Ho-Seon Park, Monika Willert-Porada, Chang-Seop Lee. Electrophoretic Deposition for the Growth of Carbon Nanofibers on Ni-Cu/C-fiber Textiles. Bulletin of the Korean Chemical Society. 2014;35(8):243–2437. DOI: http://dx.doi.org/10.5012/bkcs.2014.35.8.2431

Chapter 2

Non-Classical Crystallization of Thin Films and Nanostructures in CVD Process

Jae-soo Jung and Nong-moon Hwang

Additional information is available at the end of the chapter

http://dx.doi.org/10.5772/63926

Abstract

Non-classical crystallization, where crystals grow by the building blocks of nanoparticles, has become a significant issue not only in solution but also in the gas phase synthesis such as chemical vapor deposition (CVD). Recently, non-classical crystallization was observed in solution in-situ by transmission electron microscope (TEM) using a liquid cell technique. In various CVD processes, the generation of charged nanoparticles (CNPs) in the gas phase has been persistently reported. Many evidences supporting these CNPs to be the building blocks of thin films and nanostructures were reported. According to non-classical crystallization, many thin films and nanostructures which had been believed to grow by individual atoms or molecules turned out to grow by the building blocks of CNPs. The purpose of this paper is to review the development and the main results of non-classical crystallization in the CVD process. The concept of non-classical crystallization is briefly described. Further, it will be shown that the puzzling phenomenon of simultaneous diamond deposition and graphite etching, which violates the second law of thermodynamics when approached by classical crystallization, can be approached successfully by non-classical crystallization. Then, various aspects of non-classical crystallization in the growth of thin films and nanostructures by CVD will be described.

Keywords: chemical vapor deposition, non-classical crystallization, thin films, charged nanoparticles, gas phase nucleation

1. Introduction

The theory of classical crystal growth was established based on the concept that the building block of crystals should be individual ions, atoms, or molecules. However, there have been some experimental results, which cannot be properly explained by this classical mechanism. Rather such experimental results strongly imply that crystals should grow by the building blocks of nanoparticles, whose way of crystal growth is called 'non-classical crystallization' [1–5]. Recently, non-classical crystallization was confirmed by in-situ transmission electron microscope (TEM) observations. Although non-classical crystallization is a relatively new and revolutionary concept in crystal growth, it has now become so established that a few related books have been published and its tutorial and technical sessions had been included respectively in the spring meetings of Materials Research Society (MRS) and European Materials Research Society (EMRS) in 2014. With the establishment of non-classical crystallization, many crystals that were believed to grow by atomic, molecular, or ionic entities turn out to grow actually by nanoparticles.

Non-classical crystallization can be applied to crystal growth not only in solution but also in the gas phase synthesis of thin films and nanostructures by chemical vapor deposition (CVD) and physical vapor deposition (PVD). Hwang et al. [6–10] extensively studied non-classical crystallization in the CVD process, publishing more than 80 SCI papers. They suggested that the electric charge carried by the nanoparticles played a critical role, by which the growth of thin films and nanostructures by the building blocks of nanoparticles is made possible. This is why they called this new growth mechanism in the gas phase synthesis 'theory of charged nanoparticles (TCN)'. According to this theory, charged nanoparticles (CNPs), which are spontaneously generated in the gas phase in most CVD processes, contribute to the growth of thin films and nanostructures. If nanoparticles are neutral, they undergo random Brownian coagulation, producing a very porous structure. If nanoparticles are charged, however, they deposit as dense films without voids. This is because CNPs undergo self-assembly and are liquid-like, resulting in epitaxial recrystallization.

There seem to be two reasons why this new growth mechanism has been unknown. The first reason would be that CNPs are invisible because their size is much smaller than the wavelength of visible light. The second reason would be that it is difficult to believe that CNPs can be the building blocks for the evolution of dense films and nanostructures. The generation of CNPs in the gas phase was experimentally confirmed in many CVD processes synthesizing such as diamond [11, 12], ZrO_2 [13], Si [14], carbon nanotubes [15, 16], ZnO nanowires [17], and silicon nanowires [18]. The critical reason why these CNPs can be the building blocks of thin films and nanostructures is that the charge weakens the bond strength and makes nanoparticles liquid-like.

TCN was first suggested to explain the paradoxical experimental observation of simultaneous deposition of less stable diamond and etching of stable graphite. This phenomenon violates the second law of thermodynamics if approached by the classical concept of crystal growth by an atomic unit.

http://dx.doi.org/10.5772/63926

2. Non-classical crystallization

2.1. Theory of charged nanoparticles in CVD

Thin film growth by CVD is explained in text books as follows. Atoms or molecules are formed on the growing surface or in the gas phase as a result of chemical reactions of reactant gases. Those atoms or molecules are then adsorbed on a terrace, diffused to a ledge and become incorporated in the crystal lattice at the kink, which is called the terrace, ledge, and kink (TLK) model [19, 20]. This mechanism is called 'classical crystal growth mechanism'. Normally, a ledge of monoatomic height is regarded as a kink because the ledge is disordered or rough, consisting of lots of kinks. An atom on the terrace, which is called an adatom, has excess broken bonds but an atom at the kink has no excess broken bond. For this reason, a reversible transfer of atoms occurs only at kinks during condensation or evaporation. In other words, the interaction of atoms or molecules with the terrace is repulsive; however, their interaction with the kink is attractive. Because of this difference between the terrace and kink, atoms are only accommodated at the kink, which results in self-assembly of atoms or molecules. If the atomic interaction with the terrace should be attractive as well, there would be no atomic self-assembly, resulting in random packing of atoms and the growing film would become amorphous.

In this paradigm of thin film growth, the maximum supersaturation, which would define the maximum growth rate, would be the one which triggers the onset of gas-phase nucleation. However, according to Hwang et al. [9, 10], the supersaturation that triggers the gas-phase nucleation turns out to be so low that the film growth rate without gas-phase nucleation is negligibly low and such processing conditions would be hardly adopted in the thin film industry. In other words, under the process conditions of commercially available thin films, the gas-phase nucleation occurs in general. This means that thin films are growing inevitably under the condition of gas-phase nucleation in most CVD and PVD processes.

It was believed that the gas-phase nucleation would be harmful to the thin film growth. Gas phase-generated nanoparticles may cause killer defects, resulting in device failure, due to small feature sizes, which decrease to <100 nm [21]. However, Hwang and Lee [10] suggested that the deposition behavior of gas phase-generated nanoparticles differs drastically depending on whether they are electrically charged or not. Neutral nanoparticles produce a porous skeletal structure, usually degrading the property of films. However, CNPs tend to be liquid-like and tend to deposit epitaxially, leaving no voids behind, producing dense films. The film microstructures evolved by the deposition of liquid-like CNPs would be difficult to distinguish from those by the deposition of individual atoms or molecules.

Therefore, in order to grow a high quality film at a high deposition rate, it would be necessary to utilize the generation of CNPs in the gas phase. In accordance with this new understanding, Yoshida et al. [22] could grow high T_c superconducting ($YBa_2Cu_3O_{7-x}$) films epitaxially at a rate as high as 16 nm/s by supplying $YBa_2Cu_3O_{7-x}$ particles using the plasma flash evaporation method. Cabarrocas [23, 24], Vladimirov and Ostrikov [25], and Nunomura et al. [26] also utilize the incorporation of gas phase-nucleated nanoparticles in the plasma-enhanced CVD

(PECVD) process. During the deposition of silicon by PECVD, Cabarrocas [23, 24] deposited polymorphous films, where gas phase-generated crystalline silicon nanoparticles are incorporated into the films. Polymorphous films have better stability and electrical properties than amorphous films.

Crystal growth mechanism by the building blocks of nanoparticles has a long history. For example, more than 40 years ago, Glasner et al. [27–30] suggested that nanometer-sized nuclei were generated in the solution with Pb^{2+} during the growth of KBr and KCl. They confirmed that the crystal was grown by self-assembly of the block nuclei in the solution. The crystallinity increased with decreasing size of nuclei. Sunagawa [31, 32], made a similar suggestion that the growth unit of synthetic diamond is not an atom but a much larger unit. These suggestions were not accepted in the crystal growth community largely because the experimental tools were not available at that time to confirm the generation of nanoparticles in solution or in the gas phase. Besides, it was believed that crystal growth by the building blocks of nanoparticles would produce aggregates of nanoparticles instead of dense structures. For example, Glasner et al. [27–30]'s suggestion was doubted and criticized [33] and has been neglected in the crystal growth community.

Such a way of crystal growth by the building blocks of nanoparticles is now well-established and called 'non-classical crystallization' [34, 35]. **Figure 1** compares crystalline pathways between classical and non-classical crystallization. The building blocks of classical crystallization are atoms, ions, or molecules, which form nanoparticles (**Figure 1(a)**) [35]. As described in the classical nucleation theory, these nanoparticles may grow or shrink by the relative magnitude of surface and bulk energies. If nanoparticles reach the size of the critical nucleus, they can continue to grow into macro crystals by the attachment of an individual atom or molecule.

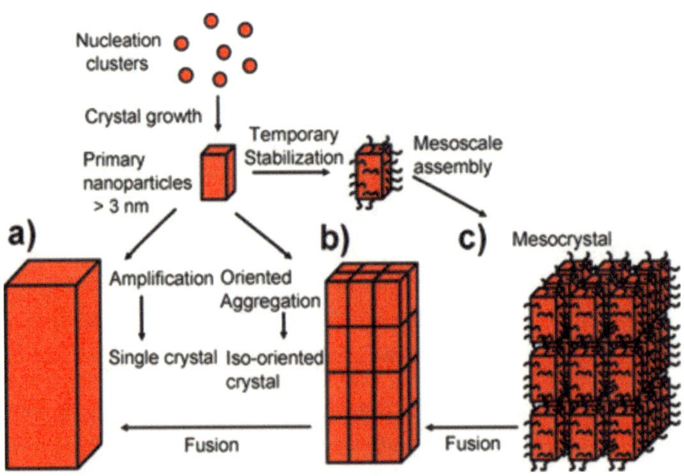

Figure 1. Schematic representation of classical and non-classical crystallization. (a) Classical crystallization. (b) oriented attachment of primary nanoparticles. (c) mesocrystal formation via self-assembly of primary nanoparticles covered with organics. Reprinted with permission from [35]. Copyright 2006 Elsevier.

http://dx.doi.org/10.5772/63926

Figure 1(b) shows the main course of non-classical crystallization, where an iso-oriented crystal grows by oriented attachment of primary nanoparticles, which can form a single crystal upon fusion of the nanoparticles. If the nanoparticles are covered by some organic components, they can form a mesocrystal by mesoscale assembly (path (c)). Cölfen and Antonietti [36] studied a mesocrystal as a superstructure of crystalline nanoparticles with external crystal faces on the scale of hundreds nanometers to micrometers. They also studied that the mesocrystal intermediates can lead to the synthesis of single crystals with included organic additives. During the synthesis process of single crystals, highly oriented nanoparticle-based intermediates could be observed as shown in **Figure 1(c)**. If these mesocrystal intermediates are heated at sufficiently high temperature, they can fuse into a single crystal.

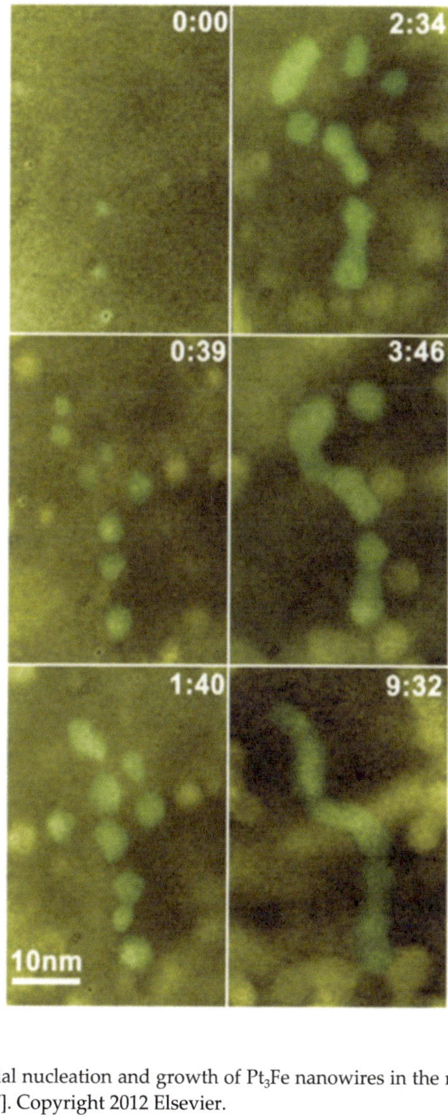

Figure 2. TEM images of the initial nucleation and growth of Pt_3Fe nanowires in the molecular precursor solution. Reprinted with permission from [37]. Copyright 2012 Elsevier.

Figure 1(c) shows that the mesocrystal intermediates clearly reveal the kinetic path of crystallization. Therefore, the mesocrystal intermediates have an important role in revealing the mechanism of non-classical crystallization. However, when the kinetics follows the path of **Figure 1(b)**, it would be difficult to distinguish from a final morphology of the crystal whether it had grown by an individual atom or nanoparticle. This is why crystal growth by nanoparticles had a great resistance in the crystal growth community in the early years.

Recently, crystal growth by nanoparticles in solution could be directly observed by TEM using a liquid cell [37, 38], which provided direction evidences for non-classical crystallization. Liao et al. [37] show detailed real-time imaging to show how Pt_3Fe nano-rods grow by nanoparticles in solution using a silicon nitride liquid cell for in situ TEM observation.

Figure 2 shows images listing the growth process of a twisted Pt_3Fe nanowire. In the initial stage of growth, many small nanoparticles are formed when the Pt and Fe precursors are reduced by electron beam illumination. Some of them grow by monomer attachment and others undergo coalescence. The nanoparticles were combined by coalescence and then relaxed into nanoparticles. Finally, the average size of these nanoparticles reached 5.3 ± 0.9 nm.

In the second stage, nanoparticles interact with each other to form nanoparticle chains. The nanoparticle chain is formed by shape-directed nanoparticle attachment with successive structural relaxation into straight Pt_3Fe nano-rods and reorientation, revealing critical mechanisms of the growth into nano-rods from nanoparticle building blocks. Therefore, even when nanoparticles attach without orientation, single-crystalline nano-rods are formed eventually. In the first stage of growth, nanoparticles meet each other, and form dimer. But if the dimer meets another nanoparticle, unlike the first stage of growth, the dimer does not coalesce into a sphere but forms a trimer by connecting the particle to the dimer end. The additional end-to-end attachments generate a nanoparticle chain.

By in situ TEM observation using graphene liquid cells, Yuk et al. [38] carried out direct atomic-resolution imaging to show how Pt crystals grow in solution. The microscope is operated at 80 kV with a beam intensity of 10^3 to 10^4 A/m^2 maintained during nanocrystal growth. Upon locating a liquid pocket on the TEM grid, the beam intensity is optimized, which reduces the Pt precursor and initiates nanocrystal growth [39]. The use of graphene liquid cells made it possible to discern colloidal Pt nanoparticles with radii as small as 0.1 nm and to track their motion, which was not possible by previous cells with silicon nitride windows [39].

Figure 3 shows the TEM images of the nanocrystals which are connected by a neck at the initial stage of coalescence. Neck growth occurs simultaneously with decreasing length (l) and thickness (t), which means that the atoms migrate to the neck region by surface diffusion [40]. After coalescence, the nanocrystal structure gradually reorganizes, evolving truncated surfaces.

Figure 3 shows the detailed process how crystalline growth occurs by the building blocks of nanocrystals. Yuk et al. [38] mentioned in the supporting information that all images were collected under ambient conditions at 23°C. It should be noted that such enhanced kinetics of liquid-like coalescence at such a low temperature can never be expected from neutral nanocrystals. Although the authors did not mention the role of charge, it should be noted that charging is unavoidable during TEM observation.

According to Hwang and Lee [10], the role of charge is critical in non-classical crystallization since it makes CNPs liquid-like. The liquid-like property of CNPs was deduced by Hwang et al. from the experimental observation that dense films are evolved by the deposition of CNPs. Considering that CNPs are liquid-like, it is expected that the bond strength should be weakened by the presence of charge. To check this possibility, we made an extensive literature survey and found the paper by Clare et al. [41], who studied the effect of charge on the bond strength in hydrogenated amorphous silicon. The main result of this paper and its implications are summarized in the following section.

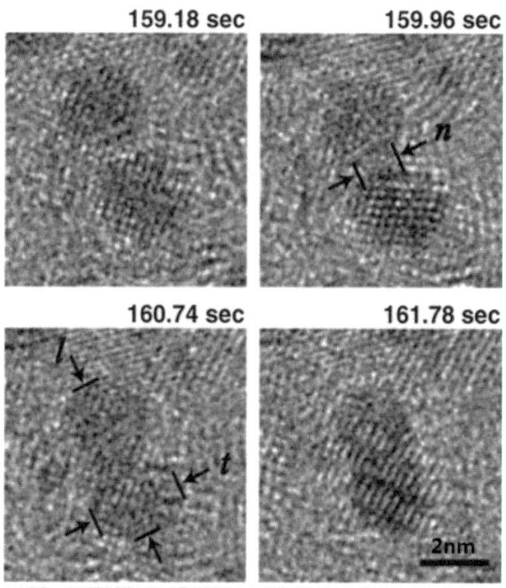

Figure 3. Pt nanocrystal dynamics of coalescence. l, t, and n in the figure represent respectively the length along the center-to-center direction, the thickness in vertical direction to the length and the neck diameter. Reprinted with permission from [38]. Copyright 2012 Elsevier.

2.2. Effect of charge on the bond strength

The effect of a single negative or positive charge on the strength of silicon-silicon and silicon-hydrogen bonds in the molecules SiH_4 and Si_2H_2 was calculated by ab initio calculations. To determine the difference in the energy to break a single Si-H bond in SiH_4, SiH_4^+ and SiH_4^- calculations were done on six species: SiH_3, SiH_4, SiH_3^-, SiH_4^-, SiH_3^+, and SiH_4^+ and the required energies were determined by comparing the bond strength of each species. Similar calculations were done with the species Si_2H_6, Si_2H_5, $Si_2H_6^-$, $Si_2H_5^-$, $Si_2H_6^+$, and $Si_2H_5^+$ to observe the effect of a lower charge/size ratio and to examine the effect of charge on the Si-Si bond energy. The results of ab initio calculations are shown in **Table 1**.

When the atoms are embedded in a lattice, they will not be free to attain geometries resembling the optimized ion geometry, although they will be able to relax to some degree. Thus, the actual effects of charge on bond strength in hydrogenated amorphous silicon will be between those indicated by the unoptimized (adiabatic) and optimized (vertical) rows of **Table 1**. They are likely to be closer to those for the unoptimized rows.

Both positive and negative charges drastically weaken the bond strength of Si-Si and Si-H. The bond strength of Si-Si is weakened from 3.2 eV to 1.11 eV when Si_2H_6 is negatively charged. It is weakened to 1.6 eV when Si_2H_6 is positively charged. The bond strength of Si-H is weakened drastically from 3.9 eV to 0.98 eV when SiH_4 is negatively charged. It is weakened to 0.3 eV when SiH_4 is positively charged.

Compound	Si–H (eV)	Si–Si (eV)
SiH_4 (optimized)	3.9	—
SiH_4^- (optimized)	0.98	—
SiH_4^+ (optimized)	0.30	—
Si_2H_6 (optimized)	3.5	3.2
$Si_2H_6^-$ (optimized)	1.02	1.11
$Si_2H_6^+$ (optimized)	1.59	1.6
SiH_6^- (unoptimized)	1.35	—
SiH_4^- (unoptimized)	0.09	—
$Si_2H_6^-$ (unoptimized)	1.34	1.3
$Si_2H_6^+$ (unoptimized)	1.49	1.6

Table 1. Calculated bond strengths of Si-H and Si-Si. Reprinted with permission from [41]. Copyright 1993 Elsevier.

The effect of charge on the bond strength can be explained by a bond order in the molecular orbital theory. A bond order, which represents the strength or stability of bond, is the number of bonding electron pairs shared by two atoms in a molecule. A bond order is defined as half the difference between the number of bonding electrons and the number of antibonding electrons as expressed by the following equation,

$$\text{Bond order} = \frac{\#of\ bonding\ electrons - \#of\ antibonding\ electrons}{2} \quad (1)$$

If a nanoparticle is charged negatively, electrons are added to the antibonding orbital. If a nanoparticle is charged positively, electrons are removed from the bonding orbital. Therefore, both positive and negative charges would decrease the bond order and thereby weaken the bond strength.

Weakening of bond strength by charge has very important implications because it means that diffusion or kinetics is enhanced. The new concept of charge-induced weakening of bond strength can explain the liquid-like property of CNPs, which was suggested by Hwang et al. [10]. This concept can also explain the rapid kinetics of coalescence in **Figures 2** and **3**. The concept of charge-enhanced kinetics can explain the enhanced chemical reactions of reactant gases even at low temperature in the PECVD process. It also explains the deposition of crystalline films at low temperature.

Moreover, there are many processes where ion or electron beams are used to enhance the kinetics at low temperature. For example, high quality films can be grown at low temperature by ion-beam-assisted deposition (IBAD) in the sputtering or evaporation processes. Also, there is a process called gas-mediated electron or ion-beam-induced deposition and etching, where beams of electrons and ions are used to modify a surface locally at micron and submicron dimensions or fabricating in three dimensions [42–44]. In all these processes, the enhanced kinetics is not clearly understood, vaguely explained by the bombarding energy of ions or electrons. However, it is highly probable that the enhanced kinetics should come from weakening of the bond strength by charge. Moreover, the effect of charge on weakening the bond strength and thereby enhancing the kinetics seems to be very general, even related with the catalytic effect and the enzyme activity in biology.

Zheng et al. [45] observed the superplastic deformation behavior of nanoscale amorphous silica near room temperature using in situ experiments inside a TEM with low beam intensities without obvious rise in sample temperature. They called the phenomenon electron-assisted superplasticity. After imaging for the positioning of the sample and the diamond flat punch, the beam was blocked with the condenser lens aperture, and the silica particle was compressed with the Hysitron Pico-indenter [46, 47]. The particle is plastically deformed because of the e-beam irradiation it had experienced during the imaging. After ~40% compression [46], where the contact pressure is estimated to be 9.2 GPa, the beam was brought on the sample to image the particle. On unloading, the total plastic percent compression was 27%. On compressive loading again with the beam on, surprisingly, the plastic flow of the glass was continuous and smooth, with no sign of shear banding or cracking. Much easier flow was observed with the ensuing beam-on deformation in a second set of in situ experiments, resulting in a pancake shape. Although the contact area kept increasing, the forces required to deform were at levels considerably lower than those in the beam-off condition. This phenomenon of electron-assisted superplasticity might also be explained by weakening of bond strength by charge.

3. Non classical crystallization in diamond deposition at low pressure

The concept that thin films grow by the building blocks of CNPs was first suggested in the low pressure synthesis of diamonds by Hwang et al. [7]. In this section, it will be briefly reviewed how the TCN was developed as the growth mechanism of diamonds in its low pressure synthesis.

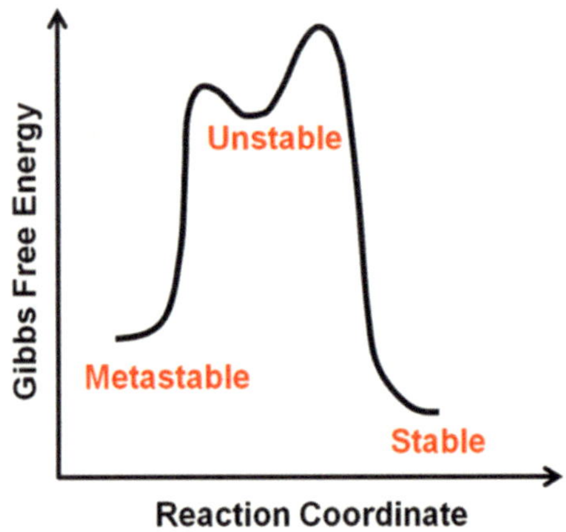

Figure 4. Thermodynamic and kinetic description of the metastable diamond formation. Reprinted with permission from [10]. Copyright 2010 Elsevier.

Low pressure deposition of diamond was reported by pioneering scientists Derjaguin and Fedoseev [48], Spitsyn et al. [49] in HWCVD and Matsumoto et al. [50] in PECVD. Although its process has been studied in depth [51–54], the underlying principle has not been clearly understood. The fundamental question is why diamond can be deposited at low pressure whereas graphite is more stable than diamond.

This question can be explained by **Figure 4** [10]. **Figure 4** describes the general thermodynamic and kinetic explanation of the case where the metastable phase can be formed more dominantly than the stable phase The Gibbs free energy of the stable phase is lower than that of the metastable phase. This difference in free energy determines the difference in the driving force of formation. In other words, the driving force for the formation of the stable phase from the unstable phase is higher than that of the metastable phase.

However, the kinetic barrier, such as a nucleation barrier, is a more important factor that determines the tendency of the formation since the formation of the stable and the metastable phases is a kinetically parallel process. The formation of metastable diamond at low pressure might be explained by this concept. To check this possibility, the nucleation barrier between graphite and diamond should be compared. The Gibbs free energy of nucleation is composed of the driving force for precipitation and the surface energy of the nucleus. Isotropic surface energies of diamond [55] and graphite [56] are respectively 3.7 J/m^2 and 3.1 J/m^2. Since the surface energy of graphite is smaller than that of diamond and the driving force for the precipitation of graphite from the gas phase is larger than that of diamond, the total Gibbs free energy of diamond is higher than that of graphite in all ranges of radius as shown in **Figure 5(a)**. Diamond is less stable than graphite in all ranges of radius in **Figure 5(a)**. Considering only **Figure 5(a)**, it appears that the formation of diamond at low pressure cannot be approached by the concept of the metastable phase formation of **Figure 4**. However, **Figure 5(a)** does not correctly compare the Gibbs free energy between diamond and graphite because the comparison should be made with respect to the number of atoms as shown in **Figure 5(b)** [57] instead of being made with respect to the radius. Although the surface energy of diamond is larger than that of graphite, the molar volume of diamond (3.41 cm^3/mole) is much smaller than that of graphite (5.405 cm^3/mole). Therefore, the surface energy multiplied by the molar area for diamond (8.38×10^{-4} J) is smaller than that for graphite (9.55×10^{-4} J).

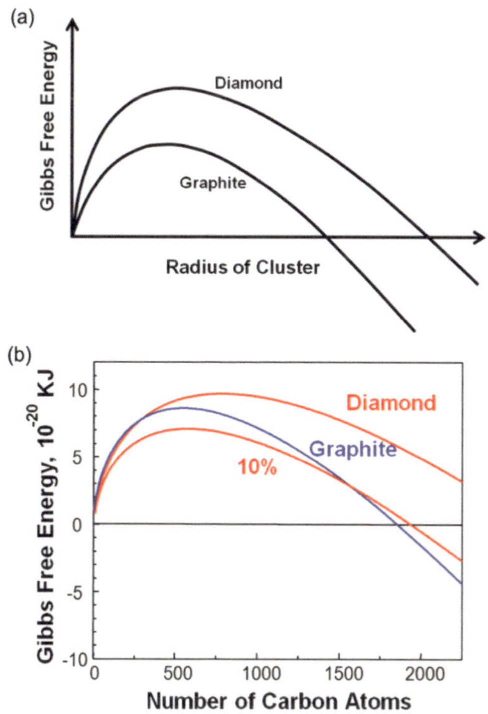

Figure 5. (a) Dependence of Gibbs free energy on the radius of diamond and graphite nanoparticle. (b) Dependence of Gibbs free energy on the number of carbon atoms for diamond and graphite. Reprinted with permission from [57]. Copyright 1997 Elsevier.

The number of atoms at which the Gibbs free energies of diamond and graphite intersect as shown in **Figure 5(b)** can be derived as,

$$n^* = 36\pi \left(\frac{\sigma_{dia} \left(\Omega_{dia}\right)^{\frac{2}{3}} - \sigma_{gra} \left(\Omega_{gra}\right)^{\frac{2}{3}}}{\Delta\mu^{dia \to gra}} \right)^3 \tag{2}$$

where σ_{dia} and σ_{gra} are respectively the surface energies of diamond and graphite, $\boldsymbol{\Omega}_{\text{dia}}$ and $\boldsymbol{\Omega}_{\text{gra}}$ respectively the atomic volumes of diamond and graphite, and $\Delta\mu_{\text{dia}\to\text{gra}}$ the free energy difference per atom between diamond and graphite. The reported surface energies of diamond and graphite are 3.7 Jm^{-2} and 3.1 Jm^{-2}, respectively. The chemical potential difference between diamond and graphite, $\Delta\mu_{dia\rightarrow gra}$, at 927°C is calculated to be −1.2101×10^{-20} J/atom. From these values, n^* for (Eq. 2) is estimated to be 351. This means that for a nucleus containing less than 351 atoms, diamond is more stable than graphite. Although the nucleation curves in **Figure 5(b)**, which are evaluated from the reported surface energy data of some uncertainty, have uncertainties, they definitely tell that diamond can be more stable than graphite for a sufficiently small size. In **Figure 5(b)**, the nucleation barrier of diamond is slightly larger than that of graphite, indicating that graphite would nucleate more dominantly than diamond. However, it should be noted that the difference of the nucleation barrier between diamond and graphite is rather small. If the nucleation barrier of diamond was slightly smaller than that of graphite, diamond synthesis at low pressure would be much easier than now and gas activation such as hot filament and plasma would not be necessary. Oxygen-deficient burning of any carbon-containing materials such as coals, fuels, and woods might have produced diamonds instead of soot.

In reality, graphitic or amorphous carbon is formed without gas activation. It is known that gas activation is essential to the low-pressure synthesis of diamond. If we assume that the role of gas activation is to reduce the surface energy of diamond, the nucleation of diamond can be more dominant than that of graphite. For example, if the gas activation reduces the surface energy of diamond by 10%, n^* in (Eq. 2) becomes 1784 at 927°C, and the nucleation barrier of diamond becomes lower than that of graphite as shown in **Figure 5(b)**.

Hwang et al. [7, 58] suggested that the essential role of gas activation is to generate an abundant amount of electric charge rather than atomic hydrogen and that the negative charge stabilizes diamond over graphite. They further suggested that charged diamond nanoparticles are generated in the gas phase, becoming the building blocks of diamond films. They reached this conclusion from the fact that the well-established experimental observation of simultaneous diamond deposition and graphite etching violates the second law of thermodynamics if diamond deposition occurs by individual atoms.

However, in the diamond CVD community, the 'atomic hydrogen hypothesis', which was suggested by Spitsyn et al. [49], has been the most popular explanation for diamond growth at low pressure. According to the hypothesis, atomic hydrogen, which is produced by gas activation such as hot filament or plasma, etches graphite much faster than diamond and therefore low-pressure synthesis of metastable diamond is possible. This hypothesis is equivalent to saying that even though diamond is less stable than graphite, diamond can grow dominantly over graphite in the presence of atomic hydrogen.

However, this statement violates the second law of thermodynamics because if stable graphite etches away, less stable diamond must etch away. Although the atomic hydrogen hypothesis has a significant drawback of violating the second law of thermodynamics as to the irreversible transfer of carbon atoms [7], it is widely accepted in the diamond CVD community largely because simultaneous diamond deposition and graphite etching are experimentally observed. In other words, the experimental observation appears to violate the second law. This is a very interesting situation. Great attention should be paid to this seeming contradiction between the experimental observation and the second law because it is expected that if such a contradiction is solved, a big discovery may be made.

Let's make a rigorous thermodynamic analysis of the simultaneous diamond deposition and graphite etching. The criterion for the irreversible transfer of atoms between the system and the surrounding, which corresponds to an open system in thermodynamics, is the chemical potential. Chemical potential is defined as the partial derivative of the Gibbs free energy with respect to the number of atoms under constant temperature and pressure [59]. The total Gibbs free energy is decreased when atoms transfer from a phase with high chemical potential to a phase with low chemical potential.

Since graphite is more stable than diamond at the given temperature and pressure of the CVD process according to the phase diagram of carbon [47, 48], the chemical potential of carbon in diamond is higher than that in graphite, which can be written as,

$$\mu_{c}^{dia} > \mu_{c}^{gra} , \tag{3}$$

where the subscript *'C'* and the superscripts *'dia'* and *'gra'* mean carbon, diamond, and graphite, respectively. Since reversible etching and deposition are not driven by the difference of the chemical potential and do not produce a net flux, the reversible etching and deposition can occur simultaneously. On the other hand, irreversible etching or deposition is driven by the chemical potential difference and produces a net flux. The irreversible etching and deposition cannot occur simultaneously.

The experimental observation of simultaneous diamond deposition and graphite etching indicates that the chemical potential of carbon in diamond is lower than that in gas and that the chemical potential of carbon in graphite is higher than that in gas. The irreversible graphite etching can be written as

$$\mu_{c}^{gra} > \mu_{c}^{gas}, \tag{4}$$

where the superscript *'gas'* represents the gas phase. Similarly, the irreversible diamond deposition can be written as

$$\mu_{c}^{gas} > \mu_{c}^{dia} \tag{5}$$

In this paper, etching and deposition refer to the irreversible process and from now on, those words will be used without the adjective 'irreversible'. The simultaneous diamond deposition and graphite etching indicate that Eqs. (4) and (5) should be simultaneously satisfied and can be written as

$$\mu_{c}^{gra} > \mu_{c}^{gas} > \mu_{c}^{dia}, \tag{6}$$

Therefore, the chemical potential of carbon in graphite is larger than that in diamond, which is identical to saying that diamond is more stable than graphite. This contradicts with Eq. (3). Therefore, it is quite evident that the simultaneous diamond deposition and graphite etching or the atomic hydrogen hypothesis violates the second law of thermodynamics.

The contradiction becomes more evident if chemical potentials of carbon for different phases are represented in diagrams as shown in **Figure 6**. **Figure 6** compares chemical potentials of carbon for graphite, diamond, and gas based on the carbon phase diagram (**Figure 6(a)**) and the atomic hydrogen hypothesis or the experimental observation of simultaneous diamond deposition and graphite etching (**Figure 6(b)**).

In the interpretation of this paradoxical observation by the atomic hydrogen hypothesis, it is implicitly assumed that the unbalanced etching rate between graphite and diamond can change the stability between graphite and diamond. This hypothesis neglects the fact that thermodynamic stability is not affected by kinetics. It should be noted that the thermodynamic stability determined by the Gibbs free energy is a state function, which depends only on the initial and final states, not on the kinetic path.

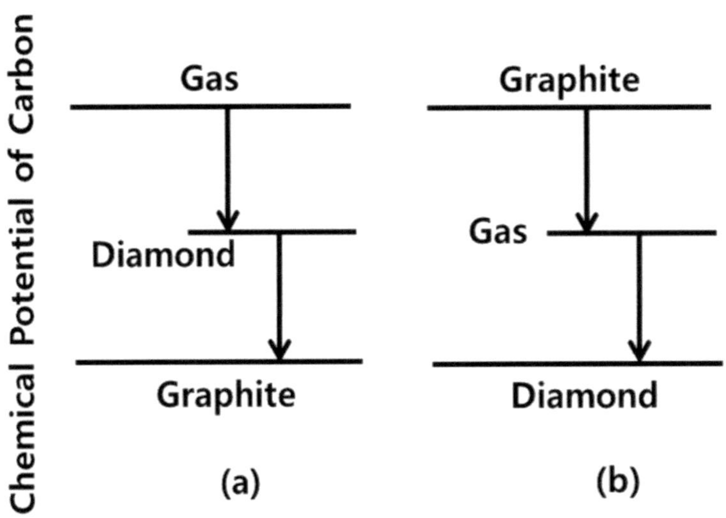

Figure 6. (a) Chemical potentials of carbon for different phases and the direction of carbon atoms indicated by the carbon phase diagram and (b) those indicated by simultaneous diamond deposition and graphite etching based on the assumption that the building block should be an individual atom. Reprinted with permission from [10]. Copyright 2010 Elsevier.

Since the second law of thermodynamics cannot be violated, there must be some logical error in interpreting the experimental observation of simultaneous diamond deposition and graphite etching. What would be the logical error? One thing that should be noted is that etching of diamond and graphite occurs under the condition where the concentration of CH_4 is high enough to deposit both phases. In order to understand this situation, let's consider the phase diagram of the C-H system as shown in **Figure 7** [6].

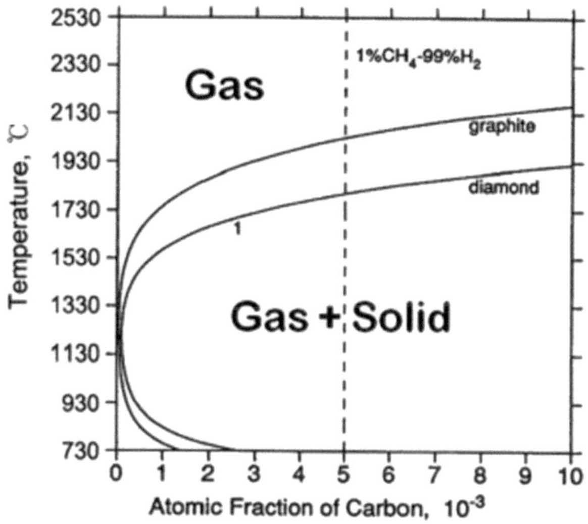

Figure 7. Phase diagram of the C–H system under 2700 Pa. The graphite and diamond lines indicate the maximum solubility of carbon in the gas phase. Reprinted with permission from [6]. Copyright 1996 Elsevier.

The vertical dashed line is for the composition of 1% CH_4-99% H_2, which is typically used in the gas-activated diamond CVD process. At the substrate temperature around 1200 K (927 °C), the vertical line is inside the two-phase region of gas + solid, which means that both diamond and graphite have a driving force for precipitation. This prediction is opposite to the experimental observation of graphite etching. This disagreement between the phase diagram and the experimental observation comes from the implicit assumption that the gas-phase nucleation does not take place. If gas-phase nucleation takes place, the equilibrium carbon content in the gas phase follows the solubility line in **Figure 7**. If diamond nucleates in the gas phase, the carbon content in the gas phase would be the solubility line for diamond. If graphite nucleates in the gas phase, the carbon content in the gas phase would be the solubility line for graphite. Carbon solubility in graphite is lower than that in diamond because graphite is more stable than diamond. This solubility tends to be minimum at ~1300 K and increases with decreasing temperature; it has a retrograde solubility. If diamond or graphite nucleates in the gas phase, carbon in the gas phase would be depleted. When this carbon-depleted gas approaches the substrate, at the temperature of which equilibrium carbon solubility is much higher, etching occurs for both diamond and graphite.

Therefore, the experimental observation of graphite etching indicates that the gas-phase nucleation of diamond or graphite occurred. Besides, in order to explain the experimental observation of simultaneous diamond deposition and graphite etching without violating the second law of thermodynamics, the nuclei should have a diamond structure and at the same time those diamond nuclei should be the building blocks of the growing diamond. According to this scenario, both diamond and graphite are etched away by the atomic unit but diamond deposition occurs by the building block of gas phase-nucleated nanoparticles. This scenario is free from the thermodynamic paradox. Therefore, the puzzling phenomenon of simultaneous diamond deposition and graphite etching provides strong evidence that diamond deposition should occur by the building blocks of gas phase-nucleated nanoparticles.

To check this scenario, the gas-phase nucleation in the diamond CVD system must be experimentally confirmed. To confirm the existence of CNPs generated in the gas phase, Jeon et al. [11] attached an energy analyzer combined with a Wien filter to the diamond hot filament CVD (HFCVD) reactor. Since the processing pressure in diamond HFCVD is higher than the required pressure for Wien filter measurements, differential pumping was done through an orifice (1.2 mmØ) and a skimmer (2 mmØ) between the measuring chamber and the CVD reactor. By such a system, they could measure the size distribution of negatively charged carbon nanoparticles in the gas phase shown in **Figure 8** [11]. The processing condition of measurement was a gas mixture of 1.5% CH_4–98.5% H_2, a reactor pressure of 800 Pa, and a wire temperature of 2100°C. The peak occurred at ~3000 atomic mass units, which corresponds to ~250 carbon atoms. Based on the negative current measured on the Faraday cup, they estimated the number density of negatively charged nuclei as ~10^6 mm^{-3}.

Figure 9 shows the paradox-free direction of carbon flux before and after gas-phase nucleation [10]. Before gas-phase nucleation, the gas phase is supersaturated and the driving force is for the deposition of both diamond and graphite (**Figure 9(a)**). On the other hand, after gas-phase nucleation, the gas phase is undersaturated and the driving force is for etching of both diamond and graphite (**Figure 9(b)**). After gas-phase nucleation, however, the gas phase is undersaturated and has the driving force for etching of both diamond and graphite (**Figure 9(b)**). Since both diamond and graphite are etched away by the atomic unit under such conditions, the atomic contribution to diamond growth is negative. Nevertheless, diamond crystals grow under such conditions. This means that the growing diamond is contributed 100% by the building blocks of gas-phase nucleated diamond nanoparticles, whereas the diamond film

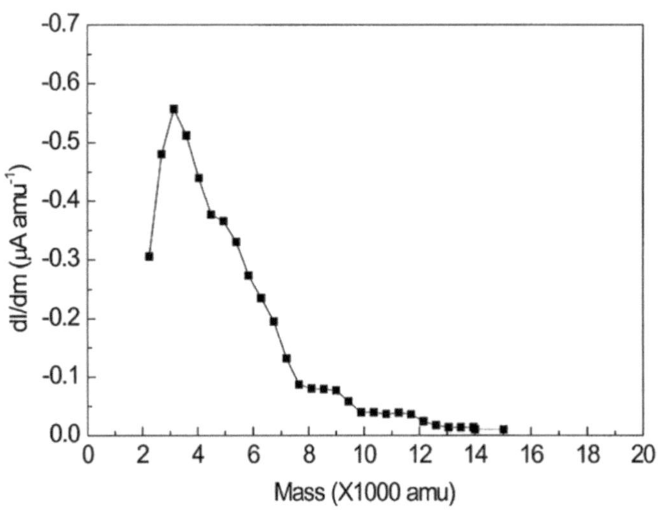

Figure 8. Mass distribution of negatively charged carbon nanoparticles generated in the HWCVD reactor. Reprinted with permission from [11]. Copyright 2000 Elsevier.

deposited on the substrate is being etched atomically. In order for diamond films to grow by the building blocks of CNPs, it must be assumed that CNPs should undergo self-assembly and epitaxial recrystallization. And in order to undergo epitaxial recrystallization, CNPs should be liquid-like. This liquid-like property would come from the weakening of bond strength by charge as explained in the previous section.

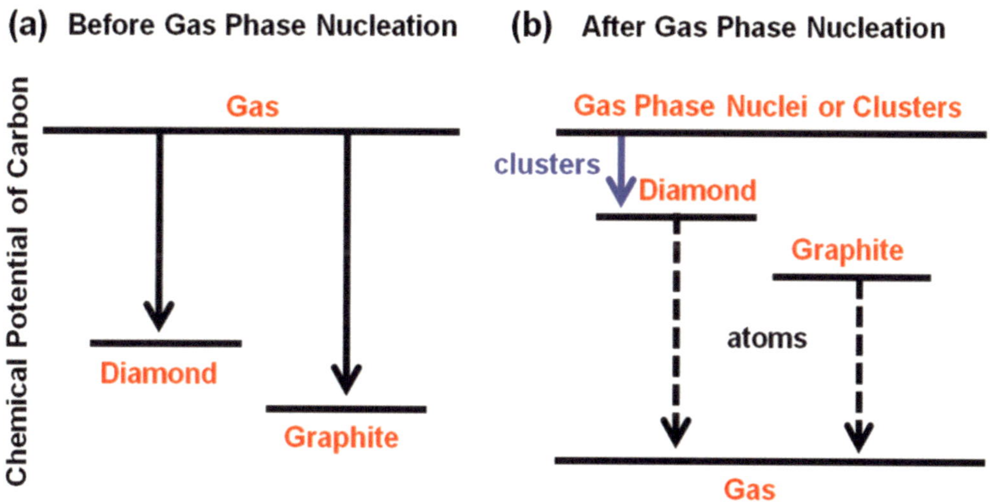

Figure 9. The chemical potentials diagram of carbon in the diamond, graphite, and gas phases before and after gas-phase nucleation. Reprinted with permission from [10]. Copyright 2010 Elsevier.

On the other hand, there is another puzzling phenomenon in the diamond CVD process: under the same deposition condition where diamond films grow on a silicon substrate, highly porous skeletal graphitic soot particles grow on an iron substrate and carbon nanotubes grow on a palladium substrate as shown in **Figure 10** [7, 61]. These results imply that the deposition mechanisms of diamond, soot, and carbon nanotubes should be closely related to one another.

From the previous analysis based on **Figures 4–9**, we know that diamond nucleates in the gas phase. Besides, **Figure 8** confirms that these nuclei are negatively charged. If this knowledge is combined with **Figure 10**, negatively charged diamond nuclei deposit as a diamond film, soot, and carbon nanotubes respectively on silicon, iron and palladium substrates. Since etching takes place atomically, the building blocks of the diamond film, soot, and carbon nanotubes must be negatively charged nanoparticles. In relation with this effect of the substrate, Huh et al. [61] studied the correlation between the deposition behavior and the charge transfer rate of the substrate and found out that there exists an almost perfect correlation that the substrate with a low charge transfer rate produces diamond whereas the substrate with a high charge transfer rate produces soot or carbon nanotubes.

According to Hwang et al. [7], when negatively charged diamond nuclei approach the silicon substrate which has a low charge transfer rate, they maintain a diamond phase and undergo self-assembly like nano-colloids and epitaxial recrystallization, producing the diamond film as shown in **Figure 10(a)**. However, when they approach the iron substrate, which has a high charge transfer rate, they lose charge to the iron substrate just before landing and transform to graphite. And the neutral graphite nanoparticles make random Brownian aggregations and produce porous skeletal soot on the iron substrate as shown in **Figure 10(b)**. In this scenario, the stability of diamond comes from charge. In other words, when carbon nanoparticles are charged, they have a diamond structure; otherwise, they have a graphite structure. In order to confirm the effect of charge on the stability of diamond, Hwang et al. [7] compared the deposition behavior between floating and grounded iron substrates after 2 hours of deposition. On the grounded iron substrate, soot particles continued to grow. On the floating iron substrate, however, diamond crystals were grown on the initially formed soot particles.

TEM observation of the soot particles by Huh et al. [61] revealed that iron nanoparticles are trapped inside each soot particle. In other words, each soot particle encloses an iron nanoparticle three dimensionally like an onion. If iron nanoparticles are not enclosed three dimensionally by carbon, it might be possible that carbon nanotubes grow. From the fact that carbon nanotubes are evolved on the palladium substrate as shown in **Figure 10(c)**, an insight can be drawn as to the growth mechanism of carbon nanotubes [62]. One possible driving force for the growth of one-dimensional carbon would be the reduction of the electrostatic energy. This possibility is supported by the bias effect on the growth of carbon nanotubes reported by Lee and Hwang [63]. They found out that carbon nanotubes grow only when the substrate is positively biased in the hot filament CVD process using a high methane concentration of 20% CH_4–80% H_2. This possibility is further supported by the experimental observation that one-dimensional carbon such as carbon nanotubes or carbon fibers grow when the catalytic metal nanoparticles are placed on the insulating substrate such as glass, but do not grow when they are placed on the conducting substrate [64].

Figure 10. (a) Diamond, (b) soot, and (c) carbon nanotubes grown respectively on Si, Fe, and Pd substrates. Reprinted with permission from [7, 61]. Copyright 1996 Elsevier.

http://dx.doi.org/10.5772/63926

4. Deposition behavior and analysis of non-classical crystallization in the CVD process

4.1. Comparison of deposition behavior between floating and grounded substrates in the silicon CVD process

The generation of CNPs in the gas phase was experimentally confirmed using a differential mobility analyzer (DMA) in many CVD systems [16, 17, 65]. Hwang et al. [9, 10] suggested that these CNPs are the building blocks of thin films and nanostructures in most CVD processes. Experimental confirmation of CNPs in the gas phase is not sufficient to say that they are the building blocks. However, the Si-Cl-H system has retrograde solubility like the C-H system [66]. Besides, in the CVD of the Si-Cl-H system, a paradoxical phenomenon of simultaneous deposition and etching of silicon was reported [67]. This phenomenon also violates the second law of thermodynamics if the silicon film grows by individual atoms. In order to avoid the violation, the charged silicon nuclei formed in the gas phase should be the building blocks of growing silicon films [9, 10].

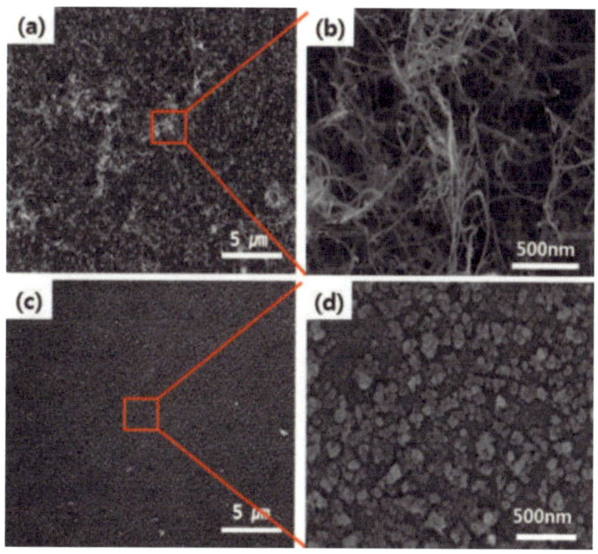

Figure 11. FESEM images: (a) low-magnification and (b) high-magnification images of a floating silicon substrate and (c) low-magnification and (d) high-magnification images of a grounded silicon substrate at a N_2 flow rate of 500 sccm. Reprinted with permission from [64]. Copyright 2014 Elsevier.

In order to show CNPs to be building blocks of thin films or nanostructures in a silicon CVD process, Youn et al. [64] compared the deposition behavior between floating and grounded silicon substrates. Charge build-up would be maximized on the floating substrate and minimized on the grounded substrate. The result is shown in **Figure 11**.

Figure 11(a) and **(b)** shows respectively low- and high-magnification FESEM images of the surface microstructure of the floating substrate after deposition for 2 hours. The substrate temperature is 900°C and the gas flow rates are 5 sccm 10% SiH_4–90% He, 50 sccm H_2, and 500 sccm N_2. **Figure 11(c)** and **(d)** shows respectively low- and high-magnification FESEM images of the surface microstructure of the grounded substrate with the deposition conditions being the same as those of **Figure 11(a)** and **(b)**. Silicon nanowires grew extensively on the floating substrate as shown in **Figure 11(a)** and **(b)**, whereas no silicon nanowire grew but only silicon nanoparticles were deposited on the grounded substrate as shown in **Figure 11(c)** and **(d)**. Diameters of the nanowires in **Figure 11(b)** are about 10–30 nm. In **Figure 11(b)**, without catalytic metals or seeds of silicon oxide, silicon nanowires were grown on the substrate. Neither vapor-liquid-solid (VLS) [68] nor oxide-assisted growth (OAG) [69] mechanism can explain this deposition behavior.

Floating and grounding of the substrate are the only differences in the processing condition between **Figure 11(b)** and **(d)**. Therefore, charge build-up would be the only difference between **Figure 11(b)** and **(d)**. Considering this, the growth of silicon nanowires would be attributed to the charge build-up. In other words, the electrostatic interaction between the CNPs and the substrate is responsible for the growth of the silicon nanowires.

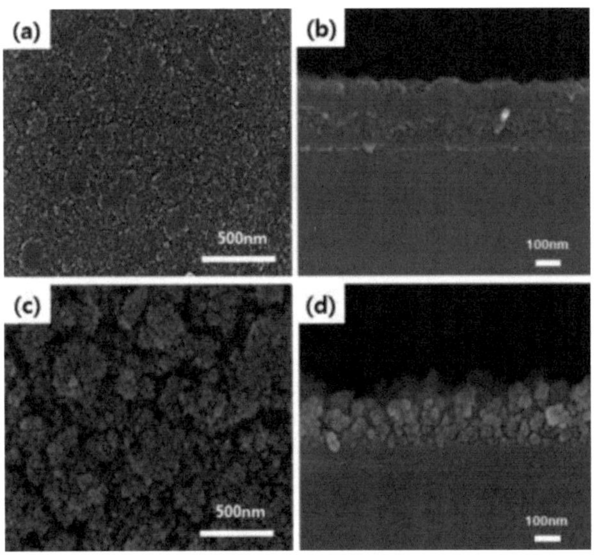

Figure 12. FESEM images for (a) plane view and (b) cross section of films deposited on a floating silicon substrate and (c) plane view and (d) cross section of films deposited on a grounded silicon substrate at a N_2 flow rate of 1000 sccm. Reprinted with permission from [64]. Copyright 2014 Elsevier.

When the N_2 flow rate was increased from 500 sccm to 1000 sccm, films instead of nanowires were grown as shown in **Figure 12**. **Figure 12(a)** and **(b)** shows respectively FESEM images of the surface and the cross section of a film deposited on a floating silicon substrate. **Figure 12(c)** and **(d)** shows those deposited on a grounded substrate. The deposition condition was the same as that of **Figure 11** except the gas flow rate of N_2. The film on the grounded substrate in **Figure 12(c)** and **(d)** was much more porous than that on the floating substrate in **Figure 12(a)** and **(b)**. Therefore, quite different microstructures are evolved between the floating and grounded substrates, indicating that the dense film in **Figure 12(a)** and **(b)** resulted from the electrostatic interaction between the CNPs and the growing surface. The film thicknesses of **Figure 12(b)** and **(d)** were respectively ~220 and ~190 nm, indicating that the growth rate of the film on the floating substrate was higher than that on the grounded substrate. Besides, the film in **Figure 12(b)** is much denser than that in **Figure 12(d)**. Therefore, the mass increase on the floating substrate in **Figure 12(b)** is much higher than that on the grounded substrate in **Figure 12(d)**. This study also indicates that the deposition behavior of charged silicon nanoparticles differs dramatically between floating and grounded substrates due to the difference in the accumulation of charge between floating and grounded substrates. These results imply that charge plays a critical role in the evolution of dense films and smooth nanowires.

4.2. Comparison of deposition behavior between conducting and non-conducting surfaces

The concept of non-classical crystallization during PECVD was suggested by Cabarrocas et al. [23, 24, 70]. They reported that silicon nanoparticles formed in the gas phase during the PECVD process are incorporated into a polymorphous structure. They also suggested that high-quality films can be deposited at a high growth rate by incorporating those nanoparticles. In order to confirm whether nanoparticles, which are generated in the gas phase, should be charged or not, Cabarrocas et al. [70] examined the deposition behavior on 5×5 cm^2 glass substrates partially coated with Cr. They also compared the deposition behavior between the glass substrates with and without hydrogen plasma pretreatment as shown in **Figure 13**.

In the case of the glass substrate without any plasma treatment, no film was deposited on the uncoated part as shown in the right of **Figure 13**. In both cases of left and right in **Figure 13**, film deposition took place only on the Cr-coated part, which is a kind of selective deposition. On the other hand, for the glass substrate exposed to hydrogen plasma, the entire surface of the substrate was deposited as shown in the left of **Figure 13**. These results strongly support their hypothesis that the building blocks of the deposition are charged.

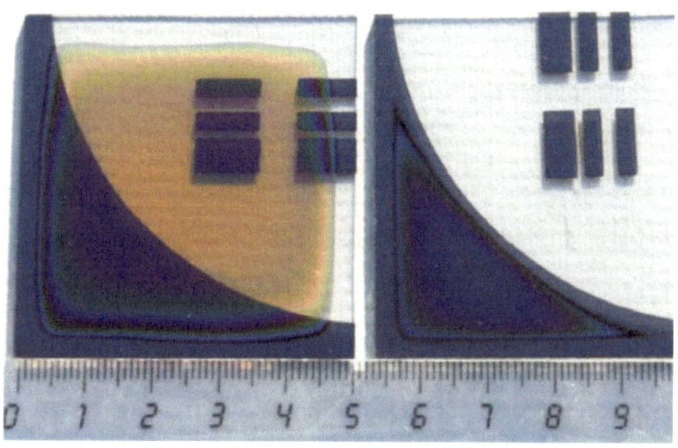

Figure 13. Image of two deposition behavior on partly Cr-coated glass substrates with a hydrogen plasma pretreatment (left) and without plasma pretreatment (right). Reprinted with permission from [70]. Copyright 2008 Elsevier.

4.3. Effect of flow rate on the deposition behavior in the silicon CVD process

The effect of various parameters in the CVD process should be newly understood based on the new paradigm of non-classical crystallization by CNPs. In this section, the effect of a carrier gas flow rate on the deposition behavior will be examined. Youn et al. [71] varied the flow rate of nitrogen carrier gas during atmospheric pressure silicon CVD and observed the microstructure evolution of films or nanostructures as shown in **Figure 14**.

Figure 14(a)–(d) shows the FESEM images of the surface microstructure of films or nanostructures synthesized on a quartz substrate respectively at nitrogen flow rates of 300, 500, 700, and 1000 sccm. At 300 sccm (**Figure 14(a)**), nanowires were grown with nanoparticles attached to nanowires. At 500 sccm (**Figure 14(b)**), nanowires were grown extensively. At 700 sccm (**Figure 14(c)**), a film was grown at first and then a porous structure evolved on the initial film. At 1000 sccm (**Figure 14(d)**), a relatively dense film was grown.

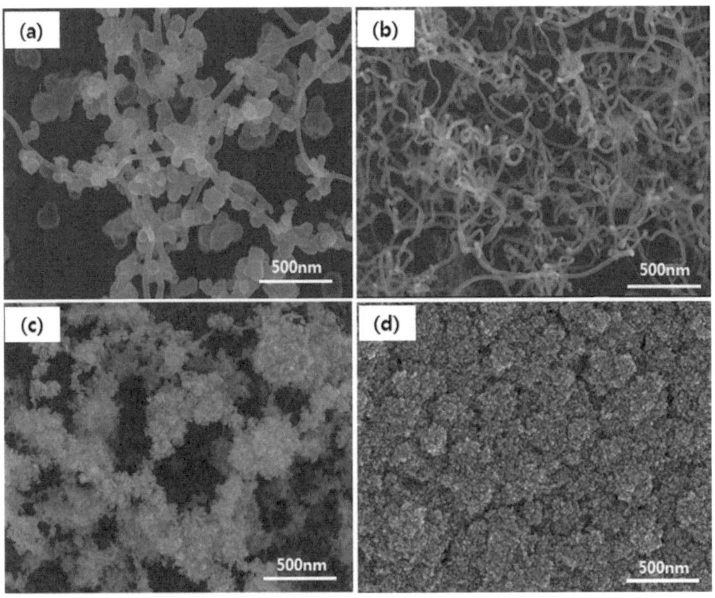

Figure 14. FESEM images of silicon films and nanostructures. The various microstructures of silicon evolved on a quartz substrate at N_2 gas flow rates of (a) 300, (b) 500, (c) 700, and (d) 1000 sccm. Reprinted with permission from [71]. Copyright 2013 Elsevier.

These results indicate that the carrier gas flow rate greatly affects microstructure evolution. To examine the effect of the carrier gas flow rate on the generation of CNPs in the gas phase, in-situ measurements of the CNPs were made using a DMA-FCE (Faraday cup electrometer) system under the same conditions as those of **Figure 14**. **Figure 15(a)** and **(b)** shows respectively the size distribution of negatively and positively charged silicon nanoparticles at various carrier flow rates. As the flow rate decreased, the number concentration of negative and positive CNPs decreased but the peak of the size distribution increased. In the case of positive CNPs in **Figure 15(b)**, the number concentration of ~40 nm nanoparticles at 1000 sccm of N_2

is roughly 100 times larger than that at 500 sccm. When the carrier flow rate is lower than 500 sccm, nanoparticles smaller than 20 nm were not detected.

Figure 14 shows that the amount of deposition on the substrate in terms of mass is decreased with decreasing flow rate of the carrier gas. This aspect would be related with **Figure 15**, which shows that the number concentration of CNPs generated in the gas phase is decreased with decreasing flow rate. The study indicated that the microstructure evolution of films, nanowires, and nanoparticles is closely related to the number concentration of CNPs, which is affected by the carrier gas flow rate.

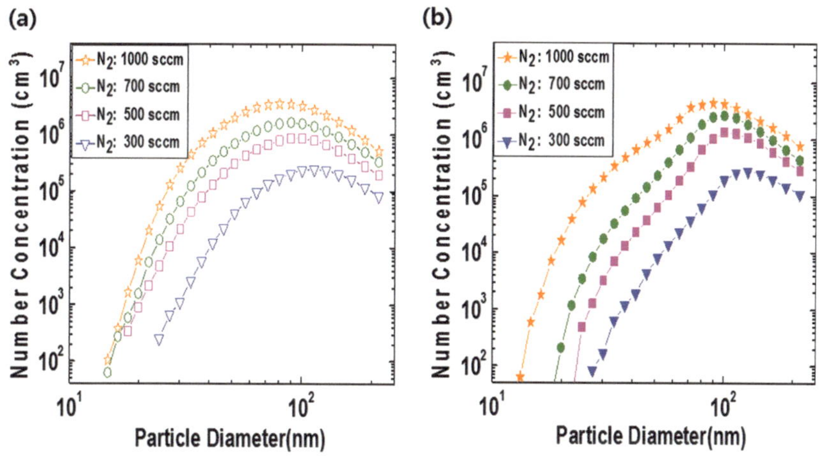

Figure 15. The number concentration and size distribution of (a) negative and (b) positive CNPs at various N_2 gas flow rates of 300, 500, 700, and 1000 sccm. Reprinted with permission from [71]. Copyright 2013 Elsevier.

Then why do the amount of the deposition and the number concentration of CNPs decrease with decreasing flow rate? What is found is that silicon tends to deposit on the tube wall near the inlet in the temperature zone of 600–800 °C, which is higher than the decomposition temperature of SiH_4. It is further observed that the amount of silicon deposited on the tube wall increases with decreasing flow rate of the carrier gas. This means that the loss of silicon on the tube wall before the hot zone at the center of the reactor is responsible for the decrease in amount of deposit on the substrate and the number concentration of charged gas-phase nuclei with decreasing flow rate.

With decreasing flow rate of the carrier gas, the mean residence time of the gas would be increased in the reactor although the total amount was decreased due to the loss on the tube wall before the hot zone. The slight increase of the particle diameter of maximum size with decreasing flow rate as shown in **Figure 15** might come from the increased mean residence time. Therefore, the deposition on the tube wall depends on the flow rate, which can be understood by the drag force imposed on CNPs by impinging molecules. When nanoparticles are the building blocks of crystal growth, the drag force should be considered in understanding the deposition behavior. Even under the condition where CNPs are generated in abundance, they might not be able to deposit because of the drag force if the flow rate in the direction parallel to the substrate is high enough. In relation to the drag force, Park et al. [72] studied the effect of the bias on the deposition behavior of ZnO nanostructures as follows.

4.4. Effect of bias on the deposition behavior of ZnO nanostructures in the CVD process

It was often found that even under the condition where CNPs are generated abundantly in the gas phase, the deposition of films or nanostructures does not occur on the substrate in specific locations of the CVD. Since non-deposition of films or nanostructures under the generation condition of CNPs would be caused by the drag force, the drag force is an important factor to be considered in the deposition of CNPs. Even when the deposition by CNPs was inhibited by the drag force, the existence of CNPs may be revealed by applying the electrical bias to the substrate. If the bias overcomes the drag force, CNPs that could not deposit on a substrate should be attracted toward the substrate and the deposition occurs. Park et al. [72] confirmed this possibility by examining the bias effect on the deposition behavior in the CVD reactor where charged ZnO nanoparticles are generated in the gas phase.

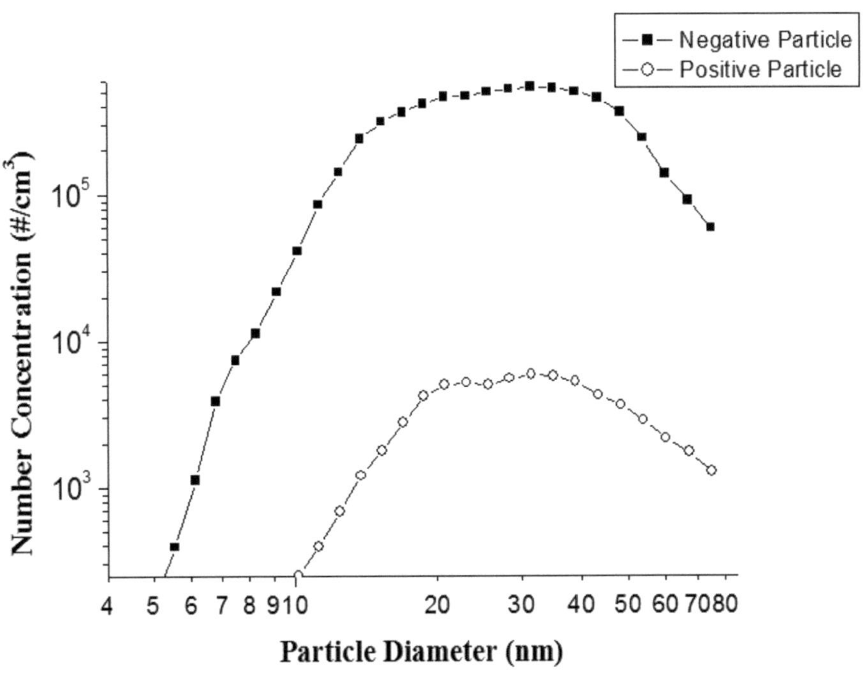

Figure 16. Size distribution of charged ZnO nanoparticles that were generated in the gas phase during the synthesis of ZnO nanostructures. Reprinted with permission from [72]. Copyright 2015 Elsevier.

Figure 16 shows the size distribution of ZnO CNPs measured by the DMA-FCE system. This result indicates that the number of CNPs in gas phase of the CVD system is large enough. However, they do not land on the substrate placed in 450 °C zone, which is rather near the outlet, being away from the center of the reactor. Then, why don't they land on the substrate placed in 450 °C zone as shown in **Figure 17(a)**?

http://dx.doi.org/10.5772/63926

One possible answer would be the drag force. The drag force is influenced by the relative velocity of particles in the medium [73]. Considering gas flow dynamics in the reactor, the hot gas that comes from the heated reactor zone would move upward over the colder gas in the unheated zone because of the density difference arising from the temperature gradient. This gas stream would exert a drag force on the nanoparticles that are generated in the hot zone so that they would move upward over the gas of 450 °C zone. To test this possibility, the electric bias was applied to the substrate holder.

The result is shown in **Figure 17**. As previously reported, charged tetrapod ZnOs as well as charged ZnO nanoparticles are generated in the gas phase during the synthesis of ZnO nanostructures [74]. This fact helps us understand the results of **Figure 17**. When +100 V was applied to the substrate holder of stainless steel, ZnO nanoparticles of ~10 nm were deposited not in a little amount on the silicon substrate as shown in **Figure 17(b)**. When the bias was increased to +300 V, slightly larger ZnO nanoparticles of ~40 nm were deposited in an appreciable amount as shown in **Figure 17(c)**. When the bias was increased even further to +600 V, tetrapod ZnOs were deposited as shown in **Figure 17(d)**. SEM observations at higher magnification revealed that tetrapod-ZnOs were deposited together with smaller ZnO nanoparticles. **Figure 17(b)–(d)** shows that as the bias voltage increased, the size of particles that could be deposited increased. This bias experiment shows that the charged ZnO nanoparticles are under the influence of the drag force and explains why CNPs measured in **Figure 16** did not land on the substrate in **Figure 17(a)**. A drag force exerting on particles by the gas flow is expressed as

$$F_D = 1/2 C_D A \rho v^2 \tag{7}$$

where F_D = drag force, C_D = drag coefficient, A = cross sectional area perpendicular to the flow, ρ = density of the medium, v = velocity of the body relative to the medium. This equation reveals that particles with a large area have a larger drag force and thereby larger resistance to move toward the substrate than those with a smaller area. Therefore, in order to deposit large particles, a high bias voltage is needed because of the high drag force arising from the upward gas stream caused by the temperature gradient. Non-deposition in **Figure 17(a)** can be understood on the basis of the drag force: under zero bias voltage, ZnO particles generated in the gas phase would just flow out with the carrier gas because the drag force inhibited the particles from moving toward the substrate. Also, under the bias of +100 V in **Figure 17(b)**, small particles of ~10 nm could be deposited, but large particles would flow out with the carrier gas. Because the drag force for small charged particles is weak, only a small electric force is needed to overcome the drag force of the small particles, which results in the deposition of small charged particles on the substrate. Under the bias of +300 V in **Figure 17(c)**, slightly larger particles of ~40 nm could be deposited with much larger particles being flowed out with the carrier gas. Under the bias of +600 V in **Figure 17(d)**, however, even larger tetrapod particles with a leg length of ~600 nm could be deposited on a substrate, indicating that the electric force is high enough to overcome the drag force. The gravity of these nanosized particles appeared to be much smaller than the electric force and could be ignored. These results indicated that the drag force becomes an important factor in deposition when nanoparticles are formed in the gas phase.

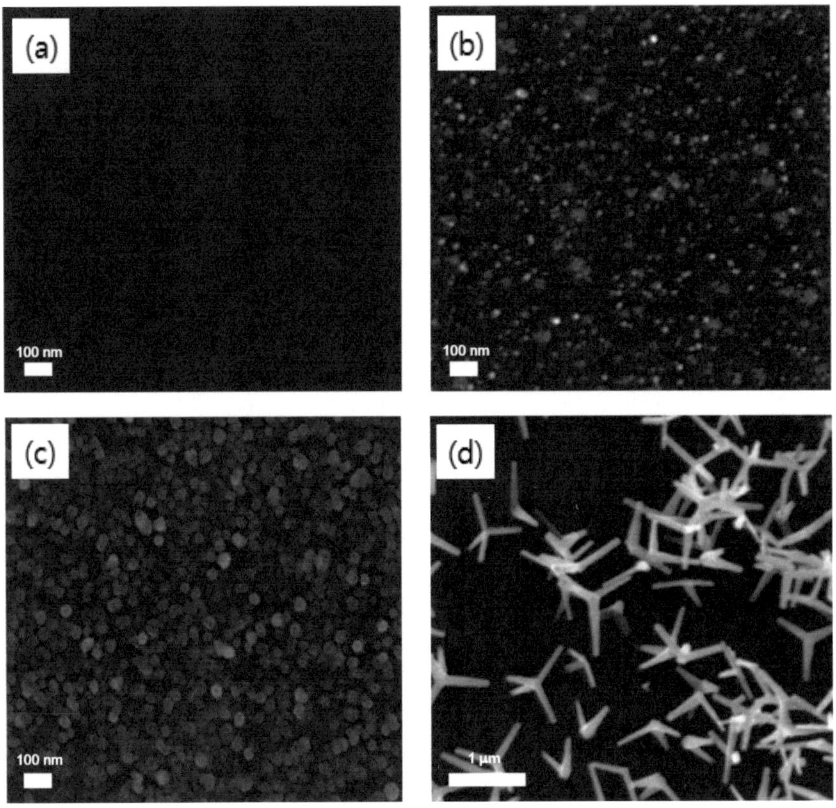

Figure 17. Deposition behavior of charged ZnO particles under the positive biases of (a) 0 V, (b) +100 V, (c) +300 V, and (d) +600 V. Reprinted with permission from [72]. Copyright 2015 Elsevier.

4.5. Effect of alternating bias on the deposition behavior of silicon in the CVD process

In the previous section, it was shown that even under the condition where CNPs are generated in abundance, they might not contribute to deposition because of the drag force. In this case, applying the DC bias would be effective to deposit those CNPs in the gas phase when CNPs are somewhat conducting as shown in **Figure 17**. When the CNPs are insulating, however, the positive bias would attract the negative CNPs and then negative charge would build up on the surface. This built-up charge would diminish the effect of the positive bias and then the positive bias would not be pronounced any more. In this case, the alternating bias would be effective because it prevents charge build up on the growing surface.

Youn at el. [75] studied the effect of alternating bias in silicon deposition on stainless substrate holder during thermal CVD. The bias frequency and voltage were varied. **Figure 18** shows the FESEM images of the surface morphology of silicon films deposited on the quartz substrates for 2 hours at reactor temperature of 900°C with gas flow rates of 5 sccm 10% SiH_4–90% He, 50 sccm H_2, and 1000 sccm N_2. **Figure 18(a)–(d)** is for the zero bias and the bias voltage of ±100 V at frequencies of 0.2, 0.5, and 5 Hz, respectively. **Figure 18(a)** shows the microstructure of the film which was deposited under the condition of the zero bias. The film has some roughness on the surface. This surface microstructure of the film is similar to that of films deposited by PECVD [76] and HWCVD [66]. **Figure 18(b)**, the microstructure of film which was deposited at a frequency of 0.2 Hz, shows the surface is covered by flake-like structures. At 0.5 Hz, the surface became smoother than that of 0.2 Hz and a flake-like structure disappeared as shown in **Figure 18(c)**. The surface of film which was deposited at 5 Hz had a very porous structure as shown in **Figure 18(d)**.

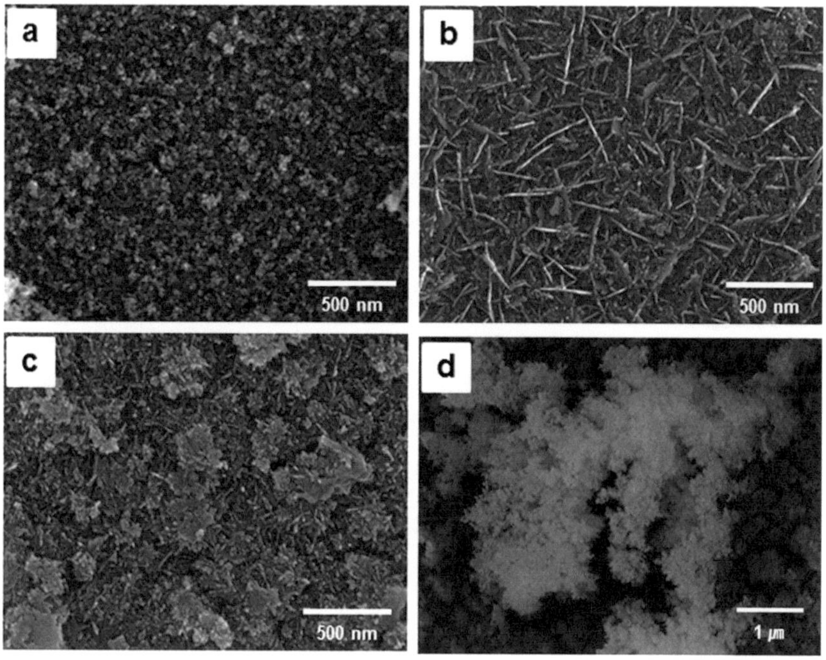

Figure 18. FESEM image of silicon deposited during 2 hour on a quartz substrate at (a) zero bias, (b) AC bias voltage of ±100 V with frequencies 0.2 Hz, (c) AC bias voltage of ±100 V with frequencies 0.5 Hz, and (d) AC bias voltage of ±100 V with frequencies 5 Hz. Reprinted with permission from [75]. Copyright 2012 Elsevier.

The effect of the bias voltage was also examined. The bias was applied to the stainless substrate holder at the biases of ±50 V, ±100 V, ±150 V, and ±200 V with a frequency of 1 Hz. **Figure 19** shows the FESEM images of the surface morphology of silicon films deposited under the same processing condition as that of **Figure 18**. Under the bias of ±50 V, the film microstructure tended to have a flake-like structure as shown in **Figure 19(a)**. When the bias was increased to ±100 V, the microstructure tended to have round nodules as shown in **Figure 19(b)**. When the bias was increased to ±150 V, the microstructure became porous as shown in **Figure 19(c)**. On applying ±200 V bias, the nanoparticles of ~100 nm were aggregated as a chain-like structure with their surface covered with numerous nano-rods as shown in **Figure 19(d)**. The nanopar-

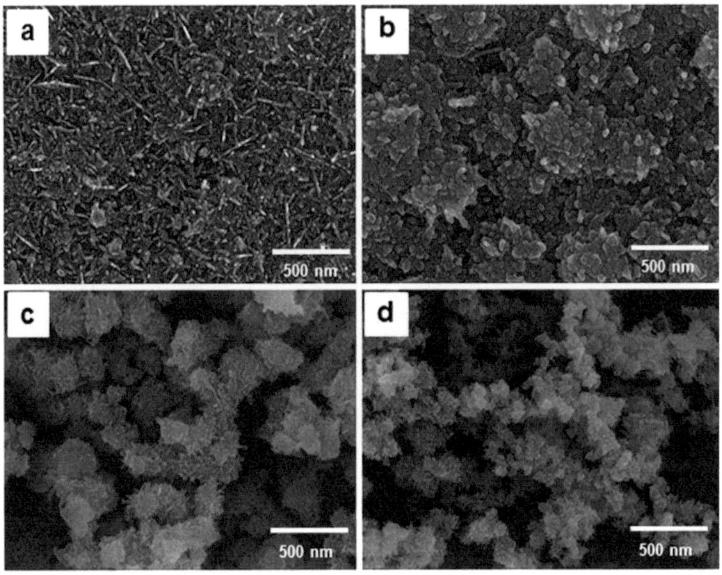

Figure 19. FESEM images of silicon films deposited on a quartz substrate at AC bias voltages of (a) ±50 V, (b) ±100 V, (c) ±150 V, and (d) ±200 V with the frequency of 1 Hz applied to the substrate holder. Reprinted with permission from [75]. Copyright 2012 Elsevier.

ticles of ~50 nm were deposited on the surface, being smaller than those deposited at ±150 V. The microstructure deposited at ±200 V was much more porous than that deposited at ±150 V.

These results imply that controlling the AC bias frequency or the magnitude of bias voltages may produce various microstructures. Besides, the AC bias technique can provide a promising method to produce highly porous microstructures, which are favorable for low dielectric materials.

Various microstructures can be synthesized by changing the bias frequency and the magnitude of bias voltages during deposition. By this technique, films of a layered structure with each layer of different microstructures can be synthesized. For example, the first layer is made dense, the second layer being porous, and the third layer being dense again. These results show clearly that the bias, either AC or DC, can be a new processing parameter in the CVD process where CNPs are generated.

4.6. Effect of size of CNPs on CVD process

The size of CNPs is also a critical factor in thin film growth by CVD. Therefore, information about the size distribution of CNPs is important in controlling the microstructure evolution of films. The size of nanoparticles increases normally as the concentration of reactant source gases increases. **Figure 20** shows the measured mass distributions of negative CNPs for four different methane concentrations during HWCVD [60]. During the mass distribution measurements, diamonds were deposited in situ on a Mo substrate placed near the orifice with a substrate temperature of 750 °C. The diamond films deposited at 1% and 1.5% CH_4, where nanoparticles of 200–300 carbon atoms were dominant as shown in **Figure 20**, showed good crystalline quality as shown in **Figure 21(a)**. For methane concentrations of 3% and 5%, where the nanoparticles in the gas phase contained more than 1000 carbon atoms and the size distribution became much broader in **Figure 20**, ball-like diamonds were deposited as shown in **Figure 21(b)**. This structure is often called a cauliflower structure because it looks similar to the vegetable cauliflower.

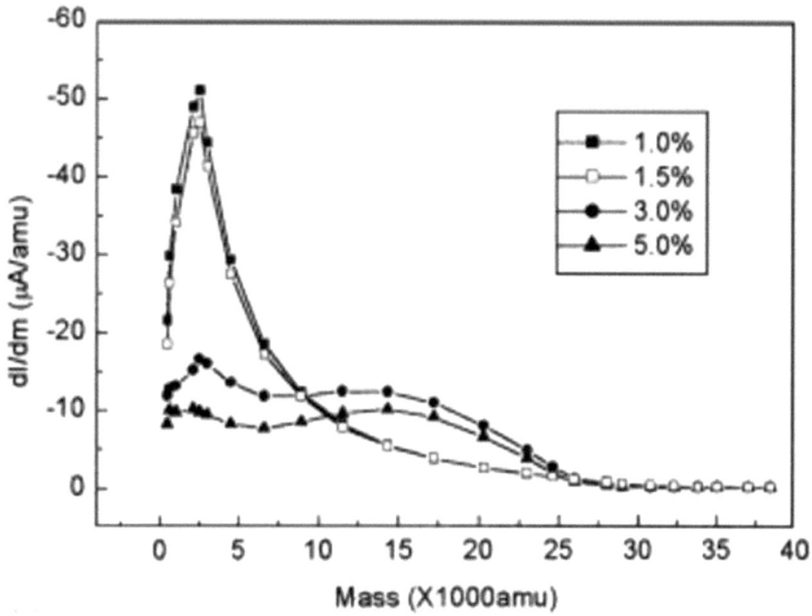

Figure 20. Measured mass distributions of negatively charged carbon clusters extracted from the hot filament reactor using gas mixtures of 1% CH_4-99% H_2, 1.5% CH_4-98.5% H_2, 3% CH_4-97% H_2, and 5% CH_4-95% H_2. Reprinted with permission from [60]. Copyright 2001 Elsevier.

To explain such an effect of size of CNPs on the microstructure evolution, weakening of bond strength by charge should be considered. In a soft charging condition such as thermal and hot wire CVD, which is in contrast with the hard charging condition in plasma CVD, most nanoparticles are expected to be singly charged. In this case, the charge effect on weakening of the bond strength would be diminished as the size increases. In other words, the smaller CNPs would be more liquid-like than the larger ones. Therefore, large CNPs generated under the high methane concentration in **Figure 20** would be less liquid-like so that CNPs frequently fail to undergo epitaxial recrystallization, producing ball-like or cauliflower diamond structures as shown in **Figure 21(b)**. On the other hand, small CNPs generated under the low methane concentration in **Figure 20** would be more liquid-like so that CNPs mostly undergo epitaxial recrystallization, producing high-quality diamonds with well-defined facets as shown in **Figure 21(a)**.

To confirm the effect of nanoparticle size on the deposition behavior, Yoshida and his colleagues [77–81] made extensive studies on the epitaxial growth of films with a building block of nanoparticles by the method called thermal plasma flash evaporation. They could deposit high-quality epitaxial $YBa_2Cu_3O_{7-x}$ films with a growth rate as high as 16 nm/s by this technique. They also could estimate the size of the nanoparticles to be about 0.3–10 nm using a micro-trench fabricated on a Si wafer [79]. They observed that small 1–2 nm nanoparticles made epitaxial spiral growth, medium size 3 nm nanoparticles produced epitaxial 2-dimensional particles, and large nanoparticles over 3 nm produced non-epitaxial island grains by scanning tunneling microscopy (STM) [80].

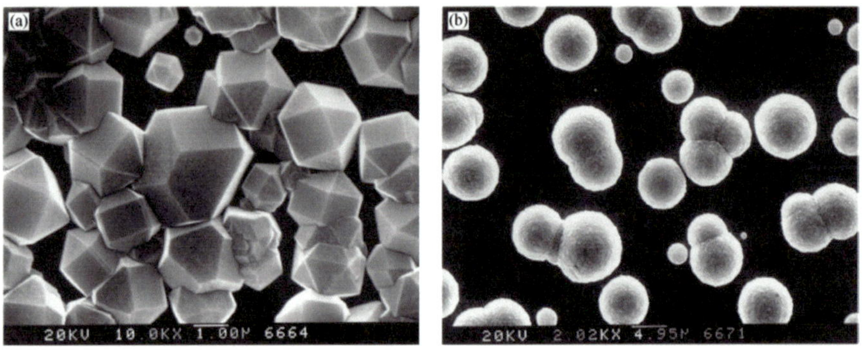

Figure 21. SEM images of diamond films deposited *in situ* during the measurement of the mass distribution of CNPs at 2100°C wire temperature and 800 Pa reactor pressure: (a) 1% CH_4-99%nH_2, (b) 3% CH_4-97%H_2. Reprinted with permission from [60]. Copyright 2001 Elsevier.

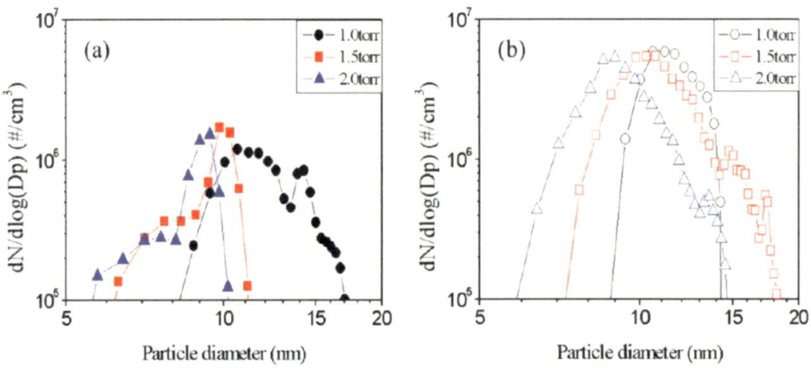

Figure 22. Size distribution of positively (a) and negatively (b) charged silicon nanoparticles measured by PBMS at a wire temperature of 1800°C with various process pressures during HWCVD. Reprinted with permission from [82]. Copyright 2014 Elsevier.

The reactor pressure also affects the size of CNPs. Yoo et al. [82] investigated the size distribution of CNPs generated in the gas phase during deposition of Si films by HWCVD through particle beam mass spectrometry (PBMS).

Figure 22 shows the size distribution of CNPs generated in the gas phase measured by PBMS in the HWCVD process with a 4% SiH_4–96% H_2 gas mixture and a wire temperature of 1800°C at 1.0, 1.5, and 2.0 Torr. The peak of the number concentration occurs at 9–12 nm at which the number concentrations of both positive and negative CNPs are 10^6–10^7 /cm^3. These nanoparticles were observed by TEM after being captured from the gas phase on a TEM grid membrane. As the pressure is increased from 0.3 to 2 Torr, not only the size and the number of captured nanoparticles are decreased, but also the rate of deposition is increased. An increase in the distance at which nanoparticles were captured from the hot wires under 1.5 Torr also reduced the size and number of nanoparticles.

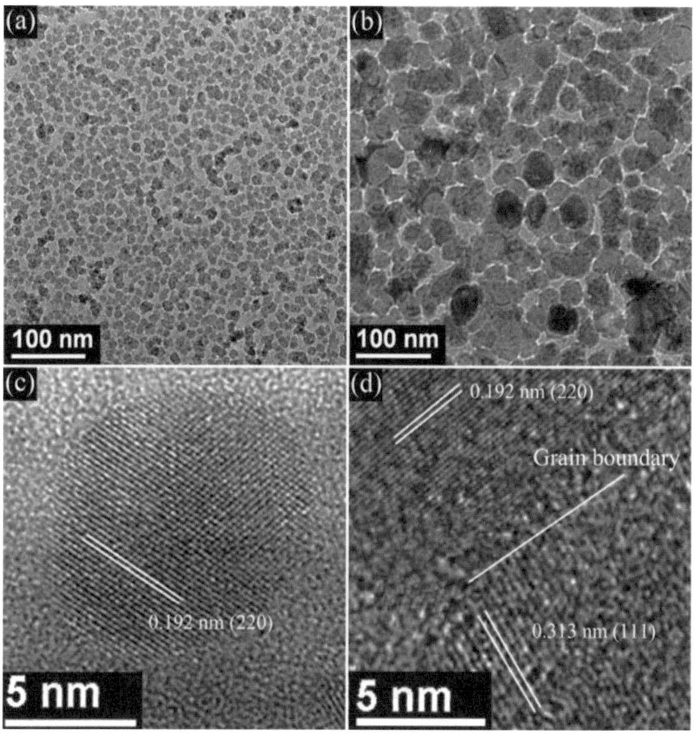

Figure 23. TEM images of initial silicon nanoparticles deposited for 20 sec after delay times of (a) 0 min and (b) 5 min and (c) and (d) showing respectively the HR images of (a) and (b). Reprinted with permission from [83]. Copyright 2012 Elsevier.

Chung et al. [83] show that the size of silicon nanoparticles generated in the gas phase during HWCVD tended to increase with increasing processing time and the size tended to be saturated above 15 min. This means that the steady state in terms of the size of CNPs is not reached in the reactor until 15 min. Therefore, in the initial stage of deposition before 15 min, small CNPs would deposit whereas in the later stage after 15 min, large CNPs would deposit. Therefore, if beginning of the deposition is delayed using the shutter above the substrate, the size of CNPs to be deposited can be controlled. Based on this observation, Chung et al. [83] used the concept of the delay time in deposition to control the microstructure. How the size of CNPs changes with the delay time can be examined by capturing CNPs on a TEM grid membrane with different delay times.

For example, **Figure 23(a)** and **(b)** shows the silicon nanoparticles captured on a Cu TEM grid membrane for 20 sec with delay times of 0 and 5 min, respectively. **Figure 23(c)** and **(d)** is the magnified view showing the lattice image of **Figure 23(a)** and **(b)**, respectively. The sizes of crystalline silicon nanoparticles in **Figure 23(a)** and **(b)** are ~10 and ~28 nm, respectively. The lattice image in **Figure 23(c)** shows a (220) plane, and that of **Figure 23(d)** consists of (220) and (111) planes, indicating that the silicon particle is a bicrystal.

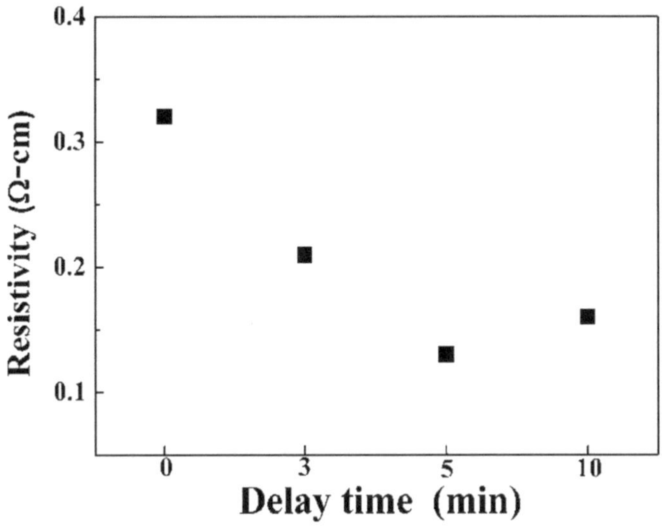

Figure 24. Resistivity of silicon films with the initial silicon layer deposited for 300 sec after delay times of 0, 3, 5, and 10 min. Reprinted with permission from [83]. Copyright 2012 Elsevier.

In order to study the effect of a different initial layer, the initial layer was deposited on the glass substrate for 20 sec after delay times of 0, 3, 5, and 10 min using the shutter. Then, these four samples were further deposited for 300 sec under the same processing conditions as that of **Figure 23**, and the resistivity of the deposited silicon films was measured. **Figure 24** shows how the resistivity of the silicon films varies with delays times. The thickness of all the films was ~170 nm. The resistivity of a silicon film deposited without a delay time (or a delay time of 0 min) is 0.32 Ω-cm. It should be noted that such low resistivity is due to HCl addition; silicon films deposited without HCl by PECVD or HWCVD have resistivity typically in the range of 5–50 Ω-cm [84, 85]. The resistivities of the silicon films prepared with delay times of 3, 5, and 10 min are 0.21, 0.13, and 0.16 Ω-cm, respectively. The resistivity tended to be saturated after a delay time of 5 min. Silicon films with a delay time show consistently lower resistivity than those without a delay time. The lowest resistivity of 0.13 Ω -cm of the film with a delay time of 5 min is 2.4 times lower than that without a delay time.

This resistivity change with delay times can be understood considering the property of CNPs. Without a delay time, the size of CNPs would be minimal and more liquid-like. These liquid-like CNPs, which would be accommodated to the substrate of an amorphous glass structure, would change to an amorphous structure, resulting in high resistivity. In contrast, with a delay time of 10 min, the size of CNPs would be maximal and less liquid-like. These CNPs would retain their crystalline silicon structure instead of changing to an amorphous structure, resulting in low resistivity.

4.7. Low-temperature deposition of crystalline silicon

There is a strong need for the low temperature synthesis of crystalline silicon films for electronics such as displays and solar cells [86–89]. The most popular method to synthesize microcrystalline films is to deposit amorphous phase silicon on a glass substrate than to crystallize it by excimer laser, thermal annealing, or metal-induced crystallization [90–92]. Since these methods increase the production cost, it is necessary to deposit the crystalline silicon film directly on a glass substrate.

From the view point of classical crystallization, silicon diffusivity below 600 °C is so low that amorphous silicon is expected on a glass substrate below 600 °C. Direct deposition of crystalline silicon at temperatures below 600 °C is not possible. Nonetheless, the direct deposition of microcrystalline silicon films on a glass substrate below 400 °C has been reported using HWCVD system [93–95] or PECVD system [96]. These results cannot be approached by the concept of classical crystallization. However, this low temperature deposition of crystalline silicon can be explained by non-classical crystallization. If crystalline silicon nanoparticles were formed in the gas phase in a high-temperature region near the hot wire and then incorporated into the films at low temperature, crystalline silicon could be deposited on a substrate at temperature below 600 °C.

If this explanation is correct, crystalline silicon nanoparticles should exist in the gas phase, which can be confirmed by capturing them on a TEM grid membrane. For this experimental confirmation, Lee et al. [84] designed the HWCVD, which consists of three chambers by two stage orifices. Each chamber was maintained in different pressures by two-stage differential pumping. The first chamber is the HWCVD reactor with the working pressure maintained at 27 Pa under a flowing gas mixture of SiH_4 and H_2, and the working pressure of the third chamber being maintained at 1.3×10^{-3} Pa during the process. The CNPs were generated in the first chamber and extracted through the first and the second orifices into the third chamber.

These CNPs were captured on the TEM grid located in the third chamber at room temperature. The wire temperature was 1560°C and the concentration of SiH_4 in the gas mixture of SiH_4-H_2 was varied as 20%, 15%, 10%, and 5%. The TEM image of the crystalline silicon CNPs captured on the TEM grid membrane is shown in **Figure 25** [84]. The CNPs which were captured on the TEM grid membrane had the crystallinity and the lattice fringes of CNPs were clearly revealed. The size of the crystalline nanoparticles decreased with decreasing SiH_4 concentration with 7–8 nm for 20% SiH_4 (**Figure 25(a)**), 6–7 nm for 15% SiH_4 (**Figure 25(b)**), 5–6 nm for 10% SiH_4 (**Figure 25(c)**), and 4–5 nm for 5% SiH_4 (**Figure 25(d)**).

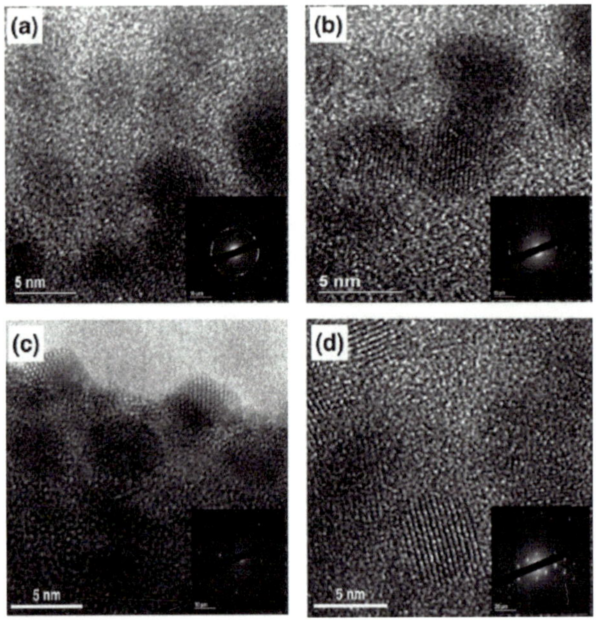

Figure 25. TEM images of silicon nanoparticles which were captured on the amorphous carbon membrane at room temperature by HWCVD with (a) 20%, (b) 15%, (c) 10% and (d) 5% flow rate of SiH_4 at a wire temperature of 1560°C. Reprinted with permission from [84]. Copyright 2008 Elsevier.

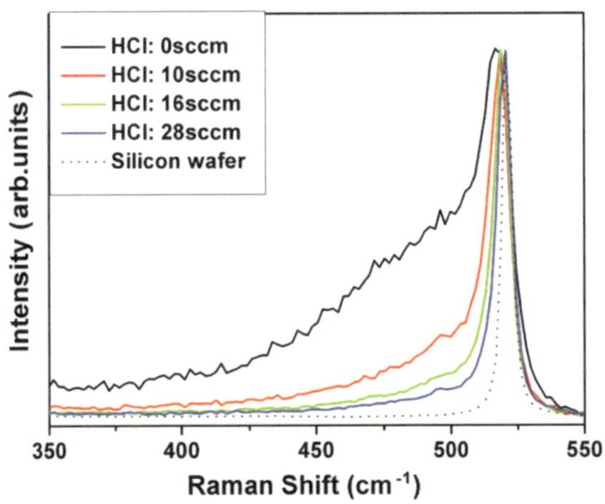

Figure 26. Raman spectra of the silicon films deposited at various HCl flow rates. The dotted line is for a single crystal silicon wafer as a reference. Reprinted with permission from [97]. Copyright 2009 Elsevier.

Classical crystallization cannot explain such formation of crystalline silicon nanoparticles at room temperature because the atomic diffusion rate is negligible. **Figure 25** strongly indicates that CNPs were formed in the gas phase of the first chamber during HWCVD and landed on the TEM grid in the third chamber.

http://dx.doi.org/10.5772/63926

There is an appreciable amount of amorphous silicon in the TEM images. This amorphous phase is expected to have been formed in the low-temperature region far from the hot wire. Therefore, to minimize the amorphous phase, low-temperature precipitation or nucleation of silicon should be inhibited. To inhibit the precipitation of silicon at low temperature, Chung et al. [97, 98] suggested HCl addition to SiH_4. Further, they confirmed that the crystallinity of silicon increased with increasing HCl addition as revealed in the Raman spectra of **Figure 26**. Since the Si-Cl-H system has retrograde solubility [99], if HCl is added to the Si-H system, the solubility of silicon in the gas phase increases with decreasing temperature, which means the equilibrium amount of silicon decreases with decreasing temperature.

This means that when gas-phase nucleation occurs, the driving force near the substrate temperature becomes for etching as in the case of the C-H system for diamond CVD. Therefore, HCl addition inhibits the precipitation of amorphous silicon at low temperature. Then, the crystalline silicon nanoparticles, which are precipitated at high temperature, will exclusively contribute to deposition without additional precipitation of amorphous silicon at low temperature, improving the crystallinity of the film.

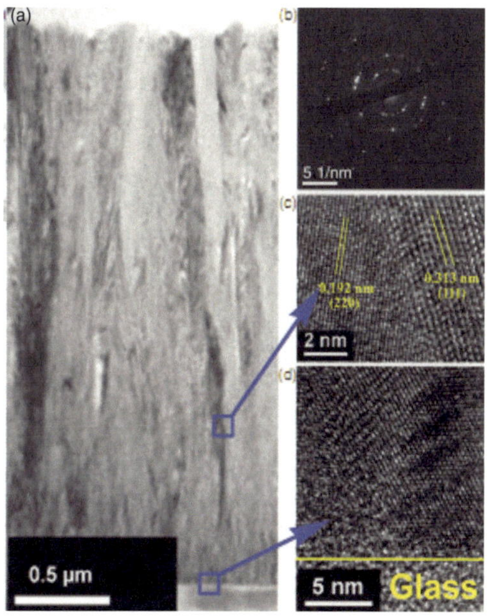

Figure 27. TEM images of the *n*-type crystalline silicon film with HCl (a) a bright-field image, and (b) image of the selective area diffraction pattern with (c) and (d) being high-resolution images on the regions indicated in (a). Reprinted with permission from [98]. Copyright 2011 Elsevier.

It is known that an amorphous silicon incubation layer, which is as thick as hundreds of nanometers, is formed on the glass substrate before crystalline silicon starts to form when the low temperature deposition of crystalline silicon is done by HWCVD or PECVD. The formation of such an amorphous incubation layer is a serious problem for thin-film transistor (TFT) applications because the thickness of the TFT should be typically less than ~200 nm.

Chung et al. [97, 98] could successfully minimize the formation of amorphous incubation layer by adding HCl as shown in **Figure 27**.

Figure 27 shows the cross section HRTEM images of the crystalline silicon film deposited at the flow rate ratios of $[HCl]/[SiH_4] = 7.5$ and $[PH_3]/[SiH_4]=0.028$. The whole deposition area of **Figure 27(a)** consists of crystalline silicon. More specifically, the high-resolution image at the middle of the film in **Figure 27(c)** reveals that the grain boundary is free from any amorphous phase between the two grains showing (111) and (220) planes. This aspect is highly in contrast with the previous result, where HCl was not added, that the crystalline silicon nanoparticles were embedded in the amorphous silicon phase. **Figure 27** showed that the amorphous incubation layer could be reduced markedly or almost removed by HCl addition. This is an example showing how new understanding of thin film growth by non-classical crystallization can be applied.

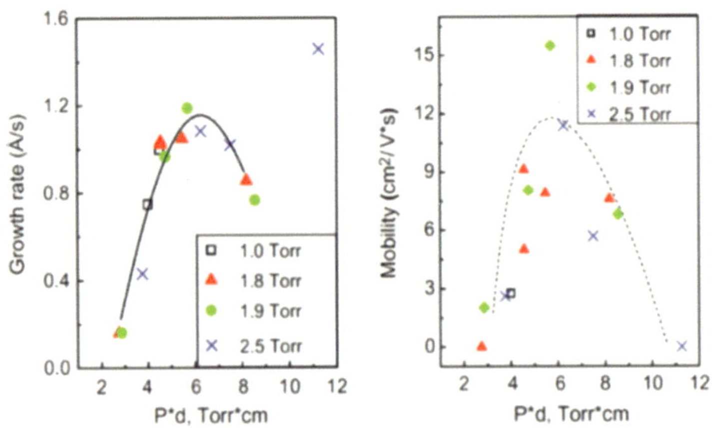

Figure 28. Growth rate and electron mobility from time-resolved microwave conductivity measurements of the total pressure and the distance of inter-electrode. Reprinted with permission from [70]. Copyright 2008 Elsevier.

As to the possibility of incorporation of particles generated in the gas phase during PECVD, Cabarrocas et al. [23, 24, 70, 96] suggested that the nanoparticles generated in the gas phase can be incorporated into films. Cabarrocas et al. [70] reported that the incorporation of crystalline nuclei generated in the gas phase increased the film deposition rate and electron mobility of films. This result is shown in **Figure 28**, where the x-axis is the product of the total pressure and inter-electrode distance. Both film growth rate and electron mobility are maximum when the process pressure is ~ 6 Torr.

They could deposit fully crystallized films, consisting of 68% small grain materials and 28% monocrystalline silicon with a 4% fraction of voids at substrate temperature of 200 °C, RF power at 20 W and gas flow rates at 1, 5, and 36 sccm respectively for SiF_4, H_2, and Ar. Considering the substrate temperature of 200 °C, the deposition of microcrystalline films also cannot be explained by classical crystallization. Therefore, the high crystallinity of films also would be attributed to the self-assembly of CNPs formed in the gas phase.

http://dx.doi.org/10.5772/63926

However, the temperature of the plasma is too low to form the crystalline silicon phase in the gas phase. This question would be explained by the role of charges produced in the plasma CVD, by which not only the kinetics of the precursor decomposition is enhanced, but also the deposition of crystalline films is made possible at such low temperatures. All these enhanced kinetics can be explained if the charge weakens the bond strength of molecules and nanoparticles.

5. Conclusion

Non-classical crystallization, where the growth units of deposition in CVD process are CNPs, was revealed by the rigorous thermodynamic analysis of the paradoxical experimental observation of simultaneous diamond deposition and graphite etching. This mechanism was further supported by the puzzling phenomenon of diamond deposition on a silicon substrate and porous skeletal graphitic soot deposition on an iron substrate. This new paradigm of crystal growth was also experimentally confirmed in many CVD processes, which means that non-classical crystallization is a general growth mechanism of thin films and nanostructures in CVD processes. Lots of fundamental and application studies should be made in this new paradigm of crystal growth.

Acknowledgements

This work was supported by the Global Frontier R&D Program (2013-073298) on Center for Hybrid Interface Materials (HIM) funded by the Ministry of Science, ICT & Future Planning and National Research Foundation of Korea(NRF) grant funded by the Ministry of Science, ICT & Future Planning(MSIP) (NO. NRF-2015R1A5A1037627).

Author details

Jae-soo Jung[1] and Nong-moon Hwang[2*]

*Address all correspondence to: nmhwang@snu.ac.kr

1 Department of Material Science and Engineering, Seoul National University, Gwanak-gu, Seoul, Republic of Korea

2 Department of Material Science and Engineering, Seoul National University and Research Institute of Advanced Materials (RIAM), Seoul National University, Gwanak-gu, Seoul, Republic of Korea

References

[1] Boufendi L, Plain A, Blondeau J P, Bouchoule A. Laure C Toogood M, Measurements of particle size kinetics from nanometer to micrometer scale in a low-pressure argon-silane radio-frequency discharge. Appl. Phys. Lett. 1992;60:169. DOI: 10.1063/1.106981

[2] Howling A A, Sansonnens L, Dorier J L, Hollenstein C. Negative hydrogenated silicon ion clusters as particle precursors in RF silane plasma deposition experiments J. Phys. D: Appl. Phys. 1993;26:1003. DOI: 10.1088/0022-3727/26/6/019

[3] Veprek S, Ambacher O, Rieger W, Schopper K, Veprek-Heijman M G J. Clusters in a silane glow discharge: mechanism of their formation and how to avoid them. Mater. Res. Soc. Symp. Proc. 1993;297:13. DOI: 10.1557/PROC-297-13

[4] Garscadden A, Ganguly B N, Haaland P D, Williams J. Overview of growth and behaviour of clusters and particles in plasmas. Plasma Sources Sci. Technol. 1994;3:239. DOI: 10.1088/0963-0252/3/3/001

[5] Stoffels E, Stoffels W W, Kroesen G M W, Hoog F J D. Dust formation and charging in an Ar/SiH_4 radio-frequency discharge. J. Vac. Sci. Technol. A. 1996;14:556. DOI: 10.1116/1.580144

[6] Hwang N M, Yoon D Y. Thermodynamic approach to the paradox of diamond formation with simultaneous graphite etching in the low pressure synthesis of diamond. J. Cryst. Growth. 1996;160:98. DOI: 10.1016/0022-0248(95)00549-8

[7] Hwang N M, Hahn J H, Yoon D Y. Charged cluster model in the low pressure synthesis of diamond. J. Cryst. Growth. 1996;162:55. DOI: 10.1016/0022-0248(95)00943-4

[8] Hwang N M, Hahn J H, Yoon D Y. Chemical potential of carbon in the low pressure synthesis of diamond. J. Cryst. Growth. 1996;160:87. DOI: 10.1016/0022-0248(95)00548-X

[9] Hwang N M, Kim D Y. Charged clusters in thin film growth. Int. Mater. Rev. 2004;49:171. DOI: 10.1179/095066004225021891

[10] Hwang N M, Lee D K. Charged nanoparticles in thin film and nanostructure growth by chemical vapour deposition. J. Phys. D: Appl. Phys. 2010;43:483001. Review Paper. DOI: 10.1088/0022-3727/43/48/483001

[11] Jeon J D, Park C J, Kim D Y, Hwang N M. Experimental confirmation of charged carbon clusters in the hot filament diamond reactor. J. Cryst. Growth. 2001;213:79. DOI: 10.1016/S0022-0248(00)00358-4

[12] Ahn H S, Park H M, Kim D Y, Hwang N M. Observation of carbon clusters of a few nanometers in the oxyacetylene diamond CVD process. J. Cryst. Growth. 2002;234:399. DOI: 10.1016/S0022-0248(01)01748-1

[13] Jeon I D, Gueroudji L, Kim D Y, Hwang N M. Temperature dependence of the deposition behavior of yttria-stabilized zirconia CVD films: approach by charged cluster

model. J. Korean Ceram. Soc. 2001;38:218. DOI: http://www.dbpia.co.kr/Journal/ArticleDetail/NODE00737510

[14] Kim J Y, Kim D Y, Hwang N M. Spontaneous generation of negatively charged clusters and their deposition as crystalline films during hot-wire silicon chemical vapor deposition. Pure Appl. Chem. 2006;78:1715. DOI: 10.1351/pac200678091715

[15] Lee J I, Hwang N M. Generation of negative-charge carriers in the gas phase and their contribution to the growth of carbon nanotubes during hot-filament chemical vapor deposition. Carbon. 2008;46:1588. DOI: 10.1016/j.carbon.2008.07.006

[16] Kim C S, Chung Y B, Youn W K, Hwang N M. Generation of charged nanoparticles during the synthesis of carbon nanotubes by chemical vapor deposition. Carbon. 2009;47(10):2511–2518. DOI: 10.1016/j.carbon.2009.05.005

[17] Kim C S, Chung Y B, Youn W K, Hwang N M. Generation of charged nanoparticles during synthesis of ZnO nanowires by carbothermal reduction. Aerosol Sci. Technol. 2009;43(2):120–125. DOI: 10.1080/02786820802499068

[18] Kim C S, Kwak I J, Choi K J, Park J G, Hwang N M. Generation of charged nanoparticles during the synthesis of silicon nanowires by chemical vapor deposition. J. Phys. Chem. C. 2010;114:3390. DOI: 10.1021/jp910242a

[19] Bennema P, Gilmer G H. 1973 Crystal Growth: An Introduction, ed. P Hartman. Wiley; 2010. 1276 p. DOI: 10.1002/bbpc.19740781138

[20] van der Eerden J P. Handbook of Crystal Growth, vol. 1. In: A Fundamentals 'Thermodynamics and Kinetics'. Amsterdam: North-Holland. DOI: 10.1107/S010876739400512X

[21] McMurry P H. Investigation of particle generation during the low-pressure chemical vapor deposition of borophosphosilicate glass films. J. Electrochem. Soc. 1998;145(6): 2051–2057. DOI: 10.1149/1.1838596

[22] Matsumoto K. Enhancement of critical current density of YBCO films by introduction of artificial pinning centers due to the distributed nano-scaled Y_2O_3 islands on substrates. Physica C. 2004;412–414:1267–1271. DOI: 10.1016/j.physc.2004.01.157

[23] Roca i Cabarrocas P. Plasma enhanced chemical vapor deposition of amorphous, polymorphous and microcrystalline silicon films. J. Non-Cryst. Solids. 2000;266–269:31–37. DOI: 10.1016/S0022-3093(99)00714-0

[24] Roca i Cabarrocas P. Plasma enhanced chemical vapor deposition of silicon thin films for large area electronics. Curr. Opin. Solid State Mater. Sci. 2002;6:439–444. DOI: 10.1016/S1359-0286(02)00112-2

[25] Vladimirov S V, Ostrikov K. Dynamic self-organization phenomena in complex ionized gas systems: new paradigms and technological aspects. Phys. Rep. 2004;393:175–380. DOI: 10.1016/j.physrep.2003.12.003

[26] Nunomura S, Kita M, Koga K, Shiratani M, Watanabe Y. Fabrication of nanoparticle composite porous films having ultralow dielectric constant. Japan. J. Appl. Phys. 2005;44:50–52. DOI: 10.1143/JJAP.44.L1509

[27] Glasner A, Kenat J. The crystallization of KCl from aqueous solutions in the presence of lead ions. I. A calorimetric study. J. Cryst. Growth. 1968;2:119–127. DOI: 10.1016/0022-0248(68)90073-0

[28] Glasner A, Skurnik S. A new mechanism for the crystallization and growth of ionic crystals, with special reference to KCl in the presence of Pb^{2+} ions. Israel J. Chem. 1968;6:69–72. DOI: 10.1002/ijch.196800011

[29] Glasner A, Tassa M. The thermal effects of nucleation and crystallization of KBr and KCl solutions. II. The heat of nucleation and the supersaturated solution. Israel J. Chem. 1974;12:799–816. DOI: 10.1002/ijch.197400069

[30] Glasner A, Tassa M. The thermal effects of nucleation and crystallization of KBr and KCL solutions. III. The heat of crystallization and the co-precipitation of lead ions. Israel J. Chem. 1974;12:817–826. DOI: 10.1002/ijch.197400070

[31] Sunagawa I. Morphology of minerals 'Morphology of Crystals'. Tokyo: Terra Sci. 26 1987:1;511–587.

[32] Sunagawa I. Growth and morphology of diamond crystals under stable and metastable conditions. J. Cryst. Growth. 1990;99:1156–1161. DOI: 10.1016/S0022-0248(08)80100-5

[33] Botsaris G D, Reid R C. Comments on the letter by Glasner and Skurnik entitled 'Growth of potassium chloride crystals from aqueous solutions. I. the effect of lead chloride'. J. Chem. Phys. 1967;47:3689. DOI: 10.1063/1.1712456

[34] Meldrum F C, Cölfen H. Controlling mineral morphologies and structures in biological and synthetic systems. Source of the Document Chemical Reviews. 2008;108:4332–4432. DOI: 10.1021/cr8002856

[35] Niederberger M, Cölfen H. Oriented attachment and mesocrystals: non-classical crystallization mechanisms based on nanoparticle assembly. Phys Chem Chem Phys. 2006;8:3271–3287. DOI: 10.1039/B604589H

[36] Cölfen H, Antonietti M. Mesocrystals: inorganic superstructures made by highly parallel crystallization and controlled alignment. Angew. Chem. – Int. Ed. 2005;44(35): 5576–5591. DOI: 10.1002/anie.200500496

[37] Liao H G, Cui L, Whitelam S, Zheng H. Real-time imaging of Pt_3Fe nanorod growth in solution. Science. 2012;336(6084):1011–1014. DOI: 10.1126/science.1219185

[38] Yuk J M, Park J, Ercius P, Kim K, Hellebusch D J. Crommie M F, Lee J Y, Zettl. High-resolution EM of colloidal nanocrystal growth using graphene liquid cells A. Science. 2012;336(6077):61–64. DOI: 10.1126/science.1217654

[39] Zheng H, Smith R K, Jun Y W, Kisielowski C, Dahmen U. Observation of single colloidal platinum nanocrystal growth trajectories. Science. 2009;324(5932):1309–1312. DOI: 10.1126/science.1172104

[40] Fang Z, Wang H. Densification and grain growth during sintering of nanosized particles. Angew. Chem. 2008;53:326–352. DOI: 10.1179/174328008X353538

[41] Clare B W, Talukder G, Jennings P J, Cornish J C L, Hefter G T. Effect of charge on bond strength in hydrogenated amorphous silicon. J. Comp. Chem. 1994;15(6):644–652. DOI: 10.1002/jcc.540150608

[42] Utke I, Hoffmann P, Melngailis J. Gas-assisted focused electron beam and ion beam processing and fabrication. J. Vac. Sci. Technol. B. 2008;26(4):1197. DOI: 10.1116/1.2955728

[43] Randolph S, Toth M, Cullen J, Chandler C, Lobo C. Kinetics of gas mediated electron beam induced etching. Appl. Phys. Lett. 2011;99(21):213103. DOI: 10.1063/1.3662928

[44] Lobo C J, Martin A, Phillips M R, Toth M. Electron beam induced chemical dry etching and imaging in gaseous NH_3 environments. Nanotechnology. 2012;23(37):375302. DOI: 10.1088/0957-4484/23/37/375302/meth

[45] Zheng H, Liu Y, Mao SX, Wang J, Huang J Y. Beam-assisted large elongation of in situ formed Li_2O nanowires. Scientif. Rep. 2012;2:542. DOI: 10.1038/srep00542

[46] Shan Z, Adesso G, Cabot A, Sherburne M, Asif S S, Warren O, Chrzan D, Minor A. Ultrahigh stress and strain in hierarchically structured hollow nanoparticles. Nat. Mater. 2008;7(12):947–952. DOI: 10.1038/nmat2295

[47] Shan Z, Li J, Cheng Y, Minor A, Asif S S, Warren O, Ma E. Plastic flow and failure resistance of metallic glass: insight from in situ compression of nanopillars. Phys. Rev. B. 2008;77(15):155419. DOI: 10.1103/PhysRevB.77.155419

[48] Derjaguin B V, Fedoseev D B. The Growth of Diamond and Graphite from the Gas Phase. Moscow: Nauka; 1989. pp. 131–248 DOI: 10.1016/0257-8972(89)90129-1

[49] Spitsyn B V, Bouilov L L, Derjaguin B V. Vapor growth of diamond on diamond and other surfaces. J. Cryst. Growth. 1981;52(1):219–226. DOI: 10.1016/0022-0248(81)90197-4

[50] Matsumoto S, Sato Y, Tsutsumi M, Setaka N. Growth of diamond particles from methane-hydrogen gas. J. Mater. Sci. 1982;17(11):3106–3112. DOI:http://dx.doi.org/10.1143/JJAP.21.L183

[51] Spear K E. Diamond—ceramic coating of the future. J. Am. Ceram. Soc. 1989;72(2):171–191. DOI: 10.1111/j.1151-2916.1989.tb06099.x

[52] Angus J C, Hayman C C. Low-pressure, metastable growth of diamond and 'diamond like' phases. Science. 1988; 241(4868):913-921. DOI : http://search.proquest.com/docview/213537217?accountid=6802

[53] Kamo M, Sato Y, Matsumoto S, Setaka N. Diamond synthesis from gas phase in microwave plasma. J. Cryst. Growth. 1983;62(3):642–644. DOI: 10.1016/0022-0248(83)90411-6

[54] Haubner R, Lux B. Diamond growth by hot-filament chemical vapor deposition: state of the art. Diamond Relat. Mater. 1993;2(9):1277–1294. DOI: 10.1016/0925-9635(93)90008-P

[55] Belokurov B V. Isotropic surface energies of diamond. Zh. Fiz. Khim. 1960;34:440

[56] Fedosayev D V, Deryagin B V, Varasavskaja I G. The growth of diamond and graphite from the gas phase. Surf. Coat. Technol. 1989;38(1–2):131–248. DOI: 10.1016/0257-8972(89)90129-1

[57] Choi K, Kang S J L, Jang H M, Hwang N M. Nucleation behavior in the presence of charge in the CVD diamond process. J. Cryst. Growth. 1997;172(3–4):416–425. DOI: 10.1016/S0022-0248(96)00759-2

[58] Hwang N M, Bahng G W, Yoon D N. Thermodynamics and kinetics for nucleation of diamond in the chemical vapor deposition process. Diamond Relat. Mater. 1992;1(2–4):191–194. DOI: 10.1016/0925-9635(92)90023-H

[59] Gaskell D R. Introduction to the Thermodynamics of Materials. Washington, DC: Taylor and Francis; 1995. DOI: 10.1016/j.msea.2008.06.016

[60] Jeon J D, Park C J, Kim D Y, Hwang N M. Effect of methane concentration on size of charged clusters in the hot filament diamond CVD process. J. Cryst. Growth. 2001;223:6. DOI: 10.1016/S0022-0248(00)00358-4

[61] Huh J M, Yoon D Y, Kim D Y, Hwang N M. Effect of substrate materials in the low-pressure synthesis of diamond: approach by theory of charged clusters. Z. Metallk. 2005;96(3):225–232. DOI: 10.3139/146.101024

[62] Yang Q, Tang Y, Yang S L, Li Y S, Hirose A. Simultaneous growth of diamond thin films and carbon nanotubes at temperatures ◎550 °C. Carbon. 2008;46(4):589–595. DOI: 10.1016/j.carbon.2008.01.005

[63] Lee J I, Hwang N M. Generation of negative-charge carriers in the gas phase and their contribution to the growth of carbon nanotubes during hot-filament chemical vapor deposition. Carbon. 2008;46(12):1588–1592. DOI: 10.1016/j.carbon.2008.07.006

[64] Youn W K, Lee S S, Lee J Y, Kim C S, Hwang N M, Iijima S. Comparison of the deposition behavior of charged silicon nanoparticles between floating and grounded substrates. J. Phys. Chem. C. 2014;118(22):11946–11953. DOI: 10.1021/jp5001144

[65] Kim C S, Youn W K, Hwang N M. Generation of charged nanoparticles and their deposition during the synthesis of silicon thin films by chemical vapor deposition. J. Appl. Phys. 2010;108(1):014313. DOI: 10.1063/1.3452352

[66] Chung Y B, Lee D K, Lim J S, Hwang N M. Reduction of amorphous incubation layer by HCl addition during deposition of microcrystalline silicon by hot-wire chemical

vapor deposition. Sol. Energy Mater. Sol. Cells. 2011;95(1):211–214. DOI: 10.1016/j.solmat.2010.04.021

[67] Kumomi H, Yonehara T, Nishigaki Y, Sato N. Selective nucleation based epitaxy (sentaxy): investigation of initial nucleation stages. Appl. Surf. Sci. 1990;41–42:638–642. DOI: 10.1016/0169-4332(89)90135-9

[68] Wagner R S, Ellis W. C. Vapor-liquid-solid mechanism of single crystal growth. Appl. Phys. 1964;4:89–90. DOI:http://dx.doi.org/10.1063/1.1753975

[69] Zhang R Q, Lifshitz Y, Lee S T. Oxide-assisted growth of semiconducting nanowires. Adv. Mater. 2003;15:635–640. DOI: 10.1002/adma.200301641

[70] Roca i Cabarrocas P, Djeridane Y, Bui V D, Bonnassieux Y, Abramov A. Critical issues in plasma deposition of microcrystalline silicon for thin film transistors. Solid-State Electron. 2008;52(3):422–426. DOI: 10.1016/j.sse.2007.10.028

[71] Youn W K, Kim C S, Hwang N M. Effect of the carrier gas flow rate on the microstructure evolution and the generation of the charged nanoparticles during silicon chemical vapor deposition. J. Nanosci. Nanotechnol. 2013;13(10):7127–7130. DOI: 10.1166/jnn.2013.7669

[72] Park S H, Park J W, Yang S M, Kim K H, Hwang N M. Effect of electric bias on the deposition behavior of ZnO nanostructures in the chemical vapor deposition process. J. Phys. Chem. C. 2015:119:25047–25052. DOI: 10.1021/acs.jpcc.5b06796

[73] Davies C N (Ed.). Aerosol Science. New York: Academic; 1966. 1548 p. DOI: 10.1126/science.157.3796.1548-a

[74] Park S H, Park J W, Kim C S, Hwang N M. Formation of tetrapod-shaped nanowires in the gas phase during the synthesis of ZnO nanostructures by carbothermal reduction. J. Nanosci. Nanotechnol. 2013;13:7198–7201. DOI: 10.1166/jnn.2013.7697

[75] Youn W K, Kim C S, Lee J Y, Lee S S, Hwang N M. Generation of charged nanoparticles and their deposition behavior under alternating electric bias during chemical vapor deposition of silicon. J. Phys. Chem. C. 2012;116:25157–25163. DOI: 10.1021/jp310705p

[76] Cheng Q, Xu S. Rapid, low-temperature synthesis of nc-Si in high-density, non-equilibrium plasmas: enabling nanocrystallinity at very low hydrogen dilution. Ostrikov, K. J. Mater. Chem. 2009;19:5134–5140. DOI: 10.1039/B904227J

[77] Hayasaki K, Takamura Y, Yamaguchi N, Terashima K, Yoshida T. Scanning tunneling microscopy of epitaxial $YBa_2Cu_3O_{7-x}$ films prepared by thermal plasma flash evaporation method. J. Appl. Phys. 1997;81:1222. DOI: 10.1063/1.364143

[78] Takamura Y, Hayasaki K, Terashima K, Yoshida T. Cluster size measurement using microtrench in a thermal plasma flash evaporation process. J. Vac. Sci. Technol. B. 1997;15:558. DOI: 10.1116/1.589292

[79] Terashima K, Yamaguchi N, Hattori T, Takamura Y, Yoshida T. High rate deposition of thick epitaxial films by thermal plasma flash evaporation. Pure Appl. Chem. 1998;70:1193–1197. DOI: 10.1351/pac199870061193

[80] Takamura Y, Hayasaki K, Terashima K, Yoshida. High-rate deposition of $YBa_2Cu_3O_{7-x}$ films by hot cluster epitaxy. J. Appl. Phys. 1998;84:5084. DOI: 10.1063/1.368757

[81] Yamaguchi N, Sasajima Y, Terashima K, Yoshida T. Molecular dynamics study of cluster deposition in thermal plasma flash evaporation. Thin Solid Films. 1999;345:34–37. DPI:10.1016/S0040-6090(99)00074-7

[82] Yoo S W, Hong J S, Lee S S, Kim C S, Kim T, Hwang N M. Nonclassical crystallization in low-temperature deposition of crystalline silicon by hot-wire chemical vapor deposition. Cryst. Growth Des. 2014;14(12):6239–6247. DOI: 10.1021/cg5008582

[83] Chung Y B, Lee S H, Bae S H, Park H K, Jung J S, Hwang N M. Effect of the initial structure on the electrical property of crystalline silicon films deposited on glass by hot-wire chemical vapor deposition. J. Nanosci. Nanotechnol. 2012;12:5947–5951. DOI: 10.1166/jnn.2012.6415

[84] Lee S S, Ko M S, Kim C S, Hwang N M. Gas phase nucleation of crystalline silicon and their role in low-temperature deposition of microcrystalline films during hot-wire chemical vapor deposition. J. Cryst. Growth. 2008;310;3659–3662. DOI: 10.1016/j.jcrysgro.2008.05.009

[85] Krishnan A T, Bae S, Fonash S J. Low temperature microcrystalline silicon thin film resistors on glass substrates. Solid State Electron. 2000;44:1163–1168. DOI: 10.1016/S0038-1101(00)00057-5

[86] Persidis E, Baur H, Pieralisi F, Schalberger P, Fruehauf N. Area laser crystallized LTPS TFTs with implanted contacts for active matrix OLED displays. Solid State Electron. 2008;52:455–461. DOI: 10.1016/j.sse.2007.10.014

[87] Shah A, Torres P, Tscharner R, Wyrsch N, Keppner H. Photovoltaic technology: the case for thin-film solar cells. Science. 1999;285:692–698. DOI: 10.1126/science.285.5428.692

[88] Uchikoga, Ibaraki N. Low temperature poly-Si TFT-LCD by excimer laser anneal. Thin Solid Films. 2001;383:19–24. DOI: 10.1016/S0040-6090(00)01644-8

[89] Kahn H, Tayebi N, Ballarini R, Mullen R, Heuer A. Sens. Fracture toughness of Polysilicon MEMS devices. Actuators A – Phys. 2000;82:274–280. DOI: 10.1016/S0924-4247(99)00366-0

[90] Shimoda T, Matsuki Y, Furusawa M, Aoki T, Yudasaka I, Tanaka H, Iwasawa H, Wang D, Miyasaka M, Takeuchi Y. Solution-processed silicon films and transistors. Nature. 2006;440:783–786. DOI: 10.1038/nature04613

[91] Jang J, Oh J Y, Kim S K, Choi Y J, Yoon S Y, Kim C O. Electric-field-enhanced crystallization of amorphous silicon. Nature. 1998;395:481–483. DOI: 10.1038/26711

[92] Morimoto R, Izumi A, Masuda A, Matsumura H. Low-resistivity phosphorus-doped polycrystalline silicon thin films formed by catalytic chemical vapor deposition and successive rapid thermal annealing. Jpn. J. Appl. Phys. 2002;41:501–506. DOI: 10.1143/JJAP.41.501

[93] Matsumura H. Summary of research in NEDO Cat-CVD project in Japan. Thin Solid Films. 2001;395:1–11. DOI: 10.1016/S0040-6090(01)01198-1

[94] Mahan A H. Status of Cat-CVD (Hot Wire CVD) research in the United States. Thin Solid Films. 2001;395:12–16. DOI: 10.1016/S0040-6090(01)01199-3

[95] Schropp R E I. Status of Cat-CVD (Hot-Wire CVD) research in Europe. Thin Solid Films. 2001;395:17–24. DOI: 10.1016/S0040-6090(01)01200-7

[96] Roca i Cabarrocas P, Fontcuberta i Morral A, Kalache B, Kasouit S. Microcrystalline silicon thin-films grown by plasma enhanced chemical vapour deposition -growth mechanisms and grain size control. Solid State Phenom. 2003;93:257–268. DOI: 10.4028/www.scientific.net/SSP.93.257.

[97] Chung Y B, Lee D K, Kim C S, Hwang N M. Effect of HCl addition on the crystalline fraction in silicon thin films prepared by hot-wire chemical vapor deposition. Vacuum. 2009;83:1431–1434. DOI: 10.1016/j.vacuum.2009.05.004

[98] Chung Y B, Park H K, Lee S H, Song J H, Hwang N M. N-type crystalline silicon films free of amorphous silicon deposited on glass by HCl addition using hot wire chemical vapour deposition. J. Nanosci. Nanotechnol. 2011;11:8242–8245. DOI: 10.1166/jnn.2011.5040

[99] Hwang N M. Deposition and simultaneous etching of Si in the chemical vapor deposition (CVD) process: approach by the charged cluster model. J. Cryst. Growth. 1999;205:59–63. DOI: 10.1016/S0022-0248(99)00247-X

Chapter 3

MOCVD Grown HgCdTe Heterostructures

Pawel Madejczyk, Waldemar Gawron,
Artur Keblowski and Adam Piotrowski

Additional information is available at the end of the chapter

http://dx.doi.org/10.5772/62952

Abstract

HgCdTe heterostructures are widely applied for IR (infrared) detector constructing. Donor- and acceptor-doping researches in (100) and (111) oriented HgCdTe layers grown by MOCVD have been studied. Fully doped HgCdTe heterostructures with acceptor concentration range between 10^{14} and 5×10^{17} cm^{-3} and donor concentration range between 10^{14} and 1×10^{18} cm^{-3} and without post-grown annealing have been reported. The electrical and chemical characterizations of HgCdTe structures grown at 360°C on GaAs substrates using DIPTe have been described. Infrared photodiodes with different composition x were constructing on the basis of obtained heterostructures enabling signal detection of any wavelength from 1 μm to above 20 μm covering SWIR, MWIR and LWIR spectral ranges. Presented experimental results show that MOCVD technology enables to grow HgCdTe structures dedicated for HOT devices.

Keywords: HgCdTe, MOCVD, Doping, Heterostructure, infrared detector

1. Introduction

Mercury cadmium telluride $Hg_{1-x}Cd_xTe$ or MCT is a pseudobinary alloy whose development was commenced by Lawson and coworkers [1]. HgCdTe ternary compound is an excellent material for infrared detectors construction. Its position is conditioned by following characteristics:

- tunable energy gap allowing to cover the 1–30 μm wavelength range,
- advantageous intrinsic recombination mechanisms that direct to higher operating temperature (HOT),

- high optical coefficients enabling high quantum efficiency.

Above-mentioned properties are simply consequence of the energy band structure of this zinc-blende semiconductor. The special advantages of HgCdTe are ability to obtain wide range of carrier concentrations, low dielectric constant, and high mobility of current carriers. The very small change of lattice constant with composition enables to grow high-quality layered and graded gap structures [2]. Thus, HgCdTe can be used for detectors operating at various modes [photodiode, photoconductor, or metal-insulator-semiconductor (MIS) detector].

This chapter reviews work from literature and some unpublished work from the authors' own laboratory that has been carried out into determining suitable extrinsic dopant species and sources for use in MOCVD growth of MCT heterostructures. The authors especially would like to present MOCVD technology with wide range of composition and donor/acceptor doping and without post-grown annealing as an excellent tool for (HOT) HgCdTe infrared photodetectors construction.

2. MOCVD grown MCT

MOCVD growth of MCT was firstly demonstrated in 1981 by Irvine and Mullin [3]. The reason for the delayed start in the growth of the mercury alloys in comparison to III–V compounds growth has been linked to the high mercury vapour pressures which needed some unconventional modifications to the MOCVD systems. MOCVD growth of MCT is determined by high vapour pressures of mercury that are necessarily to keep equilibrium conditions over the growing layer. This results from the instability of HgTe bonds in comparison with CdTe bonds and requires much lower growth temperatures than are typical for more stable compounds [4].

2.1. Thermodynamic considerations

The mercury chalcogenides are characterized by weak bonds of mercury what causes a higher equilibrium vapour pressure. If the evaporating species are Hg and Te_2, then the solid–vapour equilibrium is given by the following expression:

$$K = p_{Hg}\sqrt{p_{Te_2}} \tag{2.1}$$

where p_{Hg} is the equilibrium vapour pressure of mercury, p_{Te_2} is the equilibrium vapour pressure of Te_2, and K is an equilibrium constant which is a function of temperature. Eq. (2.1) implies that the equilibrium pressures of the component elements are linked and that there is a range of pressures over which the solid remains in equilibrium as a single phase. From the pressure–temperature phase diagram for HgTe presented in **Figure 1**, we can conclude that at MOCVD growth temperatures, the pressure can vary by three orders of magnitude and remain in equilibrium with a single phase of HgTe. However, it is clear from Eq. (2.1) that Te_2 partial pressure will be varying across this phase field in the opposite sense to the Hg partial pressure.

http://dx.doi.org/10.5772/62952

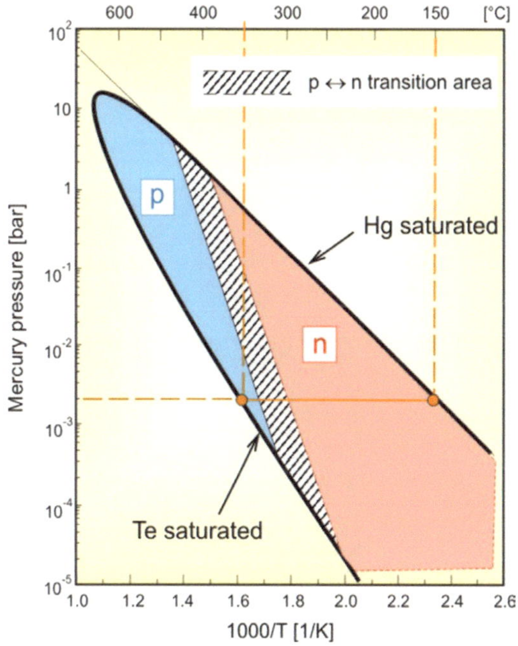

Figure 1. Mercury pressure–temperature phase diagram for HgTe. The p/n transition area indicates the transition from p-type to n-type for HgTe rich MCT annealed under equilibrium conditions [5].

The upper-phase boundary corresponds to the saturation of Hg over liquid mercury, and therefore, a horizontal line can be drawn from the equilibrium (pHg, T) to the intersection with the Hg saturated boundary in order to determine the equilibrium mercury source temperature. An example of a tie line is shown in **Figure 1** for the growth on Te rich-phase boundary at 350°C. If the HgTe layer is to be grown on Hg-rich-phase boundary, then the source and substrate temperature will be similar, which means an isothermal reaction cell. This is unacceptable for MCT MOCVD because Cd and Te organometallics will pyrolyse on the reactor cell wall before they reach the substrate.

The delivery of the vapour pressure of the Hg to the substrate for growth at 350°C must be greater than the minimum equilibrium pressure of 2×10^{-3} bar. In the author's system, the Hg source is an elemental mercury. The Hg partial pressure is about 30 mbar when the Hg zone temperature is maintained at 220°C during the IMP growth process.

2.2. MOCVD Growth approaches: IMP and DAG

There are two techniques for MOCVD growth of HgCdTe: the first based on direct growth of the ternary alloy—direct alloy growth (DAG) and the second based on interdiffused creation of the HgCdTe following initial deposition of alternating thin layers of CdTe and HgTe with total period thicknesses lower than 150 nm—interdiffused multilayer process (IMP). The latter technique takes advantage of the rapid interdiffusion rates ($D \sim 10^{-11}$–10^{-13} cm^2/s) of the cations in HgTe and CdTe at the typical growth temperatures in the range of 350–380°C. In IMP, the

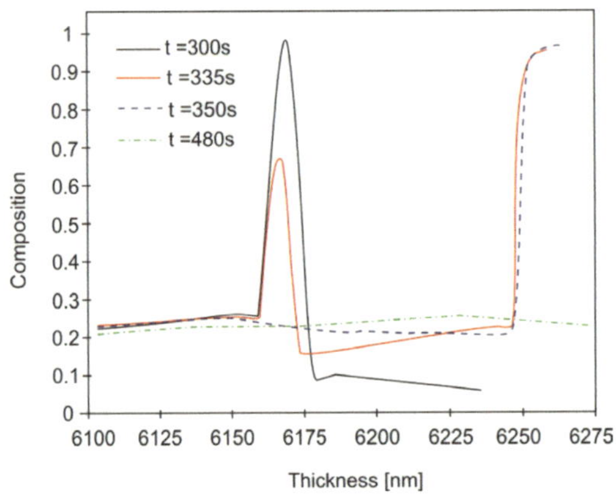

Figure 2. Composition profiles versus thickness modelled from laser reflectance data for a portion of an MOCVD-IMP growth run for x = 0.23 HgCdTe at different times (reproduced from reference [6]).

$Hg_{1-x}Cd_xTe$ final x composition is determined by the relative thickness of the HgTe and CdTe cycles which are easily monitored by controlling the times of their growth [4].

The sequence of four calculated compositional profiles for an IMP structure with 93.6 nm of HgTe and 27.6 nm of CdTe was presented on **Figure 2**. The cation interdiffusion process in IMP MOCVD growth technique of HgCdTe was demonstrated by monitoring the laser reflectance signal and modelling the IMP process. The HgCdTe was grown at 380°C. The first profile (the solid line) was taken at t = 300 s. It shows that formation of pure HgTe is inhibited by the fast diffusion of Cd atoms. The preceding CdTe layer is narrowing by interdiffusion but maintains composition x = 1 in the centre. The second period taken 35 s later is at the commence of the next CdTe growth phase, by which time the preceding CdTe layer has a maximum Cd composition of 0.6. The diffusion of Hg in CdTe is much slower than the diffusion of Cd in HgTe, and thus, the CdTe phase remains on the binary composition in the contrary to the HgTe phase. In the third time period taken 50 s after the experiment beginning, the x composition of CdTe layer has decreased below 0.3 and at after t = 180 s, entire structure is nearly completely homogeneous. The IMP growth technique is relatively ease in implementation, and its decisive advantage depends on the ability to control the Cd/Te ratio during the CdTe cycle of the growth. The dopants can be preferentially directed to the intended lattice site by controlling the Cd/Te ratio. Good uniformity in alloy composition and film thickness has been demonstrated by both techniques [7].

2.3. Growth mechanisms

A variety of different models have been used to describe the growth mechanisms of narrow bandgap II–VI semiconductors, and no clear mechanism has yet emerged from this research [4]. The haemolytic fission of alkyl radicals is a stepwise process where there is a different bond energy for the first and second radicals. In the case of DMCd, the bond energy is 193 kJ $mole^{-1}$ for the first methyl radical and 88 kJ $mole^{-1}$ for the second. The decomposition can simply be described by the following stepwise reactions where the first step is rate limiting:

$$(CH_3)_2Cd \rightarrow CH_3Cd^{\bullet} + CH_3 \quad (2.2)$$

$$CH_3Cd^{\bullet} \rightarrow CH_3^{\bullet} + Cd \quad (2.3)$$

More comprehensive study concerning HgCdTe growth mechanisms, you can find in references [4, 8].

2.4. Substrate type and orientation

The proper selection of substrate material and orientation has been a significant field for research in narrow gap II–VI semiconductors because it was recognized to be a limiting factor in the quality of the epilayers. Generally, we can divide substrates onto two categories: the first category is lattice matched II–VI substrates: CdZnTe and CdSeTe, and the second category is lattice mismatched substrates: GaAs, Si and sapphire (Al_2O_3). The CdZnTe ternary is expensive material, and it suffers severe segregation causing non-uniformity in alloy composition, and CdSeTe substrates have high impurity concentrations arising from the selenium source. Much of the MOCVD growth has concentrated on orientations close to the (100), normally with some misorientation to reduce the size of macro-defects, otherwise known as hillocks or pyramids. In a detailed analysis of the frequency and shapes of defects on different misorientations, it was concluded by Snyder et al. [9] that the optimum orientation was (100) 3°–4° towards the (111)B face. The most extensively used alternative substrate has been GaAs with a few microns of CdTe buffer layer reducing the 14% lattice mismatch between CdTe and GaAs. Most of the MOCVD growth onto silicon substrates has used a GaAs buffer layer to step the change in lattice parameter between silicon and CdTe. Substrate orientation as well as the HgCdTe orientation has been extensively reported to have fundamental influence on crystallographic defects, surface morphology, residual background concentration, and both donor- and acceptor-doping efficiency. **Figure 3** presents atomic arrangements of the (100) and (h,11)B surfaces on the (011) plane of HgCdTe structure. The HgCdTe crystallographic orientation influence on doping efficiency will be studied in next paragraphs.

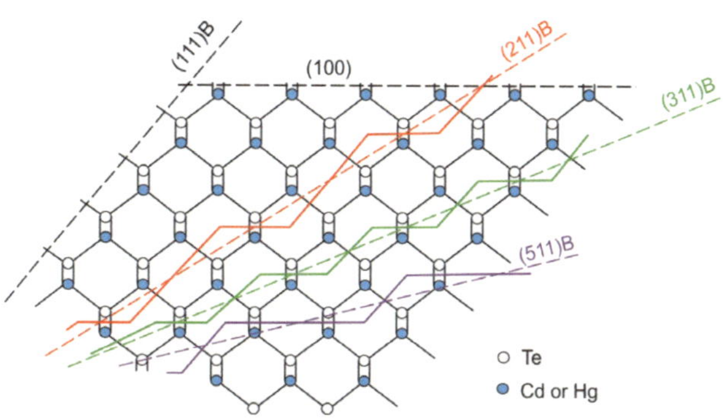

Figure 3. Atomic arrangements of the (100) and (h,11)B surfaces on the (011) plane of HgCdTe structure [10].

3. Growth of HgCdTe in Aix 200 system

HgCdTe epitaxial growth was carried out in the horizontal reactor of AIX-200 Aixtron MOCVD system (**Figure 4**). It consists of a horizontal rectangular aperture liner enclosed in an outer circular quartz tube. The system is designed to operate in the laminar flow regime with process pressures from 50 to 1000 mbar using a butterfly valve for pressure control. Reactor pressure of 500 mbar was used for all successful growth runs. Hydrogen is used as a carrier gas. Dimethylcadmium (DMCd) and diisopropyltelluride (DIPTe) are used as precursors. Ethyl-iodide (EtI) is used as a donor dopant source and TDMAAs (or AsH_3 in our previous research-es) as an acceptor dopant source. DMCd and EtI are delivered through the one channel, while DIPTe and TDMAAs are delivered through the lower channel over elemental mercury bath.

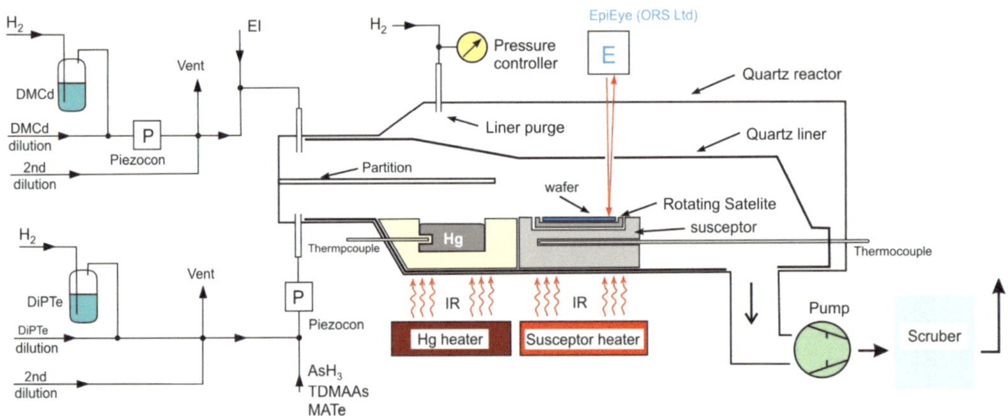

Figure 4. The scheme of the gas supply installation and the reactor in AIX 200 MOCVD system adapted to HgCdTe growth.

The Aixtron's gas foil rotation technique has been applied for better composition uniformity. There are two temperature zones in the reactor: the Mercury source zone and the growth zone with graphite susceptor controlled by external infrared heaters. High-temperature anneal was used before each growth run for reactor cleaning. Gas delivery system is additionally equipped with ultrasonic precursor concentration monitors—Piezocon and reflectometer—EpiEye. The usage of piezocons contributed to a better repeatability of the growth processes. Adaptation of EpiEye reflectometer allows for *in situ* monitoring. Typically, a 3–4-μm thick CdTe layer is used as a buffer layer reducing stress caused by crystal lattice mismatch between GaAs substrate and HgCdTe epitaxial layer structure. The buffer plays also a role of Ga diffusion barrier. The interdiffused multilayer process (IMP) technique was applied for the HgCdTe deposition [6]. HgCdTe was grown at 350°C with mercury source kept at 210°C. The II/VI mole ratio was kept in the range from 1.5 to 5 during CdTe cycles of the IMP process. An acceptor and donor doping has been examined over the wide range of compositions and doping levels of 5×10^{14}–5×10^{17} cm^{-3} have been obtained. Obtained HgCdTe heterostructures have been not annealed neither during the growth process (*in situ*) nor after the growth (*ex situ*) [11].

The set of metalorganic precursors gas supply suitable for Te and Cd delivery was presented in **Figure 5**. In our system, such a set supplies the reactor with DMCd and DIPTe. The H_2 carrier gas controlled by MFC (mass flow controller) is introduced into the bubbler and taking metalorganic precursor flows to the reactor. The bubbler is placed in the thermostatic bath. The liquid precursor temperature is one of the parameters determining precursor partial pressure in the reactor.

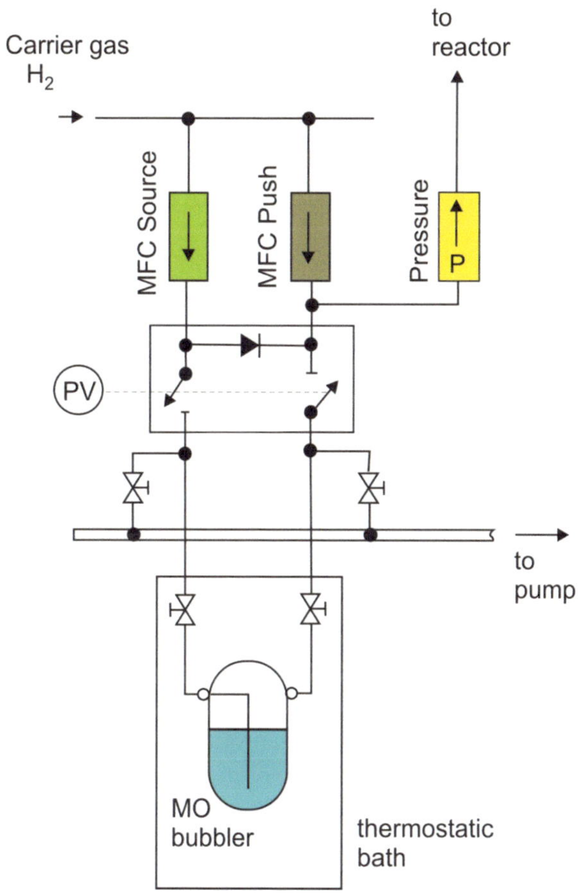

Figure 5. The set of metalorganic precursors gas supply suitable for Te and Cd delivery.

The partial pressures of cadmium and tellurium metalorganic compounds were calculated using following formulas:

$$P_p[mbar] = P_R[mbar]\frac{Q_{MO}\left[\frac{mol}{min}\right]}{Q_{Total}\left[\frac{mol}{min}\right]} \tag{3.1}$$

where Q_{Total} is the total gas flow through the reactor

$$Q_{MO}\left[\frac{mol}{min}\right]=\frac{Q_S}{22.4}\cdot\frac{p(v)}{p_c-p(v)} \tag{3.2}$$

p_c is the pressure in the reactor, *p(v)* is vapour pressure of the metalorganic in the bubbler, Q_S is the MFC source flow.

For metalorganic compounds, their vapour pressure is calculated using following formula:

$$Q_{MO}\left[\frac{mol}{min}\right]=\frac{Q_S}{22.4}\cdot\frac{p(v)}{p_c-p(v)} \tag{3.3}$$

where T is bubbler temperature in Kelvins, and A and B are material constants shown in **Table 1**.

Material constant	**Metalorganic sources**			
	Cd	Te	EtI	Zn
A	1850	2309	1715	2109
B	7764	8288	7877	8280

Table 1. Material constant examples of the metalorganic sources.

In the author's system, the Hg source is an elemental mercury. The Hg partial pressure can be read from **Figure 1** and is about 30 mbar when the Hg zone temperature is maintained at 220°C during the IMP growth process.

The examples of surface morphology of obtained (111) and (100) oriented HgCdTe layers are presented in **Figures 6** and **7**, respectively. Both layers were not doped intentionally and are about 10 μm thick. The pictures were obtained using optical microscope with Nomarski contrast with ×1000 magnification. Both layers were grown on (100) GaAs substrate with 3° misoriented towards the (111)B face. The (111) CdTe buffer growth on (100) GaAs was obtained using Te flush during nucleation. The (100) CdTe buffer growth on (100) GaAs was obtained using Cd flush during nucleation. Both obtained layer are characterized by high uniformity in the composition and the thickness as well due to the Aixtron's gas foil rotation technique. It is clearly seen that the smoothness of (100) HgCdTe is much better in comparison with (111) HgCdTe what is attributable to microtwins on (111) orientation. The surface roughness coefficient Rq is equal 70 and 6 nm for (111) and (100) orientation, respectively. Rq coefficient was measured on Veeco optical profiler type Wyko NT 1100.

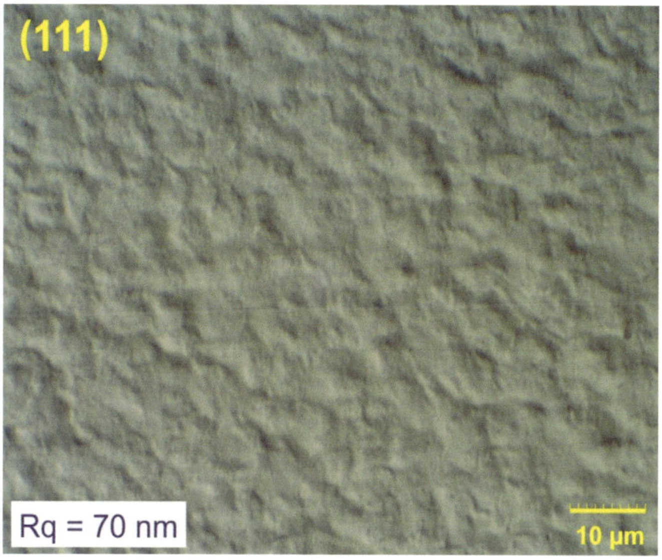

Figure 6. The example of surface morphology of obtained (111) oriented HgCdTe layer.

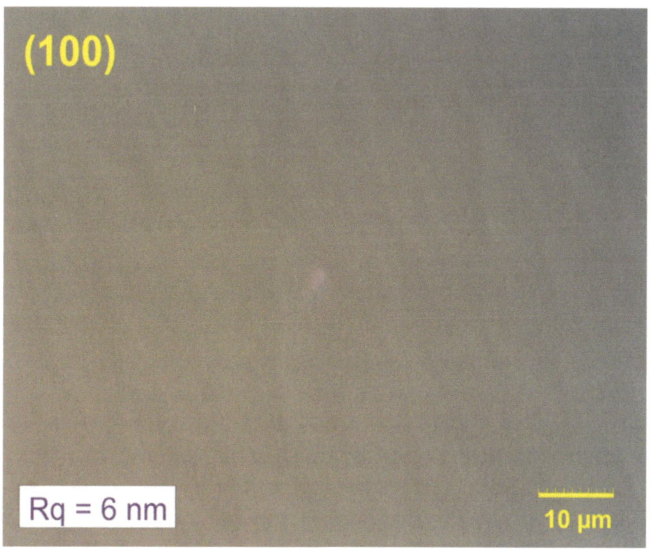

Figure 7. The example of surface morphology of obtained (100) oriented HgCdTe layer.

3.1. Undoped electrical properties

Indispensable in understanding the electrical properties of doped layers is the ability in interpretation of undoped layers results. The electrical properties of undoped MOCVD-grown MCT are dependent on the different parameters such as contamination from alkyls or substrates [3, 12], crystallographic defects such as microtwins on (111) orientation, quench cooling conditions and native defects like mercury vacancies and others [13]. Although such defects do not affect interpretation of highly doped layers like contact layers, they may confuse interpretation of low-doped layers like absorbing layers in photodiodes structure. **Figure 8** presents residual background concentration of undoped HgCdTe layers versus consecutive number of growth processes carried out within 3 years in author's lab.

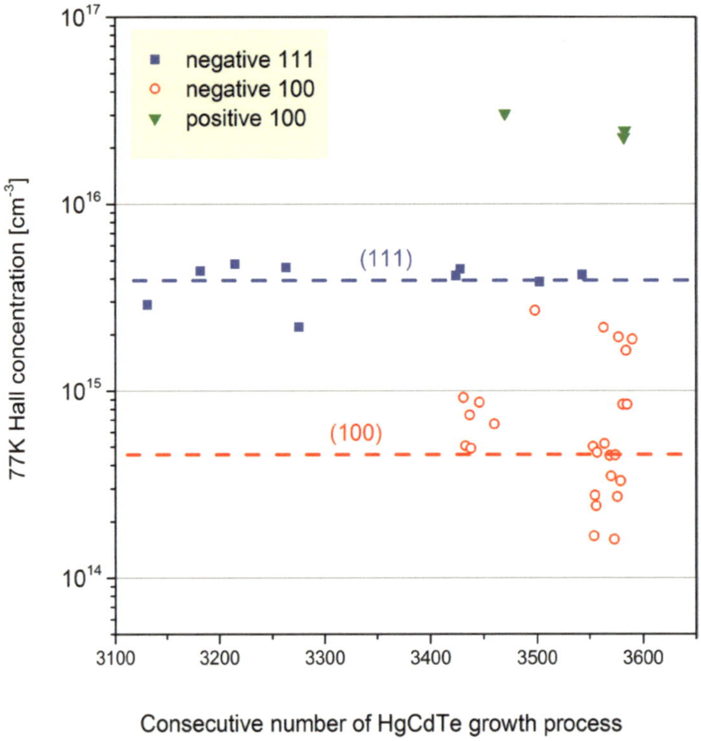

Figure 8. Residual background concentrations of undoped HgCdTe layers versus the numbers of growth processes.

The residual background concentration is a matter of huge concern of engineers form each semiconductor laboratory because it is the serious test of the equipment and applied materials purity as well as the result of the efforts of the staff. The mean residual concentration maintained in our laboratory is about 4×10^{15} cm^{-3} for (111) orientation and 4.5×10^{14} cm^{-3} for (100) orientation. An order of magnitude higher background concentration for (111) concentration is attributable to the microtwins presence. In the **Figure 8**, we can observe single points with positive (hole) concentrations above level of 10^{16} cm^{-3} for (100) orientations. These happened during experiments with lowering the mercury zone temperature in order to reduce the hillock density. Then, the mercury vacancies were created. The lowest background donor concentration obtained in our system is about 2×10^{14} cm^{-3} what is similar with the best results reported from other labs. The matter of residual donor concentration is particularly important during p-type doping at the low level.

http://dx.doi.org/10.5772/62952

4. Doping study

Improved performance IR photodiodes which require the growth of doped heterostructures have been theoretically modelled [14, 15]. Such structures require control of the dopant concentrations and sharp transitions between either dopant or different concentrations within a structure combined with bandgap changes induced by varying the alloy composition. To realize these doped junction structures, the dopants chosen need to have low diffusion coefficients at the growth temperature. Another considerable issue during doped heterostructures deposition is the control of any dopant memory effects. Some dopant sources may demonstrate 'run-to-run' memory effects in which the dopant is found to contaminate the run subsequent to the one in which the dopant was used. The dopant may induce memory effects within a single run which can limit the abruptness of the dopant transition. Then, the order of the layers within the structure can influence the sharpness of the electrical junction. Vast majority of papers concerning the influence of substrate orientation on impurity incorporation have pointed the complexity of the phenomena in play and have only given qualitative interpretations.

4.1. Donor-doping control

The selection for donor dopant in MCT is between the group III elements (Al, Ga, In) on metal sites and the group VII elements (Cl, Br I) on the Te sites. Despite some concerns over the diffusion coefficient of the group III species, these have been the most widely studied with most work concentrating on indium as the slowest diffuser of the group III elements [16, 17]. Although the reactive nature of the halogen elements, iodine doping has been studied at PRL [18] for the first time. The real aim of donor doping for the most device applications is to obtain control of the donor concentrations at $\approx 10^{15}$ cm^{-3} for absorbing layers and at $\approx 10^{17}$ cm^{-3} for n^+ contact layers.

4.1.1. Indium- and aluminium-doping studies

Trimethyl indium (TMIn) has been studied most widely as a doping source [13]. Despite its low vapour pressure, this source yields very high dopant incorporation. In incorporated preferentially into HgTe (20–50 times higher than in CdTe). This effect was use to attempt to lower the doping range by only injecting the TMIn during the CdTe cycles. However, the controllable doping range with a bubbler temperature of 2°C was still only from 2×10^{17} to 3×10^{18} cm^{-3}. Lower bubbler temperatures (-10°C) have been applied to obtain minimum concentration of 5×10^{16} cm^{-3} in MCT grown at 370°C using DAG. 'Effuser' mode bubbler operation with TMIn was investigated to dope at low level, but results were not promising. Although this did enable layers with $N_{D-A} = 2 \times 10^{16}$ cm^{-3} to be grown, the rate of increase in the donor concentration with bubbler temperature was very abrupt, making reproducible control very difficult. Similar rapid doping transitions were observed using a lower vapour pressure source [ethyldimethyl indium (EDMIn)] in bubbler mode meant that control of the donor concentration was only possible down to $\approx 10^{17}$ cm^{-3} [19]. This threshold effect was assigned to alkyl adsorption effects which were dependent on reactor design. Adsorption in the gas lines and manifold can cause 'run-to-run' memory effects extended over several growth runs. The electrical properties of as-grown In-doped layers, which contain significant Hg vacancy concentrations, suggest that only 30% of the In is electrically active and that the In introduces additional Hg vacancies causing autocompensation in these layers [13].

4.1.2. Iodine-doping studies

The group VII elements, substituting on Te lattice sites, were expected to be slower diffusers. Preliminary researches of iodine doping in MOCVD were carried out with elemental iodine I_2. It has a vapour pressure of 0.3 Torr at 25°C. Electrically active donor incorporation was observed at low—10^{15} cm^{-3} doping levels with high mobilities. The proper control with abrupt doping profiles was not demonstrated, however, because I_2 reacts with DMCd. Improved doping efficiency has been expected when alkyl iodides were applied. In DAG deposition, iodine doping from isopropyl iodide has been reported to levels as low as 5×10^{15} cm^{-3}. However, isopropyl iodide introduced a memory effect. The initial choice for a MO iodide source was ethyliodide (EtI) due to its commercial availability. EtI is a highly effective precursor for doping without any memory effects [20]. Controlled iodine doping has been achieved in the range of 3×10^{14}–2×10^{18} cm^{-3} with 100% electrical activation following a standard Hg-rich stoichiometric anneal at 235°C. However, some works present similar donor-doping results without annealing. A double-dilution bubbler configuration is presented in **Figure 9**, and it has been used to obtain controlled low-level doping.

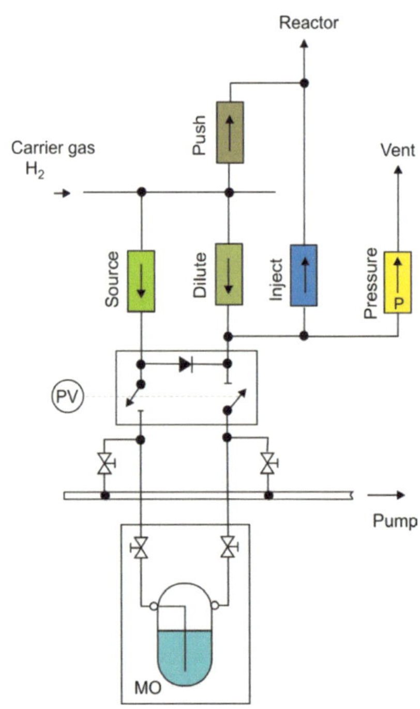

Figure 9. The double-dilution bubbler configuration for EtI doping.

http://dx.doi.org/10.5772/62952

The special doping structures contribute to reduce the number of growth experiments required to assess a doping source. Multilevel 'staircase' structures quickly yielded dopant source calibration curves and growth of a highly doped layer between two low-doped regions in a 'sandwich' structure yielded information on diffusion and reactor memory effects. **Figure 10** shows a 'staircase' (111) HgCdTe structure obtained using EtI which was assessed by SIMS and differential 77 K Hall profiling. Presented SIMS profile is shown from the growth perspective, while vast majority of such curves published in the literature are referenced from the layer surface. The errors in differential Hall concentrations become larger when the etch step size is around 2 μm. Consequently, it is very difficult to align the step positions with the grown layer interfaces. The correlation obtained between the chemical and electrical results implied the iodine incorporated on the correct Te lattice site; and that within the accuracy of the techniques, the activation efficiency is high. However, when we look at the **Figure 10**, the EtI-doping efficiency (the adjacency between chemical and electrical concentration) is decreasing with increasing dose level. This was corroborated by further layers, and **Figure 11** shows the resultant EtI calibration curve showing the linear control of both iodine and 77 K Hall concentration as a function of the injected EtI concentration in the reactor. For the comparison calibration curve for (100), HgCdTe layers is shown. We can see above one order of magnitude higher EtI-doping efficiency: incorporation and activation for (111)HgCdTe than for (100). The iodine should be located in the tellurium sublattice sites to act as a donor. The Te sides on the (111)B surface provide a more stable adsorption site with three bonds from the underlying group II atoms. On the contrary, the (100) surface provides a weaker adsorption site with two bonds (**Figure 3**).

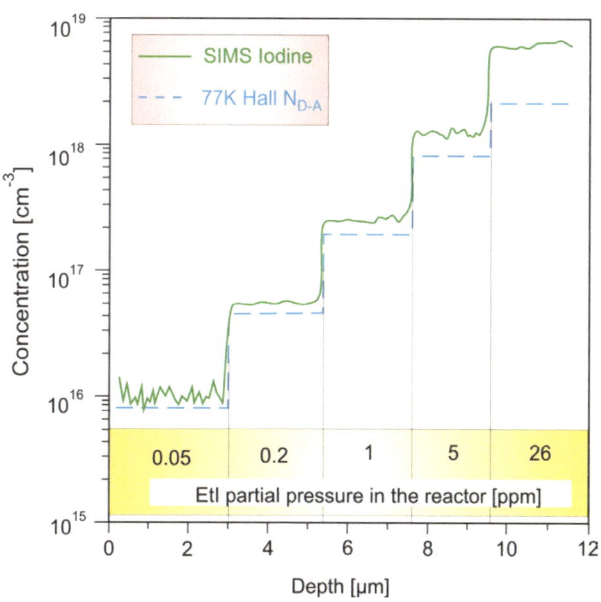

Figure 10. SIMS and differential excess donor concentration profiles of a 'staircase' (111) HgCdTe structure, doped using EtI [21].

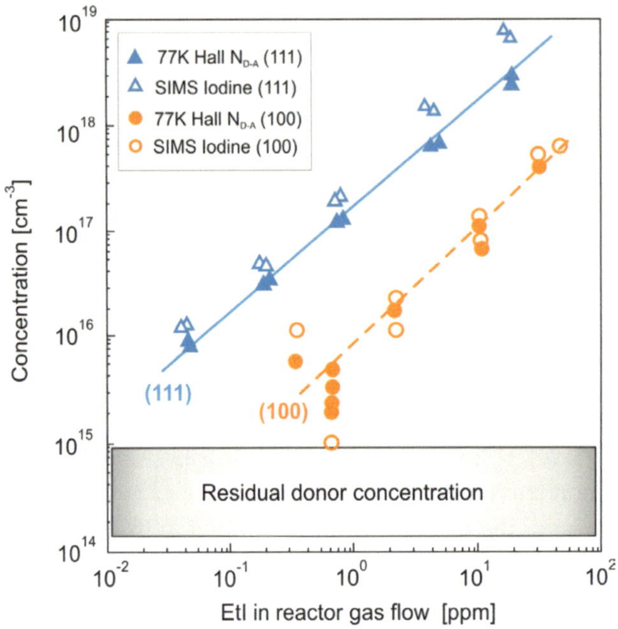

Figure 11. Calibration curve obtained with EtI for (111) and (100) oriented HgCdTe layers.

4.2. Acceptor doping

Acceptor doping should be provided by applying elements from the group IA (Li, Na, K), IB (Cu, Ag, Au) or group VB (P, As, Sb, Bi). The group I elements should be positioned on the metal lattice sites in order to provide theirs activity as the acceptors. The elements from group V need to be located on the tellurium sublattice sites. The group V elements are slower-diffusing species in HgCdTe and are versatile dopant capable of producing a variety of abrupt, doped junctions. If the group V elements were located on metal sites then they are likely to reveal donor behaviour. The most widely used element for acceptor doping is arsenic [22–28], although researches involving phosphorus and antimony have also been reported [13].

Figure 12 presents incorporation and activation of As from TDMAAs and AsH_3 versus precursor partial pressure for (111)HgCdTe. The chemical As concentration was determined by SIMS measurements. There is also comparison with 77K N_{A-D} (100)HgCdTe doped from TDMAAs. The As chemical concentration increases roughly proportionally with As precursor partial pressure up to 10 ppm. The levels of As incorporation from AsH_3 and TDMAAs into growing (111)HgCdTe samples are equal considering the measurement uncertainty. Arsenic precursors were introduced to the reactor during CdTe cycles of IMP growth process. The arsenic atoms appearance accelerates DMCd pyrolysis in the reactor that causes high shift in x composition. Chemical concentration of arsenic atoms measured by SIMS does not comply with acceptor electrical concentration determined by 77-K Hall measurements so we do not observe full arsenic activation. Also, for the same dose of arsenic in the reactor, the holes concentration in the (100)HgCdTe is about one order of magnitude higher than in

(111)HgCdTe. The arsenic incorporation considerably depends on crystallographic orientation. This dependence can be explained by considering that the two Cd and As atoms are readily active in the incorporation process. The created Cd–As species can break into on the surface if proper conditions are realized. In this respect, the (100) plane offers the best configuration because this surface has two double dangling bonds available for both Cd and Te (or substituting As) surface atoms (**Figure 3**). This mechanism can thus account for the highest doping observed with (100) substrates. Going away from (100) plane towards (111)A or (111)B leads to reduced As incorporation, and this corresponds to the fact that the relative fraction of (100) terraces decreases at the expense of (111) steps.

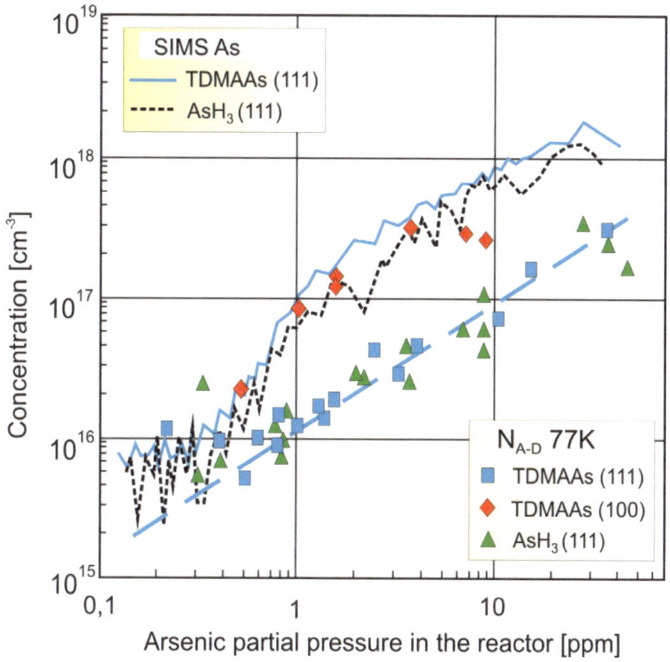

Figure 12. Incorporation and activation of As from TDMAAs and AsH_3 versus partial pressure of precursors.

5. Heterostructures and devices

After achieving device-quality MCT layers with doping control over the ranges required, the growth of doped junctions can be studied. When both donor and acceptor doping is required, for example in $P^+/\pi/N^+$ photodiodes, we prefer MCT with (111) crystallographic orientation because of limits with high-concentration doping control in (100)MCT. The capital letters P and N in device description means acceptor- and donor-doped layers with wider gap material. The upper index '+' denote high doping. If the particular device requires p-type doping only then is constructed on the basis of (100)HgCdTe. As an example, the MWIR photoresistors are presented in the Section 5.3. Cryogenic cooling of detectors has always been the disadvantage of sensitive infrared (IR) systems. Conventional HgCdTe IR photodetectors should be cooled down below ambient temperature to reduce noise and leakage currents resulting from thermal generation processes. Presented in previous section progress in MOCVD technology let construct IR detectors operating without cryogenic cooling. Uncooled—operating in room temperature or thermoelectrically cooled devices—operating typically at 230 K is named as higher operating temperature (HOT) devices. The examples of IR HOT detectors operating in different IR spectral ranges are shown in the following sections.

5.1. LWIR photodiodes

The dark current in LWIR HgCdTe photodiodes is primary determined by Auger generation processes at elevated temperatures. The low-doped absorber layer becomes intrinsic, and the carrier concentration is higher than the doping level. The device structures with a combination of exclusion and extraction junctions in $N^+/\pi/P^+$ configurations have demonstrated suppression of Auger mechanisms by reducing the absorber carrier density below thermal equilibrium in a reverse bias condition. Classical $N^+/\pi/P^+$ structure has been expanded with graded interface layers denoted as 'G'. Graded doping and composition x layers represent the real structure which profile is shaped by interdiffusion processes during $Hg_{1-x}Cd_xTe$ growth at 350°C. Thus, $N^+/G/\pi/G/P^+/G/n^+$ HgCdTe photodiode structure has been obtained as it is shown in **Figure 13**. In order to improve electrical contact properties of metallization to P^+ layer, the structure was upgraded with p^+/n^+ tunnelling junction on the top.

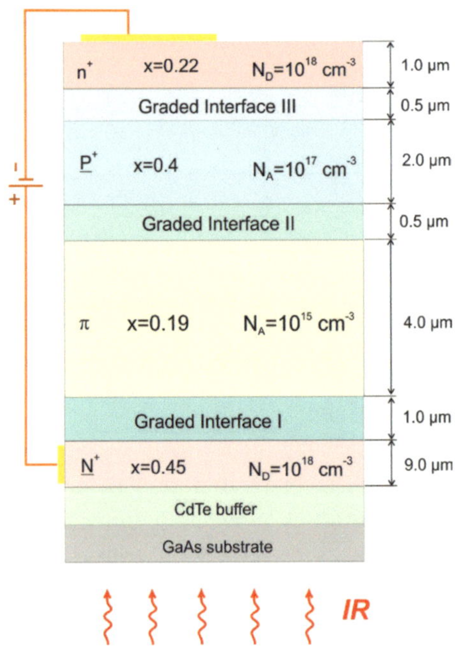

Figure 13. The LWIR $N^+/G/\pi/G/P^+/G/n^+$ HgCdTe photodiode diagram. x is the alloy composition, N_A—the acceptor concentration, and N_D—the donor concentration [29].

The cladding N^+-layer should be thicker than minority carrier diffusion length and doped at possible highest level providing reduction of series resistance. Analogously parameters of cladding P^+-layer are defined. The thickness of active π-type region (absorbing layer) should be shorter than minority carrier diffusion length. Generally, the thickness of absorber layer is a compromise between requirements of high absorption efficiency and low thermal generation. In our experiments, the thickness of absorber layer is varied from 3 to 6 μm. The acceptor doping of the absorber should be at possible low level just to overcompensate the donor background concentration of the material.

The SIMS profiles of $N^+/G/\pi/G/P^+/G/n^+$ heterostructure measured by CAMECA IMS 6F using positive and negative Cs ions are presented in **Figures 14** and **15**, respectively. The *x*-composition represented by the solid green lines has been calculated taking into account the measured SIMS points for Cd, Hg and Te elements. There is quite good convergence between projected and measured values in the absorber region. There are discrepancies between positive and negative ion measurements which result from incorrect SIMS ions calibration; about half of the order of magnitude differences between positive and negative ions for the arsenic-doping profile. The diffusion processes during the epitaxial growth cause gradient profile in the interface layers adjacent to the absorber region.

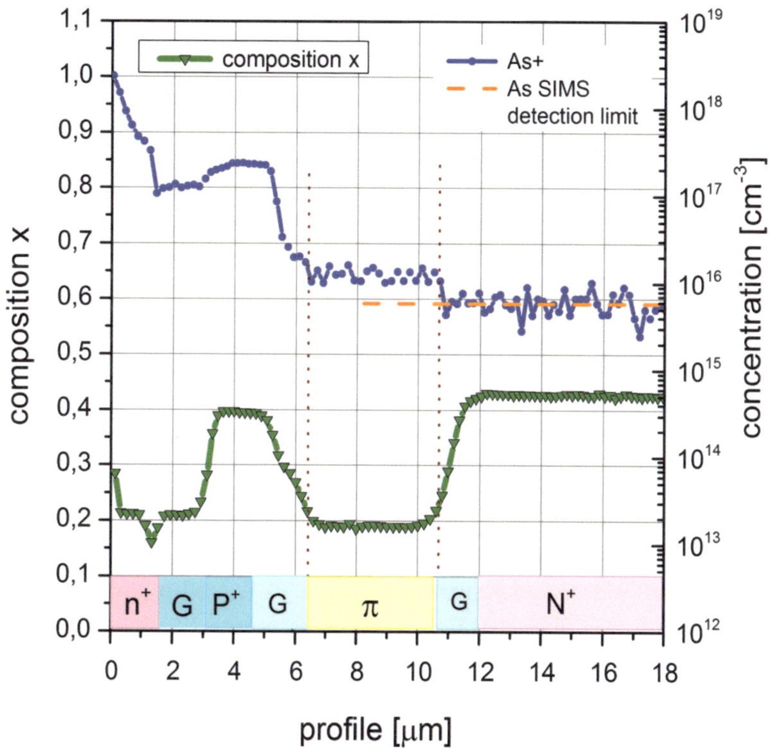

Figure 14. SIMS profile through $N^+/G/\pi/G/P^+/G/n^+$ (111)HgCdTe heterostructure measured by using positive caesium ions.

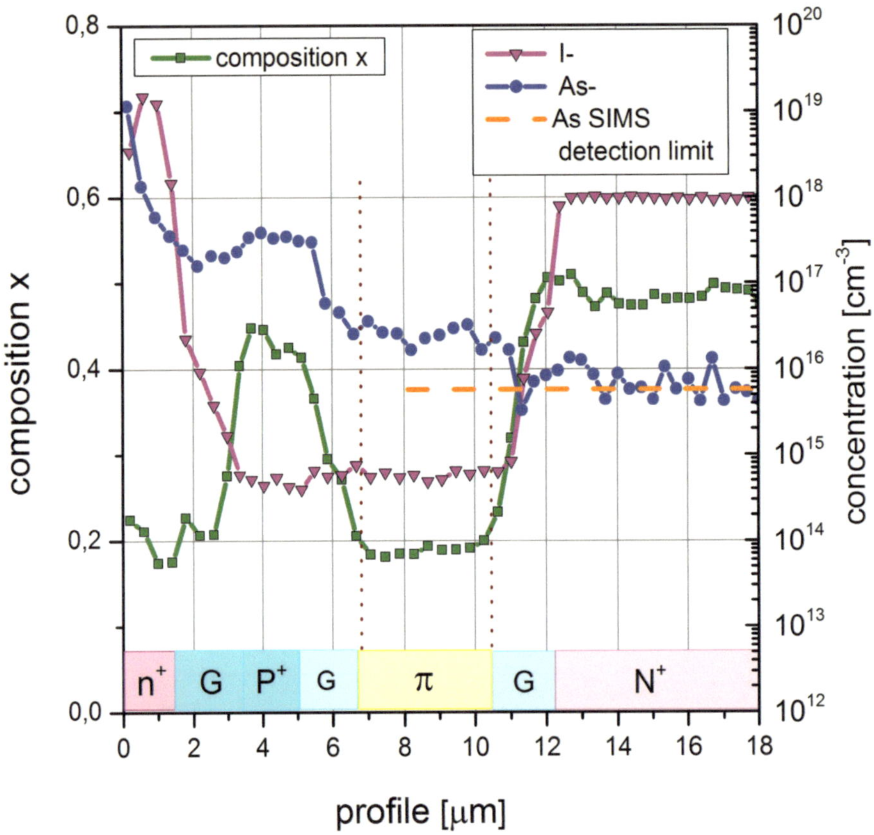

Figure 15. SIMS profile through N^+/G/π/G/P^+/G/n^+ (111)HgCdTe structure measured by using negative caesium ions.

The gradient slope of arsenic line between P^+ and absorber π-region seen in **Figures 14** and **15** indicates arsenic diffusion from P^+-region to absorber region what is dangerous especially in thin, 3-μm thick, absorber layers. The iodine profile (the solid pink line with triangles) seen in **Figure 15** indicates expected donor-doping levels: 1×10^{19} cm^{-3} for n^+-layer and 1×10^{18} cm^{-3} for N^+-layer with EI doses 90 and 19 ppms, respectively. Unwanted iodine presence in the absorber region is at the level of 1×10^{15} cm^{-3}, what is close to the SIMS detection limit for this element.

The $Hg_{0.81}Cd_{0.19}Te$ N^+/G/π/G/P^+/G/n^+ photodiode's current–voltage characteristic is presented in **Figure 16** for device with active region doped with 7 ppms of TDMAAs dose during the active layer deposition. Measurements were taken at 300 K (without any cooling) using the Keithley 2400 sourcemeter. The electrical area of devices is 8.1×10^{-9} m^2. The solid line denotes theoretically calculated results applying the *APSYS* simulation platform. The negative differential resistance between -150and -250 mV is attributable to the suppression of Auger processes due to the exclusion and extraction phenomena. The dark current is determined by both band-to-band (BTB) and trap-assisted (TAT) tunnelling mechanisms which evidences the quality of the material, for example the point defects like mercury vacancies. More comprehensive studies concerning fitting procedure have been reported in references [14, 15].

http://dx.doi.org/10.5772/62952

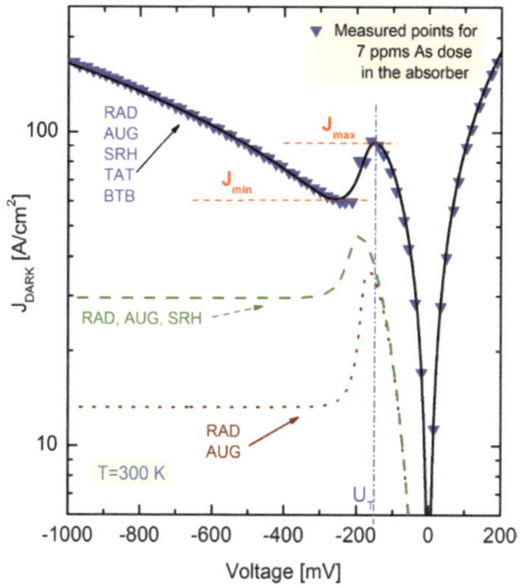

Figure 16. Modelled and measured characteristics of dark current density versus bias voltage for LWIR N⁺/G/π/G/P⁺/G/n⁺ HgCdTe photodiode at room temperature [29].

The spectral characteristics have been measured at zero and reverse bias polarization using the Perkin Elmer FT-IR Spectrometer type Spectrum 2000. **Figure 17** presents the current responsivity versus wavelength for long wavelength $Hg_{0.81}Cd_{0.19}Te$ detector measured at 300 K. TDMAAs dose of 3.6 ppms was applied during the growth of the absorbing region. The values of the reverse biases are typically chosen considering current–voltage analysis and are typically beyond the threshold voltage U_T (designated in **Figure 16**). We can increase the current responsivity up to 50 times by the use of the reverse bias and thus causing Auger generation process suppression.

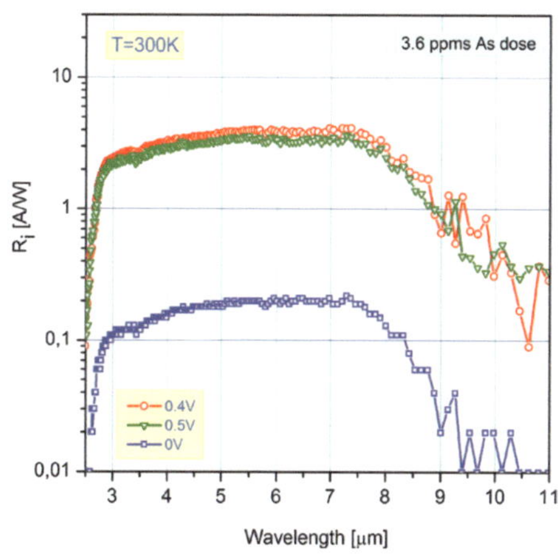

Figure 17. Current responsivity versus wavelength for LWIR HgCdTe photodiodes measured at room temperature [29].

5.2. MWIR photoresistors

The voltage responsivity versus wavelength for (100) and (111) HgCdTe MWIR photoresistors is presented in **Figure 18**. Photoresistors constructed from (100)HgCdTe have higher responsivity than these constructed from (111)HgCdTe. As it was discussed in the Section 4.2, there is higher acceptor-doping efficiency in (100)HgCdTe in comparison with (111)HgCdTe. However, presented values of the voltage responsivities are not fully comparable, because of the differences in $Hg_{1-x}Cd_xTe$ compositions in presented devices. The absorber composition in (100)HgCdTe was x = 0.337 whereas in (111)HgCdTe was x = 0.331. The cut-off wavelength λ_{CO} of measured photoresistors moves to longer waves with decreasing temperature because the energy gap is narrower.

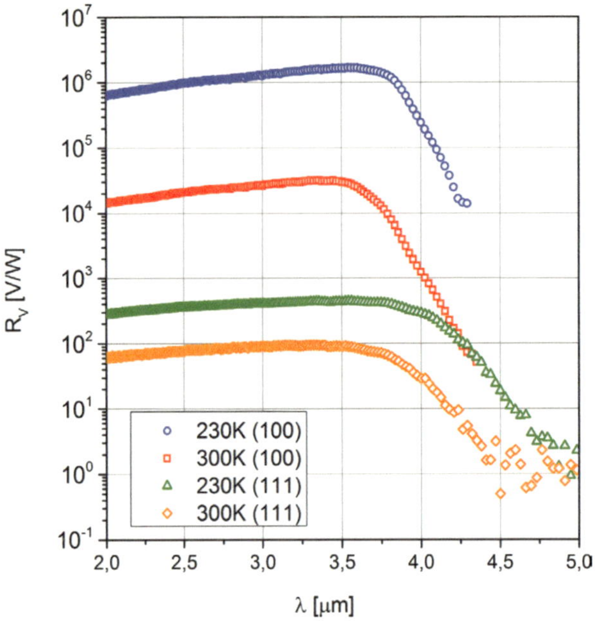

Figure 18. The voltage responsivity versus wavelength for (100) and (111) HgCdTe MWIR photoresistors.

6. Summary

In spite of other competitive technologies like InAs/GaSb superlattices, MCT material takes leading position in infrared detector industry at the time of writing this chapter. Donor- and acceptor-doping researches in (100) and (111) oriented HgCdTe layers grown by MOCVD have been studied. Fully doped HgCdTe heterostructures with acceptor concentration range between 10^{14} and 5×10^{17} cm^{-3} and donor concentration range between 10^{14} and 1×10^{18} cm^{-3} and without post-grown annealing have been reported. The electrical and chemical characterizations of HgCdTe structures grown at 360°C on GaAs substrates using DIPTe have been described. Infrared photodiodes with different composition x were constructing on the basis of obtained heterostructures enabling signal detection of any wavelength from 1 μm to above 20 μm covering SWIR, MWIR and LWIR spectral ranges. Presented experimental results show that MOCVD technology enables to grow HgCdTe structures dedicated for HOT devices.

Acknowledgements

This study was supported by Polish National Science Centre as the Research Project No. UMO-2013/08/A/ST5/00773 and Polish National Centre for Research and Development as the Project No. TANGO1/2665576/NCBR/2015

Author details

Pawel Madejczyk[1*], Waldemar Gawron[1], Artur Keblowski[2] and Adam Piotrowski[2]

*Address all correspondence to: pmadejczyk@wat.edu.pl

1 Military University of Technology, Warsaw, Poland

2 VIGO System S.A., Ozarow Mazowiecki, Poland

References

[1] Lawson WD, Nielson S, Putley EH, Young AS. Preparation and properties of HgTe and mixed crystals of HgTe–CdTe. Journal of Physics and Chemistry of Solids. 1959;9:325–329. doi:10.1016/0022-3697(59)90110-6

[2] Rogalski A. Infrared Detectors 2nd ed. Boca Raton: CRC; 2011. 876 p.

[3] Irvine SJC, Mullin JB. The growth by MOVPE and characterization of $Cd_xHg_{1-x}Te$. Journal of Crystal Growth. 1981;55:107–115. doi:10.1016/0022-0248(81)90277-3

[4] Irvine SJC. Metal-organic vapour phase epitaxy. In: Capper P, editor. Narrow-gap II–VI Compounds for Optoelectronic and Electromagnetic Applications. 1st ed. London: Chapman & Hall; 1997. pp. 71–96.

[5] Jones CL, Quelch MJT, Capper P, Gosney JJ. Effects of annealing on the electrical properties of $Cd_xHg_{1-x}Te$. Journal of Applied Physics. 1982;53:9080–9092. doi: 10.1063/1.330419

[6] Svoronos SA, Woo WW, Irvine SJC, Sankur HO, Bajaj J. Model of the interdiffused multilayer process. Journal of Electronic Materials. 1996;25:1561–1571. doi:10.1007/BF02655400

[7] Mitra P, Case FC, Reine MB. Progress in MOVPE of HgCdTe for advanced infrared detectors. Journal of Electronic Materials. 1998;27:510–520. doi:10.1007/s11664-998-0007-5

[8] Bell W, Stevenson J, Cole-Hamilton DJ, Halis JE. Evidence for a surface-bound free radical mechanism during the decomposition of $^{i}Pr_2Te$ in the presence or absence of mercury and/or Me_2Cd under MOVPE conditions obtained from deuterium-labelled precursors. Polyhedron. 1994;13:1253–1265. doi:10.1016/S0277-5387(00)80259-4

[9] Snyder DW, Mahajan S, Ko EI, Sides PJ. Effect of substrate misorientation on surface morphology of homoepitaxial CdTe films grown by organometallic vapor phase epitaxy. Applied Physics Letters. 1991;58(8):848–850. doi:10.1063/1.104509

[10] Svob L, Cheye I, Lusson A, Ballutaud D, Rommeluere JF, Marfaing Y. Crystallographic orientation dependence of As incorporation in MOVPE-grown CdTe and corresponding acceptor electrical state activation. Journal of Crystal Growth. 1998;184/185:459–464. doi:10.1016/S0022-0248(98)80096-1

[11] Madejczyk P, Gawron W, Piotrowski A, Klos K, Rutkowski J, Rogalski A. Improvement in performance of high-operating temperature HgCdTe photodiodes. Infrared Physics and Technology. 2011;54:310–315. doi:10.1016/j.infrared.2010.12.036

[12] Madejczyk P, Gawron W, Piotrowski A, Klos K, Rutkowski J, Rogalski A. Influence of TDMAAs acceptor precursor on performance improvement of HgCdTe photodiodes. Acta Physica Polonica A. 2010;118(6):1199–1204.

[13] Maxey CD, GaleIG, Clegg JB, Whiffin. Doping studies in MOVPE-grown HgCdTe. Semiconductor Science and Technology. 1993;8:S183–S196. doi:10.1088/0268-1242/8/1S/042

[14] Martyniuk P, Gawron W, Madejczyk P, Rogalski A, Piotrowski J. Modelling of HgCdTe LWIR detector for high operations temperature conditions. Metrology and Measurement Systems. 2013;XX(2):159–170.

[15] Martyniuk P, Gawron W, Pawluczyk J, Keblowski A, Madejczyk P, Rogalski A. Dark current suppression in HOT LWIR HgCdTe heterostructures operating in non-equilibrium mode. Journal of Infrared and Millimeter Waves. 2015;34(4):385–390.

[16] Gough JS, Houlton MR, Irvine SJC, Shaw N, Young ML, Astles MG. The growth and properties of In-doped metalorganic vapor phase epitaxy interdiffused multilayer process (HgCd)Te. Journal of Vacuum Science and Technology. 1991;B9(3):1687–1690. doi:10.1116/1.585400

[17] Irvine SJC, Bajaj J, Bubulac LO, Lin WP, Gedridge RW, Higa KT. A new n-type doping precursor for MOCVD-IMP growth of detector quality MCT. Journal of Electronic Materials. 1993;22(8):859–864. doi:10.1007/BF02817498

[18] Easton BC, Maxey CD, Whiffin PAC, Roberts JA, Gale IG, Grainger F, Capper P. Impurities and metal organic chemical-vapor deposition growth of mercury cadmium telluride. Journal of Vacuum Science and Technology. 1991;B9(3):1682–1686. doi: 10.1116/1.585399

[19] Edwall DD. Comparison of spatial compositional uniformity and dislocation density for organometallic vapor phase epitaxial HgCdTe grown by DAG and IMP. Journal of Electronic Materials. 1993;22(8):847–851. doi:10.1007/BF02817496

[20] Maxey CD, Jones CL, Metcalfe NE, Catchpole R, Houlton MR, White AM, Gordon NT, Elliott CT. Growth of fully doped HgCdTe heterostructures using a novel iodine doping source to achieve improved device performance at elevated temperatures. Journal of Electronic Materials. 1996;25(8):1276–1285. doi:10.1007/BF02655020

[21] Madejczyk P, Piotrowski A, Gawron W, Klos K, Pawluczyk J, Rutkowski J, Piotrowski J, Rogalski A. Growth and properties of MOCVD HgCdTe epilayers on GaAs substrates. Optoelectronic Review. 2005;13(3):59–71.

[22] Vydyanath HR. Mechanisms of incorporation of donor and acceptor dopants in (Hg,Cd)Te alloys. Journal of Vacuum Science and Technology. 1991;B9(3): 1716–1723. doi:10.1116/1.585405

[23] Tascar NR, Bhat IB, Parat KK, Ghandhi SK, Scilla GJ. Extrinsic p-doped HgCdTe grown by direct alloy growth organometallic epitaxy. Journal of Vacuum Science and Technology. 1991;B9(3):1705–1708. doi:10.1116/1.585403

[24] Bevan MJ, Chen MC, Shih HD. High-quality p-type HgCdTe prepared by metalorganic chemical vapor deposition. Applied Physic Letters. 1995;67(23):3450–3452. doi: 10.1063/1.115276

[25] Song JH, Kim JW, Park MJ, Kim JS, Jung KU, Suh SH. Iodine and arsenic doping of (100)HgCdTe/GaAs grown by metalorganic vapor phase epitaxy using isopropyl iodide and tris-dimethylaminoarsenic. Journal of Crystal Growth. 1998;184/185:1232–1236. doi:10.1016/S0022-0248(98)80258-3

[26] Edwall DD, Bubulac LO, Gertner. p-type doping of metalorganic chemical vapor deposition—grown HgCdTe by arsenic and antimony. Journal of Vacuum Science and Technology. 1992;B10(4):1423–1427. doi:10.1116/1.586265

[27] Elliott J, Kreismanis VG. Orientation dependence of arsenic incorporation in metalorganic chemical-vapor deposition-grown HgCdTe. Journal of Vacuum Science and Technology. 1992;B10(4):1428–1431. doi:10.1116/1.586266

[28] Mitra P, Schimert TR, Case FC, Barnes, Reine MB, Starr R, Weiler MH, Kestigian. Metalorganic chemical vapor deposition of HgCdTe p/n junctions using arsenic and

iodine doping. Journal of Electronic Materials. 1995;24(9):1077–1085. doi:10.1007/BF02653056

[29] Madejczyk P, Gawron W, Martyniuk P, Keblowski A, Piotrowski A, Pawluczyk J, Pusz W, Kowalewski A, Piotriwski J, Rogalski A. MOCVD grown HgCdTe device structure for ambient temperature LWIR detectors. Semiconductor Science and Technology. 2013;28:105017–105023. doi:10.1088/0268-1242/28/10/105017

Chapter 4

Hot Filament Chemical Vapor Deposition: Enabling the Scalable Synthesis of Bilayer Graphene and Other Carbon Materials

Frank Mendoza, Tej B. Limbu, Brad R. Weiner and Gerardo Morell

Additional information is available at the end of the chapter

http://dx.doi.org/10.5772/63921

Abstract

The hot filament chemical vapor deposition (HFCVD) technique is limited only by the size of the reactor and lends itself to be incorporated into continuous roll-to-roll industrial fabrication processes. We discuss the HFCVD reactor design and the interplay between the reactor parameters, such as filament and substrate temperatures, filament-to-substrate distance, and total pressure. Special attention is given to the large-area synthesis of bilayer graphene on copper, which is successfully grown by HFCVD with transmittance greater than 90% in the visible region and no gaps. We also discuss the HFCVD synthesis of carbon nanotubes, microcrystalline diamond, and nanocrystalline diamond.

Keywords: hot filament chemical vapor deposition, bilayer graphene, diamond, carbon nanotubes, 2D material

1. Introduction

Graphene can be obtained by a variety of techniques, including mechanical and chemical exfoliation of graphite, and chemical vapor deposition (CVD) methods, such as hot wall and hot filament CVD. Each technique has its own advantages and disadvantages that have to be weighed according to the intended application. The mechanical exfoliation of graphite produces high-quality monolayer graphene flakes [1, 2], whereas chemical exfoliation produces graphene flakes with significant structural defects [3]. Although the individual micron-scale graphene flakes can be used to make miniature test devices that provide useful insight into the physical properties and potential applications of graphene [3], the reality is that graphene powders obtained by exfoliation methods are not useful for the mass production of reliable and reproducible electronic devices. However, hot-wall CVD appears to be the most straightforward method to produce monolayer graphene films with low defect density, typically on Cu substrates [4]. Nonetheless, the graphene films obtained by this technique are polycrystalline and discontinuous [5]. **Table 1** compares the physical properties of graphene obtained by different methods.

Method	Crystallite size (μm)	Sample size (mm)	Charge carrier mobility (room temperature) ($cm^2V^{-1}s^{-1}$)	Sheet resistance (Ω/sqr) / 2L
Mechanical exfoliation [1, 2]	~1 [5]	~1 [8]	(2L) >10^4 (at low temperature) [9]	10^2 [9]
Chemical exfoliation [3]	≤0.1 [3]	Large area of overlapping flakes [3]	100 (for a layer of overlapping flakes) [5]	30 kΩ Drop-cast film [3]
Chemical exfoliation through graphene oxide [3]	~100 [3]	Large area of overlapping flakes [3]	200–300 (for a layer of overlapping flakes) [8]	8–150 kΩ Multilayer Langmuir-Blodgett film [3]
CVD [9]	0.5 [5]	~5 [10]	5000 [9]	900 [11]
Epitaxial growth (SiC) [12, 13]	50 [5]	50 [5]	15,000 [14]	150–300 [15]
HFCVD [7]	≤0.1 [7] 0.1–0.2 [16]	30 [7]	~2500 [This work]	~800 [This work] ~650 [16]

Table 1. Properties of graphene films obtained by different methods.

An alternative approach to advance the potential applications of graphene is to focus the attention on bilayer graphene, which has a zero band gap like monolayer graphene. Furthermore, the band gap of bilayer graphene has been shown to be tunable, and its domain walls have shown ballistic electron-conducting channels [6]. These exciting physical properties expand the potential applications of bilayer graphene beyond those of monolayer graphenes.

Mendoza et al. [7] reported that large-area bilayer graphene can be grown by hot filament chemical vapor deposition (HFCVD) with transmittance greater than 90% in the visible region and no gaps. The size of the graphene wafers that can be obtained by HFCVD is only limited by the size of the reactor and lends itself to be incorporated into a continuous roll-to-roll fabrication process, thus lowering the cost per unit area and enabling the integration of bilayer graphene in large-scale industrial production lines. The HFCVD method has the potential to enable large-area applications of graphene, such as transparent electrodes for flat panel displays and solar cells. In this chapter, we review the HFCVD state of the art for the synthesis of bilayer graphene and provide insight into how to tune the synthesis process to obtain optimum bilayer graphene films. We also discuss the HFCVD synthesis of diamond and carbon nanotubes.

2. The hot filament chemical vapor deposition reactor

The hot filament chemical vapor deposition (HFCVD) technique relies on a heated coiled wire to decompose the precursor reactants present in the gas mixture and deposit a film on the substrate surface kept near the filament at lower temperatures. HFCVD was introduced in 1979 for the synthesis of amorphous silicon films from silane gas at low substrate temperatures and high deposition rate [17]. For the synthesis of carbon materials (e.g., diamond, carbon nanotubes, graphene), a gas mixture consisting of a hydrocarbon (e.g., methane, acetylene) diluted in hydrogen is decomposed utilizing a refractory metal filament (such as tungsten, tantalum, or rhenium) that is resistively heated and maintained to 1800–2300°C, as measured with a dual wavelength pyrometer. Heterogeneous reactions at the hot filament surface and the emissive properties of the incandescent filament [18–21] initiate the decomposition of the hydrocarbon gases and the subsequent cascade of chemical vapor reactions. The filament temperature plays an essential role and influences the type and quality of the carbon materials that are grown. For example, at filament temperatures below 1800°C, little or no diamond film is obtained. Higher filament temperatures up to 2300°C lead to higher growth rates and higher quality polycrystalline diamonds. However, the higher the filament temperature, the faster the filaments become carburized. The metal carbides are brittle, resulting in relatively short

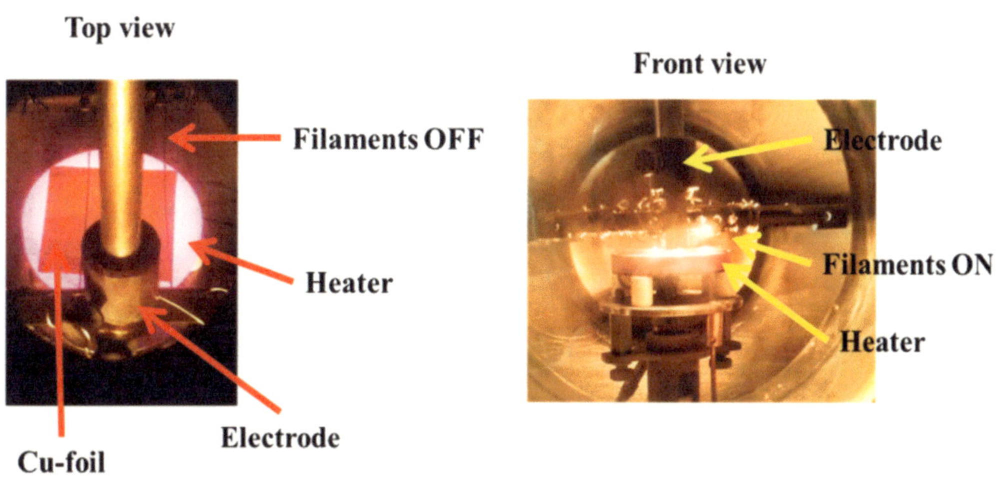

Figure 1. Internal details of the HFCVD setup for large-area graphene film deposition.

filament lifetimes and unsteady growth conditions. For stability and reproducibility, the filament material of choice is rhenium, which carburizes very slowly and lasts many cycles before requiring replacement.

For a 5-cm wafer, the HFCVD reactor typically consists of a 20-liter six-way cross stainless steel chamber. The substrate holder sits on the top of the substrate heater at the center of the chamber. The filament assembly is kept 5–10 mm above the substrate holder. **Figure 1** shows an actual picture of the main internal components in an HFCVD system. The filament wire diameter, length, and geometry must be optimized according to the intended film size, as well as the filament temperature and gas pressure required by the specific chemical vapor deposition reaction. It is necessary to assess the filament material's resistivity as a function of temperature and pressure. For example, the empirical resistivity of rhenium is approximately $2 \times 10^{-6}\ \Omega\cdot m$ at 2000°C and 3×10^{3} Pa. The filament is held fixed by two parallel electrodes made of a refractory metal, typically molybdenum, to avoid evaporation of the electrode material onto the growing film. Coil-shaped filaments are often used because they enhance the total surface area available for initiating the heterogeneous decomposition reactions, but they tend to bend downward over time during film growth causing unsteady growth conditions. A filament arrangement consisting of multiple parallel stretched filaments of 0.25–0.50 mm diameter, as shown in **Figure 1**, is usually a better choice because it enhances the uniformity and total area of the resulting films while also providing considerable surface area for the heterogeneous decomposition reactions. However, the multiple filament arrangement requires very high-current power supplies.

The substrate temperature required for the synthesis of a particular material is determined by the residence time and mobility of reactant molecules and radicals that adsorb at the substrate surface. In the HFCVD reactor, the substrate temperature is intrinsically affected by the filament temperature and power, filament-to-substrate distance, and total pressure. Therefore, active substrate temperature control is needed in most cases in order to be able to optimize the quality of the films grown. While the filament temperature is dictated by the energy required to break the precursor gas molecules to initiate the chemical vapor reactions, the filament-to-substrate distance and total pressure are determined by the mean free path of critical radicals that form at or near the filament and must travel to the substrate surface before they get scavenged through other chemical reactions that do not lead to film deposition.

Besides determining the required filament and substrate temperatures, filament-to-substrate distance, and total pressure, the choice of substrate material and gas composition are also critical parameters to obtain the intended film type and quality. For example, under the same conditions, diamond can grow on molybdenum but not on quartz. The right gas-phase chemical environment and substrate temperature are conditions necessary but not sufficient to grow the intended film. The heterogeneous reactions that take place at the substrate surface are critically important in any chemical vapor deposition process. Therefore, the appropriate substrate must be provided to adsorb the reactive chemical species from the gas phase and catalyze the film synthesis reactions. After the right substrate is provided, the relative abundance of reactive chemical species will determine the film quality and growth rate and is primarily determined by gas composition. Priming or preparation of the substrate surface is usually needed because it is covered with all sorts of adsorbates from the ambient environment; seeding may also be needed in some cases. The priming may be done before insertion in the HFCVD chamber or during the initial HFCVD process. For example, diamond growth on silicon must be preceded by diamond seeding prior to insertion in the HFCVD and by the removal of the native silicon oxide layer, which can be accomplished by attack of the hydrogen radicals in the HFCVD [Solid State Communications 116, 217, 2000]. In this case, there is an induction period during which the intended film does not grow until the substrate surface is ready to adsorb the reactive species. In the case of graphene grown on copper foil, the copper oxide layer is first removed by exposure to hydrogen at 1000°C.

3. Reports on HFCVD graphene

Singh et al. [16] reported the growth of large sp^2 domain size (117–279 nm) single, bilayer, and multilayer graphene films on thick Ni foils using HFCVD at a relatively low substrate temperature of 700°C and low pressure (<200 Pa) at relatively short deposition time (10–30 min). The strategy used in this work is based on the mechanism by which hydrocarbon molecules decompose at the metal surface and diffuse into the bulk. Carbon diffuses into the bulk of nickel while it is kept hot and precipitates on the nickel surface forming nonuniform and discontinuous graphene layers during the fast cooling phase. This work presents the HFCVD as the advantage to dissociate at high temperature on filaments (tungsten filament at 1800–2200°C) source gases while temperature on heater is relatively low (700°C) compared with other CVD process obtaining high-quality films. Accordingly, with the different characterization (i.e., Raman spectra and SAED-TEM) presented in his report, the graphene samples (1–6 layers) obtained on nickel shows good crystallinity and also low sheet resistance values (630 Ω/sqr).

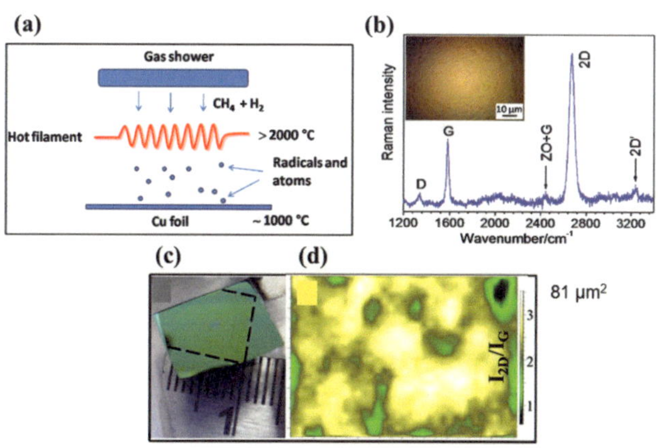

Figure 2. (a) Schematic diagram of the HFCVD used, (b) Raman spectra of the graphene obtained, (c) picture of the graphene transferred on SiO_2 substrate, and (d) Raman mapping of a selected area of the hydrogenated graphene. (Image adapted with permission from Kataria et al. [22]).

Using Cu instead of Ni and higher substrate temperature, Kataria et al. [22] reported the growth of relatively large discontinuous patches (~100 μm^2) of high-quality monolayer graphene at 1000°C by HFCVD, see **Figure 2a** the schematic diagram about HFCVD used. This report is summarized in **Figure 2**. Through **Figure 2b**, we can see that the graphene obtained is monolayer, which is corroborated by Raman area map in **Figure 2c**. The Raman spectra in these samples showed an additional band termed 2D′ peak (2946 cm^{-1}) in combination with D (~1353 cm^{-1}) and G (1622 cm^{-1}) bands, indicating hydrogenated graphene. Kataria's group analyzed the possibility to remove hydrogen from the graphene samples by applying temperatures up to 600°C in inert atmosphere. These procedures and Raman spectroscopy examination confirm the desorption of hydrogen from the graphene samples.

Behura et al. [23] reported the synthesis of nonuniform large-domain graphene films on Cu substrates inside a HFCVD reactor [24, 25]. From the report, it is not clear that they actually used the filament during the synthesis process. The material obtained shows high number of defects in the film. This is observable through the large intensity of D-band accordingly with the Raman spectra. Also, the intensity ratio between the G and 2D bands is ~0.7, and the FWHM of the 2D band is ~55 cm^{-1} indicating the growth of few-layer graphene. This report is summarized in **Figure 3**.

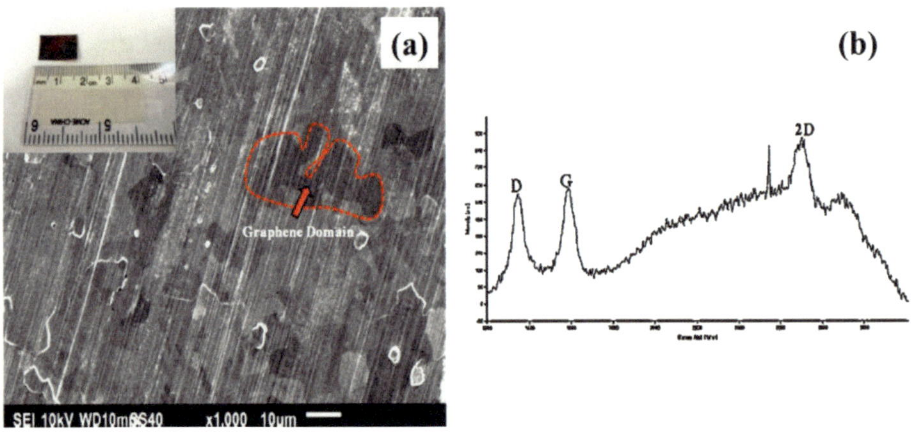

Figure 3. (a) SEM micrograph of graphene films on Cu surface with optical photograph shown in the inset, (b) Raman spectrum of graphene film on Cu surface. (Image adapted with permission from Behura et al. [23]).

4. Obtaining uniform bilayer graphene by HFCVD

The general process to grow graphene by HFCVD is represented in **Figure 4**. The first step is the removal of the copper oxide layer from the copper substrate by exposure to hydrogen at 1000°C. We carry out this step at 4.7 × 10^3 Pa for 30 min. The second step is the actual growth of graphene by adding methane to the gas mixture and heating the filament to 1800°C.

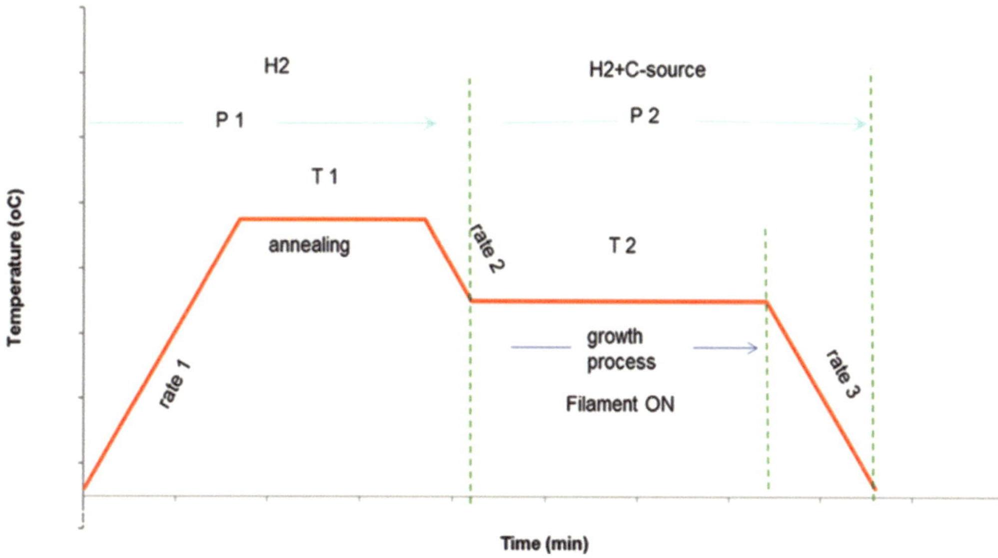

Figure 4. Performance of the temperature during the cleaning process and deposition time.

Mendoza et al. [7] performed a parametric study that included the variation of methane concentration, total pressure, and substrate temperature. The filament-to-substrate distance and filament temperature were kept at 5 mm and 1800°C, respectively. The methane concentration was varied in the 1–50% range in 5% steps, while keeping the pressure at 4.7×10^3 Pa and the substrate temperature at 850°C. The ratio of the intensity of the 2D band to that of the G band is used to determine the number of graphene layers, while the ratio of the intensity of the D band to that of the G is used to determine the degree of disorder in the trigonal carbon network.

Graphene growth took place only in the 15–25% methane concentration range. The pressure was then changed in the 2.7–4.7×10^3 Pa range, and the quality of the graphene films deteriorated as indicated by the D band. Increasing the pressure to 5.3×10^3 Pa yielded amorphous carbon. The substrate temperature was then lowered from 850 to 750°C in steps of 50°C, resulting in lower graphene quality as indicated by the D band. At temperatures higher than 850°C, the graphene quality improves and monolayer is achieved, but it grows in discontinuous patches. In HFCVD, there seems to be a fundamental trade-off between continuous-discontinuous coverage and bilayer-monolayer graphene [26].

Based on the above-described results, we extended the deposition time from 30 to 100 min for the HFCVD graphene growth in the temperature range from 750 to 1000°C, while keeping the gas mixture of 15% methane in hydrogen at 4.7×10^3 Pa. As shown in **Figure 5**, multilayer graphene is obtained below 850°C, and turbostratic bilayer graphene is obtained between 850 and 1000°C. The most uniform bilayer graphene is obtained at 950°C; the graphene film grown at 1000°C is discontinuous.

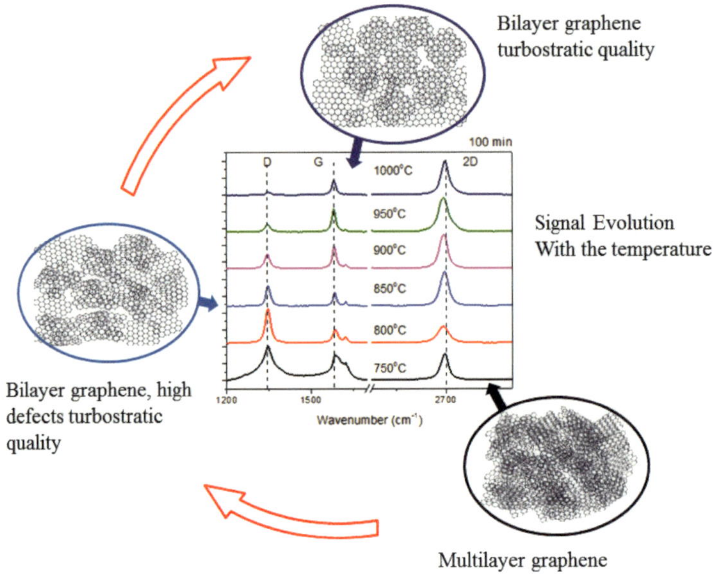

Figure 5. Graphene film evolution as a function of substrate temperature.

We then carried out graphene synthesis experiments as a function of time, keeping the substrate temperature at 950°C and the other parameters constant as well. **Figure 6** shows the Raman spectra of the graphene films as a function of deposition time. The optimum bilayer graphene quality and continuous surface is obtained for the 100-min deposition time. As the

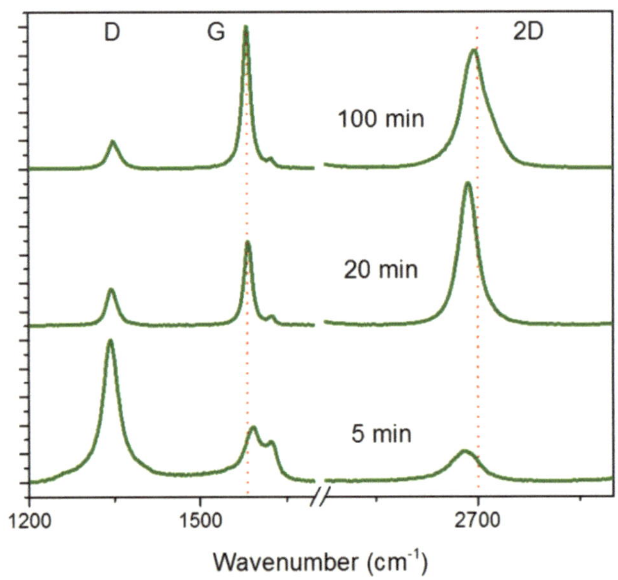

Figure 6. Effect of growth time on bilayer graphene synthesis at 950°C.

http://dx.doi.org/10.5772/63921

deposition time is decreased in 5-min interval down to 20 min, the film continuity is compromised, and gaps remain open on the substrate surface as seen under the optical microscope. The graphene that is grown for less than 10 min shows a very large D band, indicating very small crystallite sizes (<10 nm). Hence, it appears that 100 min allows enough time for the graphene crystallites to grow, defects to heal, and gaps to close. Above 100 min, multilayer graphene starts to dominate the film composition.

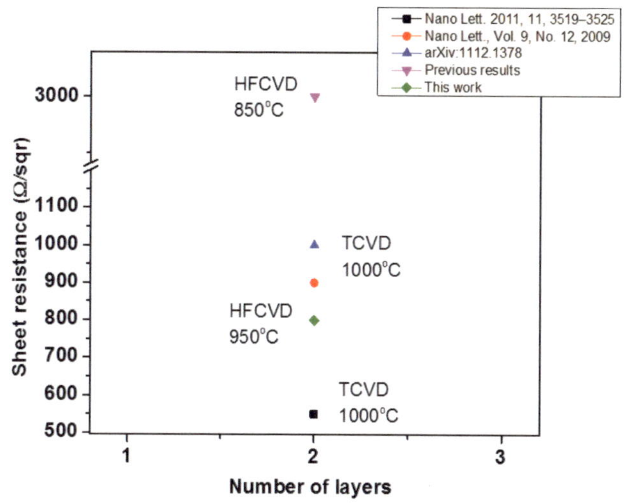

Figure 7. Comparison of sheet resistance values reported for graphene.

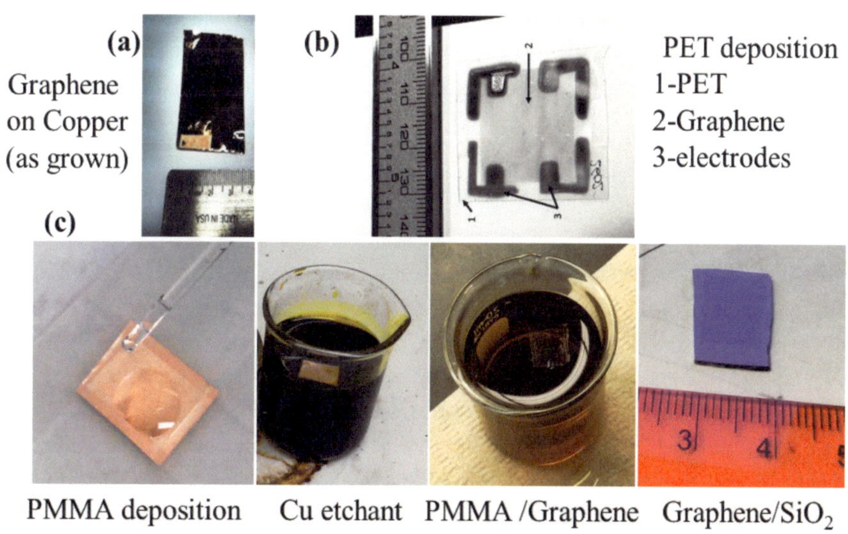

Figure 8. (a) Graphene on copper foil, (b) graphene transferred onto PET, and (c) graphene transfer process from Cu to SiO_2/Si.

The bilayer graphene grown by HFCVD should also be compared with other graphene reports in terms of transmittance and sheet resistance (Ω/sqr) [27–31]. **Figure 7** shows the sheet resistance values for HFCVD and hot wall or thermal chemical vapor deposition (TCVD) graphene with visible transmittance in the 90–95% range. The HFCVD bilayer graphene films grown at 850°C show an average sheet resistance in the order of 3×10^3 Ω/, whereas those grown at 950°C show an average sheet resistance value in the order of 8×10^2 Ω/sqr. The latter value is close to values reported for monolayer graphene grown by TCVD techniques, as seen in **Figure 8**. Moreover, the indium tin oxide (ITO) transparent electrode that is deposited on glass for flat panel systems has sheet resistance values in the order of 1.5×10^3 Ω/sqr [32].

5. Graphene transfer

For transparent electrode applications, the bilayer graphene grown on copper substrates (**Figure 8a**) must be transferred onto the transparent substrate of choice (e.g., PET, **Figure 8b**). The transfer process is another critical step that can preserve or damage the integrity of the graphene films. First, a polymer support is attached to the graphene film, such as polymethyl methacrylate (PMMA), polydimethylsiloxane (PDMS), polyethylene terephthalate (PET). The polymer support is designed to hold the graphene film when the underlying Cu foil is etched away in iron chloride ($FeCl_3$) at 70°C, as shown in **Figure 8c**. Once the Cu is removed, the graphene on polymer support is thoroughly cleaned in HCl to remove residual $FeCl_3$ and Fe^{3+}. Then, the graphene on polymer support is rinsed with deionized water several times and let to dry. The graphene is then ready to be transferred onto any substrate according to the intended application. In the example shown in **Figure 8**, the HFCVD bilayer graphene film was transferred onto a SiO_2/Si wafer by pressing them firmly together and slowly peeling off the polymer support. The integrity of resulting graphene film depends greatly on carefully carrying out the transfer process. The electrical properties of graphene are particularly sensitive to remnant $FeCl_3$ impurities and cracks produced in the transfer process.

6. HFCVD diamond

Microcrystalline and nanocrystalline diamond films are grown by HFCVD on Si, Mo, and other carbide-forming substrates. The substrates need to be seeded by polishing the surface with <0.1-μm synthetic diamond powder. Before the clean reactive gas mixture is introduced into the HFCVD chamber, it is evacuated to 10^{-3} Pa or lower. The gas flow of methane and hydrogen is controlled to obtain fixed methane concentrations of 0.3 and 2.0% for microcrystalline and nanocrystalline diamond, respectively. The gas mixture enters at a rate of 100 sccm and is activated by a heated Re filament positioned at 8 mm above the substrate. The choice of Re for the filament material has the advantage that it does not react with carbon and therefore is not consumed during the diamond reaction. The filament is resistively heated in the range of 2300–2500°C, as measured by an optical pyrometer. The total gas pressure of the chamber is kept at approximately 2.7 ± 0.1 Pa. The substrate temperature can be varied in the 600–900°C range.

http://dx.doi.org/10.5772/63921

The growth rate varies in the 0.1–0.5 μm/h range depending on the substrate temperature [33]. The SEM images in **Figure 9** illustrate the differences in morphology and grain size between microcrystalline and nanocrystalline diamond.

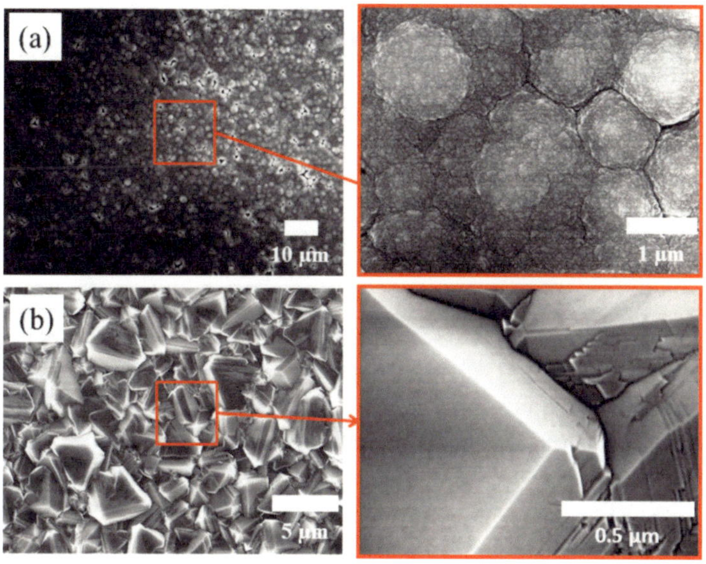

Figure 9. Micrographs of (a) nanocrystalline and (b) microcrystalline diamond.

7. HFCVD carbon nanotubes

Carbon nanotubes (CNTs) are grown by HFCVD on Cu at a substrate temperature of 900°C using a concentration of 2.0% methane in hydrogen. The growth parameters are similar to those required to grow nanocrystalline diamond; the crucial difference is the choice of substrate material, which should act as catalyst for the growth of CNTs. Before the clean reactive gas mixture is introduced into the HFCVD chamber, it is evacuated to 10^{-3} Pa or lower. The gas mixture enters at a rate of 100 sccm and is activated by a heated Re filament positioned at 8 mm above the substrate. The filament is resistively heated in the range of 2300–2500 °C, as measured by an optical pyrometer. The total gas pressure of the chamber is kept at approximately 2.7 ± 0.1 Pa. A thick layer of carbon nanotubes is readily obtained in about 15 min [34, 35].

The SEM images (**Figure 10a** and **b**) show that the films are composed of entangled clusters of nanotubes or nanofiber structures with the catalyst material present in the form of spherical tips. A more detailed by TEM (**Figure 10c**) shows that the fibers are bamboo-like carbon nanotubes (BCNTs) with diameters ranging from 50–100 nm and variable lengths. They contain nanocavities that are stacked one over the other with closed walls consisting of a number of graphene layers.

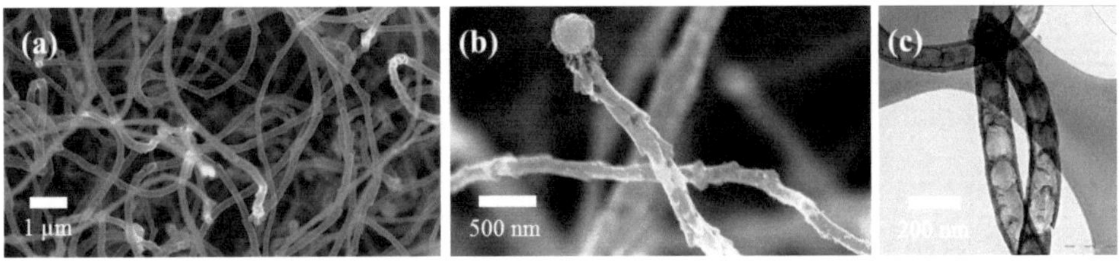

Figure 10. Scanning electron microscopy of (a) HFCVD carbon nanotubes, (b) Bamboo-like carbon nanotubes (BCNTs), and (c) transmission electron microscopy of BCNTs.

8. Summary

In this chapter, we discussed the HFCVD reactor design and the interplay between the reactor parameters, such as filament and substrate temperatures, filament-to-substrate distance, and total pressure. Special attention was given to the large-area synthesis of bilayer graphene on copper, but we also discussed the HFCVD synthesis of microcrystalline diamond, nanocrystalline diamond, and carbon nanotubes. Large-area bilayer graphene grown by HFCVD has transmittance greater than 90% in the visible region and no gaps. The size of the graphene wafers that can be obtained by HFCVD is only limited by the size of the reactor and lends itself to be incorporated into a continuous roll-to-roll fabrication process, thus lowering the cost per unit area and enabling the integration of bilayer graphene in large-scale industrial production lines. The HFCVD method has the potential to enable large-area applications of graphene, such as transparent electrodes for flat panel displays and solar cells.

Acknowledgements

This work is funded by the Institute for Functional Nanomaterials (NSF Grant No. 1002410) and in part by PR NASA EPSCoR (NNX15AK43A).

Author details

Frank Mendoza[1,2*], Tej B. Limbu[1,2], Brad R. Weiner[1,3] and Gerardo Morell[1,2]

*Address all correspondence to: fran.mendoza@upr.edu

1 Institute for Functional Nanomaterials, University of Puerto Rico, San Juan, USA

2 Department of Physics, University of Puerto Rico at Rio Piedras, San Juan, USA

3 Department of Chemistry, University of Puerto Rico at Rio Piedras, San Juan, USA

References

[1] Kusmartsev FV, Wu WM, Pierpoint MP, and Yung KC.; Application of Graphene within Optoelectronic Devices and Transistors. arXiv:1406.0809 [cond-mat.mtrl-sci].

[2] Novoselov KS, Geim AK, Morozov SV, Jiang D, Zhang Y, Dubonos SV, Grigorieva IV, Firsov AA. Electric Field Effect in Atomically Thin Carbon Films. Science 2004;306:666–669

[3] Park S and Ruoff RS. Chemical methods for the production of graphenes. Nat. Nanotechnol. 2009;4:217–224.

[4] Li X., Cai W., An J., Kim S., Nah J., Yang D., Piner R., Velamakanni A., Jung I., Tutuc E., Banerjee SK, Colombo L and Ruoff RS. Large-area synthesis of High Quality and Uniform Graphene films on Copper Foils. Science, 2009, 324, 1312–1314

[5] Novoselov KS, Fal'ko VI, Colombo L, Gellert PR, Schwab MG, and Kim K. A road map for graphene. Nature. 2012;490(7419):192–200.

[6] Zhang Y, Tsung T-T, Girit C, Hao Z, Martin MC, Zettl A, Crommie MF, Shen YR, and Wang F. Direct observation of widely tunable bandgap in bilayer graphene. Nature. 2009;459:820–823.

[7] Mendoza F, Limbu TB, Weiner BR, and Morell G. Large-area bilayer graphene synthesis in the hot filament chemical vapor deposition reactor. Diamond Relat. Mater. 2015;51:34.

[8] Eigler S, Enzelberger-Heim M, Grimm S, Hofmann P, Kroener W, Geworski A, Dotzer C, Röckert M, Xiao J, Papp C, Lytken O, Steinrück H-P, Müller P, and Hirsch A. Wet chemical synthesis of graphene. Adv. Mater. 2013;25:3583–3587.

[9] Mattevi C, Kim H, and Chhowalla M. J. A review of chemical vapour deposition of graphene on copper. Mat. Chem. 2011;21(10): 3324–3334.

[10] Kim KS, Zhao Y, Jang H, Lee SY, Kim JM, Kim KS, Ahn JH, Kim P, Choi J-Y, and Hong BH. Large-scale pattern growth of graphene films for stretchable transparent electrodes. Nature. 2009;457:706–710.

[11] Li X, Zhu Y, Cai W, Borysiak M, Han B, Chen D, Piner RD, Colombo L, and Ruoff RS. Transfer of Large-Area Graphene Films for High-Performance Transparent Conductive Electrodes. Nano. Lett. 2009;9:4359–4363.

[12] Berger C, Song Z, Li X, Wu X, Brown N, Naud C, Mayou D, Li T, Hass J, Marchenkov AN, Conrad EH, First PN, and Heer WA. Electronic Confinement and Coherence in Patterned Epitaxial Graphene. Science. 2006;312:1191–1196.

[13] Sutter P. Epitaxial Graphene: How silicon leaves the scene. Nat. Mater. 2009;8(3):171.

[14] Wu X, Hu Y, Ruan M, Madiomanana NK, Hankinson J, Sprinkle M, Berger C, and Heer WA. Half integer quantum Hall effect in high mobility single layer epitaxial graphene. Appl. Phys. Lett. 2009;95:223108.

[15] Krupka J and Strupinski W. Measurements of the sheet resistance and conductivity of thin epitaxial graphene and SiC films. Appl. Phys. Lett. 2010;96:082101.

[16] Singh M, Jha HS, and Agarwal P. Growth of Large Sp2 domain size single and multilayer graphene films at low substrate temperature using hot filament chemical vapor deposition. Mat. Lett. 2014;126:249–252.

[17] Wiesmann H, Ghosh AK, McMahon T, and Strongin M. J. a-Si:H produced by high-temperature thermal decomposition of silane. Appl. Phys. 1979;50:3752.

[18] Dupuie JL and Gulari E. Hot Filament enhanced chemical vapor deposition of AIN thin films. Appl. Phys. Lett. 1991;59(5):549–551.

[19] Jackson MJ, editor. Microfabrication and nanofacturing. Microfabrication and Nanomanofacturing. 1st ed. Boca Raton, FL: International Standard Book. 9780824724313; 2010. 401 p.

[20] Wei QP, Ashfold MNR, Mankelevich YA, Yu ZM, Liu PZ, and Ma L. Diamond growth on WC-Co substrates by hot filament chemical vapor deposition: Effect of filament-substrate separation. Diamond Relat. Mater. 2011;20:641–650.

[21] Ali M, Ürgen M, and Atta MA. Tantalum carbide films synthesized by hot filament chemical vapor deposition technique. Surf. Coat. Technol. 2012;206:2833–2838.

[22] Kataria S, Patsha A, Dhara S, Tyagi AK, and Barshilia HC. J. Raman, Raman imaging on high-quality graphene grown by hot-filament chemical vapor deposition. Spectrosc. 2012;43:1864–1867.

[23] Behura SK, Mukhopadhyay I, Jani O, Yang Q, and Hirose A. Synthesis and characterization of graphene films by hot filament chemical vapor deposition. Proceedings of the 24th CANCAM Saskatoon; Saskatchewan, Canada. 2013.

[24] Yang Q., Xiao C., Chen W., Singh AK, Asai T., and Hirose A, Growth mechanism and orientation control of well-aligned carbon nanotubes. Diam. Relat. Mater. 2003;12:1482–1487.

[25] Behura SK, Nayak S, Yang Q, Hirose A, Jani O. J. Chemical Vapor Deposited Few-Layer Graphene as an Electron Emitter. Nanosci. and Nanotechnol. 2016;16:287–295

[26] Hawaldar R, Merino P, Correia MR, Bdikin I, Grácio J, Méndez J, Martín-Gago J-A, and Singh MK. Large-area high-throughput synthesis of monolayer graphene sheet by Hot Filament Thermal Chemical Vapor Deposition. Sci. Rep. 2012;2:682.

[27] Chen S, Cai W, Piner RD, Suk JW, Wu Y, Ren Y, Kang J, and Ruoff R. Synthesis and Characterization of Large-Area Graphene and Graphite Films on Commercial Cu-Ni Alloy Foils. Nano. Lett. 2011;11:3519–3525.

[28] Buijnsters JG, Shankar P, van Enckevort WJP, Schermer JJ, and Ter Meulen JJ. The Adhesion of Hot-Filament CVD Diamond Films on AISI Type 316 Austenitic Stainless Steel. Diamond Relat. Mater. 2004;13:848–857.

[29] Chen H, Jang C, Xiao S, Ishigami M, and Fuhrer MS. Intrinsic and extrinsic performance limits of graphene devices on SiO_2. Nat. Nanotechnol. 2008;3:206.

[30] Li X, Zhu Y, Cai W, Borysiak M, Han B, Chen D, Piner RD, Colombo L, and Ruoff RS. Transfer of large-area graphene films for high-performance transparent conductive electrodes. Nano. Lett. 2009;9:124359.

[31] Lee S, Lee K, Liu C-H, and Zhong Z.; Homogeneous bilayer graphene film based flexible transparent conductor. arXiv:1112.1378 [cond-mat.mes-hall]

[32] Hong S-J, Kim M-S, Kim J-W, and Shin M. Characteristics of indium-tin-oxide (ITO) glass re-used from old TFT-LCD panel. Mater. Trans. 2012;53(5):968–972.

[33] Gupta S, Weiner BR, and Morell G. J. Synthesis and characterization of sulfur-incorporated microcrystalline diamond and nanocrystalline carbon thin films by hot filament chemical vapor deposition. Mater. Res. 2003;18:363.

[34] Katar SL, Hernandez D, Labiosa AB, Mosquera-Vargas E, Fonseca L, Weiner B, and Morell G. SiN/bamboo like carbon nanotube composite electrodes for lithium ion rechargeable batteries. Electrochimica Acta 2010;55:2269.

[35] Room temperature gas sensor based on tin dioxide-carbon nanotubes composite films. Mendoza F, Hernández DM, Makarov V, Febus E, Weiner BR, Morell G Sensors and Actuators B: Chemical; 190:227–233

Chapter 5

In Situ Observation of Chemical Vapour Deposition Using Langasite Crystal Microbalance

Hitoshi Habuka

Additional information is available at the end of the chapter

http://dx.doi.org/10.5772/62389

Abstract

A method of *in situ* observation using langasite crystal microbalance (LCM) is described for chemical vapour deposition (CVD). First, the frequency behaviour of the LCM is expressed using the equation having the optimized coefficients in a wide range of gas-phase conditions for the CVD. Next, by the LCM frequency behaviour, the existence of surface chemical reactions in a CVD reactor is determined. Additionally, the LCM can determine the lowest temperature for initiating the film deposition. In the last part, the temperature change related to the film formation process is described.

Keywords: Chemical vapour deposition, *In situ* observation, Langasite crystal microbalance, Surface reaction, Gas properties

1. Introduction

Chemical vapour deposition (CVD) is currently a fascinating technology for producing thin films in various advanced industries [1]. The CVD process is a complicated one having transport phenomena linked with the gas-phase and surface chemical reactions. For clarifying and designing the CVD process, the phenomena in the CVD reactor should be understood and optimized. For this purpose, the computational fluid dynamics (CFD) has been significantly advanced [2] and is actually very useful. In contrast, an experimental approach is still not easy [3], because the sensors seriously suffer from thermal, mechanical and chemical damage by the reactive and high-temperature ambient.

The *in situ* monitoring technique by means of the langasite crystal microbalance (LCM) [4–6] can be an appropriate experimental solution, because the LCM can sensitively detect various changes in the CVD reactor, such as heat, flow and film deposition [5, 6]. The information obtained by the LCM should be used for understanding the CVD phenomena. Here, the LCM frequency nonlinearly influenced by various parameters should be clarified. Additionally, a practical process for the measurement and analysis should be developed. In this chapter, the process to *in situ* observe the CVD is thus explained using the LCM.

In Section 1, the frequency behaviour of the LCM is discussed using ambient gas mixtures at atmospheric pressure and at various temperatures. In order to express the LCM frequency decrease with the increasing concentrations of various gases in ambient hydrogen, the LCM frequency difference between the gas mixture and the carrier gas is practically expressed by optimizing the coefficients accounting for the gas properties. In Section 2, a method for determining the existence of surface chemical reactions is explained. The parameter, $C(T)$, of the equation is used for expressing the LCM frequency change in a hydrogen–monomethylsilane system. In Section 3, the lowest temperature for initiating the film deposition is evaluated using trichlorosilane gas and boron trichloride gas. In Section 4, the way to obtain the temperature change related to the film formation in a trichlorosilane–hydrogen system is described. First, the time constants for the LCM frequency change due to the surface and gas-phase temperature change are determined. Next, the continuous LCM frequency decrease is assigned to the weight increase by the film formation in a steady state. Based on an evaluation of the difference in the LCM frequency between those with and without the heat change related to the film formation, the surface temperature decrease caused by changing the precursor concentration is obtained.

2. Langasite crystal microbalance frequency behaviour over wide gas-phase conditions for chemical vapour deposition

In this section, a relationship between the LCM frequency and the gas properties is expressed [7] as a practical equation applicable to the CVD conditions using several gases at various temperatures and gas concentrations.

2.1. Experimental procedure

Figure 1 shows the horizontal cold wall CVD reactor containing the langasite ($La_3Ga_5SiO_{14}$) crystal microbalance (LCM) [4, 8]. This reactor consists of a gas supply system, a rectangular-shaped quartz chamber and five infrared lamps. Hydrogen, nitrogen, trichlorosilane, monomethylsilane and boron trichloride gases are used. The carrier gas is hydrogen. The LCM (Halloran Electronics, Tokyo, Japan) having a fundamental frequency of 10 MHz is placed 5 mm above the silicon wafer surface. The LCM frequency decreases with the increasing temperature [5, 6]. The silicon wafer and the LCM are simultaneously heated by infrared light from halogen lamps through the quartz chamber.

A typical process is shown in **Figure 2**. First, the LCM is heated to 160–600°C in hydrogen at atmospheric pressure. After waiting until the LCM frequency becomes stable, various gases are introduced at atmospheric pressure into the reactor chamber. The total gas flow rate is adjusted to 1 slm.

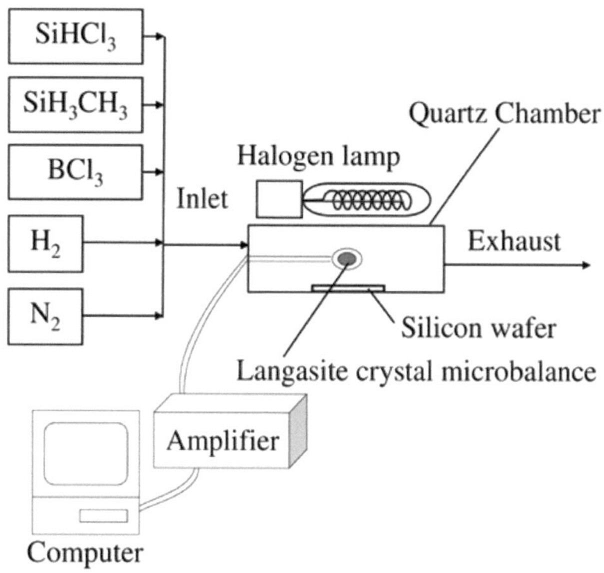

Figure 1. Chemical vapour deposition reactor containing a langasite crystal microbalance.

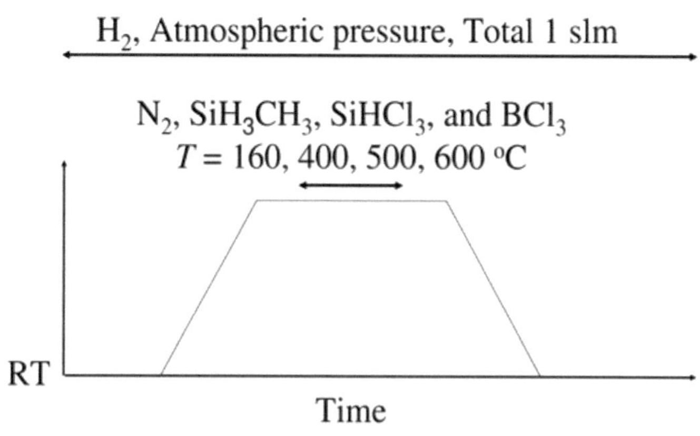

Figure 2. Process for measuring the frequency of the langasite crystal microbalance under various conditions.

2.2. LCM frequency and fluid property

The relationship between the LCM frequency and the gas properties is described, accounting for the various relationships shown in **Figure 3**. Following a previous paper [9], the LCM frequency change, Δf_{Gas}, at a fixed temperature, T (K), is expressed using the product of the gas density, ρ_{gas}, and gas viscosity, η_{Gas} as follows:

$$\Delta f_{\mathrm{Gas}} = -C(T)\rho^{y}_{Gas}\eta^{z}_{Gas}, \tag{1}$$

$$C(T) = \left(\frac{f_C^x}{\pi \rho^y_c \mu^z_c} \right) \tag{2}$$

where f_c is the resonant frequency of the fundamental mode of the langasite crystal. ρ_c and μ_c are the density and shear modulus, respectively, of the langasite crystal. Based on Kanazawa et al. [9], the x, y and z values are 1.5, 0.5 and 0.5, respectively. In a vacuum, the Δf value becomes zero.

The LCM frequency change from the vacuum condition to the hydrogen and to the gas mixtures are shown in **Figure 3** and are expressed by Eqs. (3) and (4).

$$\Delta f_{\mathrm{H_2}} = -C(T)\rho^{y}_{\mathrm{H_2}}\eta^{z}_{\mathrm{H_2}} \quad \text{(in hydrogen).} \tag{3}$$

$$\Delta f_{\mathrm{Mix}} = -C(T)\rho^{y}_{\mathrm{Mix}}\eta^{z}_{\mathrm{Mix}} \quad \text{(in gas mixture).} \tag{4}$$

Although the LCM frequency in the vacuum, as a initial value, should be determined, experiment by experiment, the LCM frequency measurement in the vacuum is often not easy in the atmospheric and low-pressure CVD system. Thus, the accurate measurement of the LCM frequency difference from the vacuum condition, such as $\Delta f_{\mathrm{H_2}}$ and Δf_{Mix}, is difficult.

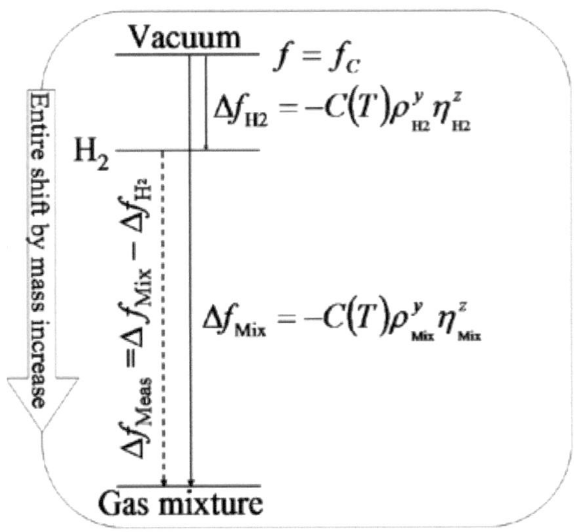

Figure 3. Frequency of langasite crystal microbalance for hydrogen and gas mixtures.

In contrast, the LCM frequency change at the CVD condition from the carrier gas ambient, such as hydrogen ambient, is easily measured by simple operation. The LCM frequency change from that in the hydrogen (100%), Δf_{Meas}, is expressed by Eq. (5).

$$\Delta f_{Meas} = \Delta f_{Mix} - \Delta f_{H_2}. \tag{5}$$

Following this, the Δf_{Meas} value can be accurately determined, while the mass change of the LCM occurs by means of the multiple use of the LCM.

By evaluating the ratio of the measured LCM frequency in the gas mixture to that in hydrogen, Δf_{Meas} is shown to have a relationship with the properties as described by Eq. (6).

$$\Delta f_{Meas} = \frac{\Delta f_{H_2}}{\rho_{H_2}^{y}\eta_{H_2}^{z}} \rho_{Mix}^{y}\eta_{Mix}^{z} - \Delta f_{H_2}. \tag{6}$$

The Δf_{H_2} value can be sufficiently smaller than the first term in Eq. (6).

$$\ln \Delta f_{Meas} = y \ln \rho_{Mix} + z \ln \eta_{Mix} + A, \tag{7}$$

where A is a constant. Using Eq. (7) accounting for various LCM frequencies and the fluid properties, the y and z values can be optimized.

In Eq. (7), the gas density and the gas viscosity are obtained following the ideal gas law and the Chapman–Enskog equation [10], respectively. The viscosity of the gas mixture is calculated following Pollard and Newman [11].

2.3. Influenced of gas density and viscosity

Figure 4 shows the Δf_{Meas} values obtained at 160, 400, 500 and 600°C using various gas mixtures, such as hydrogen–nitrogen, hydrogen–trichlorosilane, hydrogen–monomethylsilane. In this figure, the Δf_{Meas} value is plotted as a function of $\rho^y\mu^z$ assuming that both the y and z values are 0.5 following a previous study [9]. As shown in this figure, the Δf_{Meas} value follows a single trend and does not depend on the gas species. However, the behaviour shown in this figure is not linear.

Next, using Eq. (7), the y and z values are obtained. Using the measurement shown in **Figure 4**, both the y and z values are determined to be 1.3. As shown in **Figure 5**, the Δf_{Meas} value has a linear relationship with $\rho^{1.3}\mu^{1.3}$. Similar to **Figure 4**, the LCM frequency change in **Figure 5** does not depend on the gas species.

Using Eqs. (1) and (2), the $C(T)$ values at 160, 400, 500 and 600°C are obtained as shown in **Figure 6**. This figure shows that the $C(T)$ value is expressed following the Arrhenius-type behaviour.

$$C(T) = 5.4 \times 10^{11} e^{(-2100/T)} \tag{8}$$

Following Eqs. (1), (2), (5) and (8) using the y and z values of 1.3, the Δf_{Meas} value is calculated and plotted versus the measurement from 160 to 600°C, as shown in **Figure 7**. This figure shows that the calculation could sufficiently reproduce the measurement.

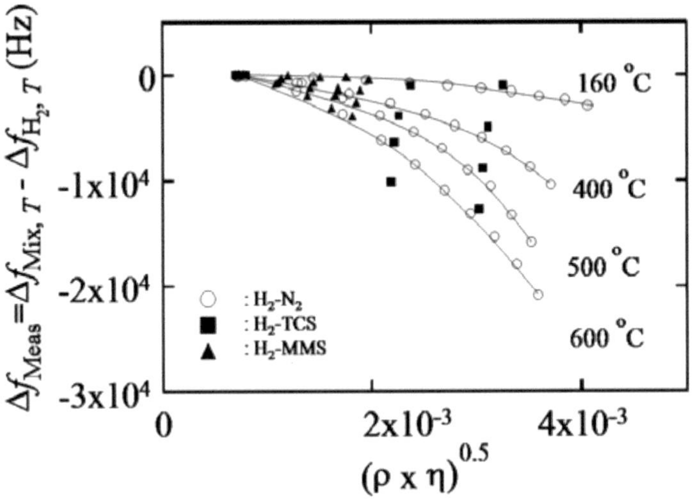

Figure 4. The Δf_{Meas} value obtained for the gas mixture containing nitrogen, trichlorosilane and monomethylsilane in the hydrogen carrier gas at 160–600°C, plotted as a function of $\rho^{0.5}\mu^{0.5}$.

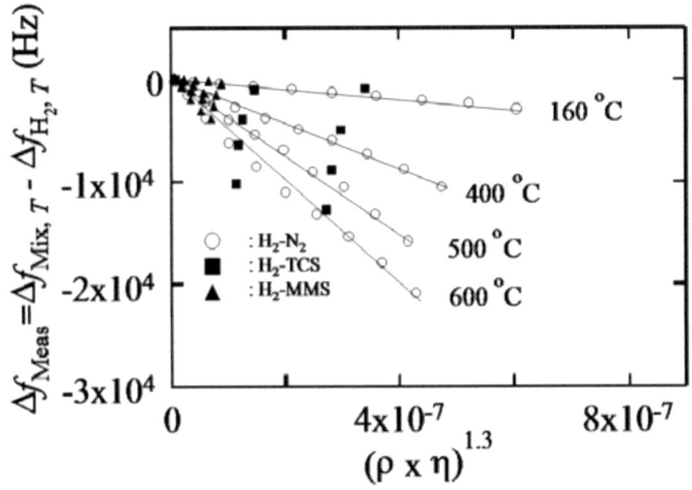

Figure 5. The Δf_{Meas} value obtained for the gas mixture containing nitrogen, trichlorosilane and monomethylsilane in the hydrogen carrier gas at 160–600°C, plotted as a function of $\rho^{1.3}\mu^{1.3}$.

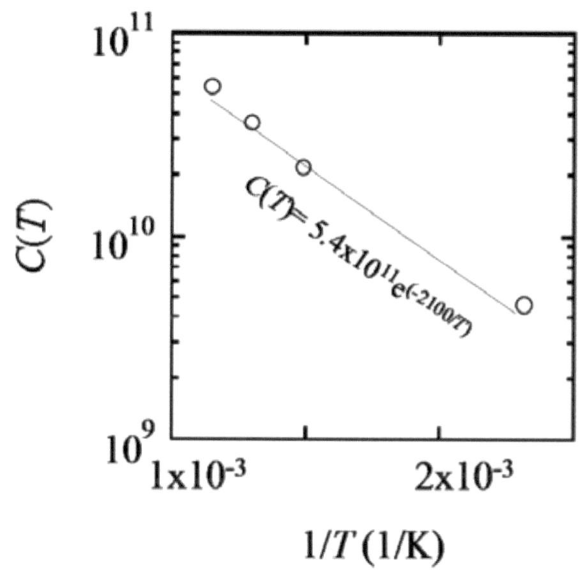

Figure 6. $C(T)$ values changing with $1/T$.

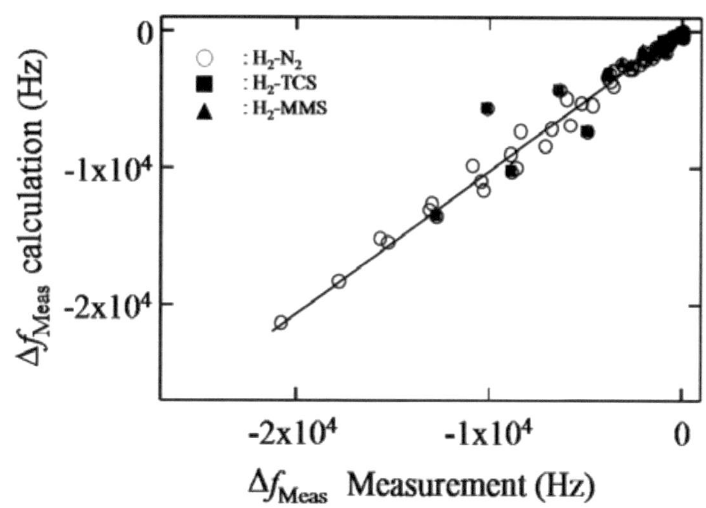

Figure 7. Correlation of the Δf_{Meas} values between the measurement and the calculation by Eqs. (1), (2), (5) and (8) using y and z of 1.3 at the temperatures of 160–600°C.

3. Method for determining chemical vapour deposition occurrence

In this section, the method for determining the surface chemical reaction occurrence [12] is explained. The silicon carbide film formation from monomethylsilane gas is discussed.

3.1. LCM frequency

The LCM frequency differences from ambient hydrogen in monomethylsilane–hydrogen system are shown in **Figure 8**. Immediately after opening the gas valve at 0 s, a high concentration of monomethylsilane gas, remained at 100% between the gas valve and the mass flow controller, is introduced into the reactor. During the high concentration gas passing through the reactor, the LCM frequency significantly decreases. Thereafter, the diluted monomethylsilane gas reaches the LCM in the reactor. Thus, the fluctuation of the LCM frequency becomes quite small.

As shown in **Figure 8**, the LCM frequency difference measured at 300°C is very stable after 20 s. This indicates that the gas temperature and the fluid properties, such as the gas density and gas viscosity, are in a steady state. This simultaneously indicates that there is no chemical reaction at 300°C, due to no thermal change caused by no reaction heat. The behaviour at 400 and 500°C is similar to that at 300°C. The trend in the LCM frequency difference at 300, 400 and 500°C is the same and parallel to each other. Thus, the MMS-H_2 system below 500°C is concluded to undergo no chemical reaction.

Although the LCM frequency difference at 550°C seems to be relatively stable, it very gradually increases after 20 s. Similar to this, the LCM frequency difference at 600°C also slightly increases with a larger gradient than that at 550°C. This gradual increase in the LCM frequency difference indicates that any transient change related to a chemical reaction, such as temperature, gas density and gas viscosity, occurred and continued in the reactor during the introduction of the monomethylsilane gas.

As shown in **Figure 8**, the LCM frequency increases with the decreasing temperature [13]. Additionally, the surface chemical reaction for the silicon carbide (SiC) film deposition from the monomethylsilane gas is endothermic [13]. The frequency decrease due to the weight

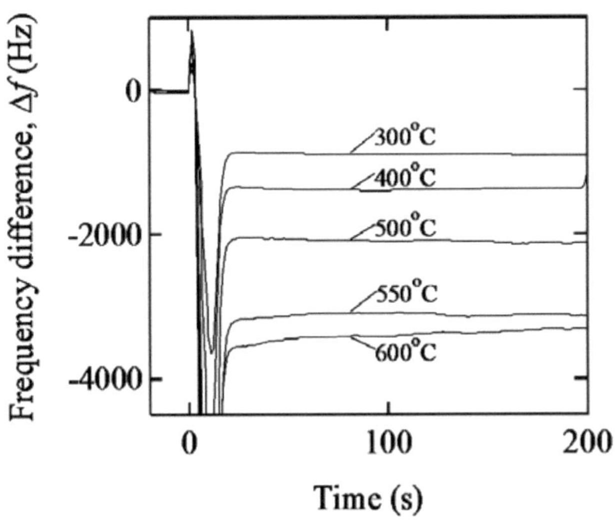

Figure 8. The LCM frequency change immediately after the introduction of the monomethylsilane gas to the hydrogen ambient atmosphere.

http://dx.doi.org/10.5772/62389

increase by the film deposition is overcompensated by the frequency increase due to the temperature decrease by the endothermic reaction. Thus, the LCM frequency continues to increase till reaching a steady state. From **Figure 8**, the chemical reaction occurring at the LCM surface is considered to continue after 200 s.

The temperature change should be enhanced by the increasing reaction rate, due to the greater reaction heat. In order to clearly show this trend, the LCM frequency gradient at various temperatures for the monomethylsilane–hydrogen system is evaluated, as shown in **Figure 9**. The LCM frequency gradient in the low-temperature range between 300–500°C is <0 Hz/s. This value is recognized to show the state with no chemical reaction. In contrast, the LCM frequency gradient increases at 550°C from a value <0 to the positive value of 2 Hz/s. It further increases to a value >4 Hz/s at 600°C. Because the LCM frequency gradient increases with the increasing temperature, the surface chemical reaction at the LCM surface is initiated in the temperature range between 500 and 600°C.

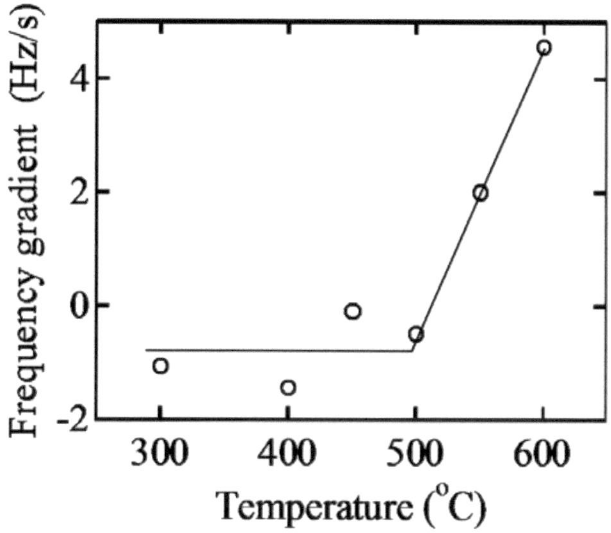

Figure 9. LCM frequency gradient immediately after introducing the MMS gas at various temperatures.

3.2. *C*(*T*) parameter

Figure 10 shows the LCM frequency change with the increase in $\rho_{\mathrm{Mix}}^{1.3}\eta_{\mathrm{Mix}}^{1.3}$ of the gas mixtures of monomethylsilane, nitrogen and hydrogen at 400, 500 and 600°C. In this figure, the frequency difference of the nitrogen–hydrogen system (white circles) follows Eq. (1) (x and y = 1.3) and shows a linear relationship. The monomethylsilane–hydrogen system (triangles) at 400 and 500°C coincides with the behaviour of the nitrogen–hydrogen system. In contrast, the monomethylsilane–hydrogen system showed a slightly higher gradient in the low $\rho^{1.3}\eta^{1.3}$range. Thus, the $C(T)$ value of Eq. (1) is expected to indicate the occurrence of a surface chemical reaction.

The *C*(*T*) values obtained using Eq. (1) are plotted versus $1/T$. as shown in **Figure 11**. The *C*(*T*) values for the nitrogen–hydrogen system have a linear relationship following Arrhenius-type equation at temperatures from 300 to 600°C. In contrast, the monomethylsilane–hydrogen system shows an increase in the *C*(*T*) value at the higher temperatures.

In order to clearly recognize the difference, the *C*(*T*) value difference of the monomethylsilane–hydrogen system from the nitrogen–hydrogen system are shown in **Figure 12**. The *C*(*T*) value difference at temperatures lower than 500°C is around zero. Thus, the monomethylsilane gas at <500°C behaves the same as that of nitrogen. The monomethylsilane–hydrogen system shows an increased *C*(*T*) value to $>1 \times 10^{10}$ at temperatures between 500 and 550°C. Because these values are obtained separate from the fluid properties, the *C*(*T*) value change is understood as the surface chemical reaction for the CVD occurring at the higher temperatures. The obtained temperature is consistent with that from our previous study [6].

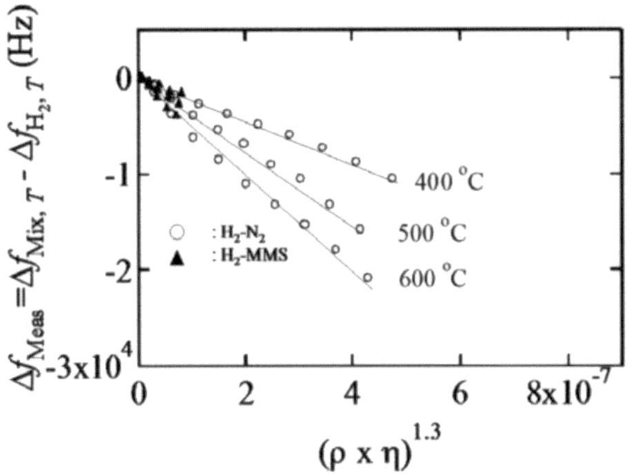

Figure 10. LCM frequency difference for monomethylsilane gas (dark triangle) and nitrogen gas (circle) from ambient hydrogen.

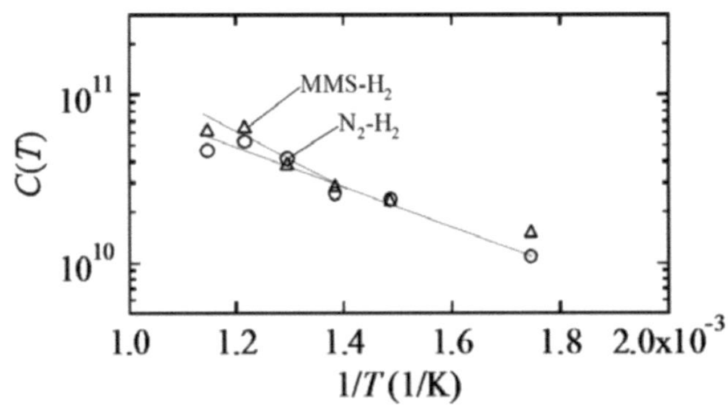

Figure 11. *C*(*T*) values of monomethylsilane–hydrogen system (triangles) and nitrogen–hydrogen system (circles).

http://dx.doi.org/10.5772/62389

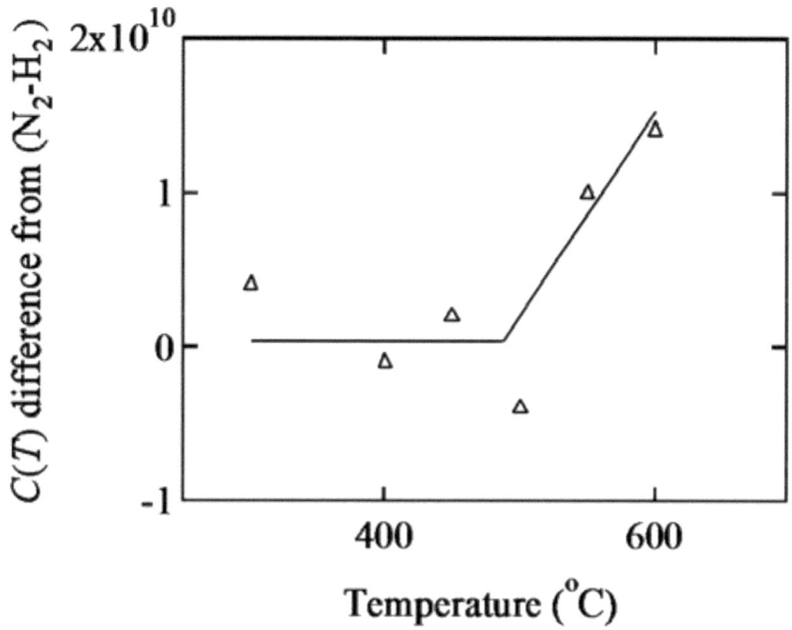

Figure 12. $C(T)$ value difference of MMS–H_2 system from N_2–H_2 system.

4. *In situ* observation of chemical vapour deposition using $SiHCl_3$ and BCl_3 gases

The film deposition behaviour using multiple precursors is explained [14]. The boron-doped silicon film is formed using the trichlorosilane ($SiHCl_3$, TCS) gas and the boron trichloride (BCl_3) gas for the film deposition and boron doping, respectively.

4.1. Chemical reaction occurrence

The chemical reaction behaviour of boron trichloride gas is measured, as shown in **Figure 13**, at various boron trichloride concentrations and temperatures. At 400°C, the LCM frequency quickly decreases corresponding to the change in the fluid properties by an increase in the boron trichloride concentration. The LCM frequency is kept constant at each boron trichloride concentration. Thus, the boron trichloride gas does not have any chemical reaction and film deposition at this temperature. At 470°C, the LCM frequency sometimes shows flat and other times decrease. The film deposition occurrence is not obvious. In contrast, at 570°C, the LCM frequency decrease is clear at each boron trichloride concentration, as the typical behaviour showing the film deposition.

Similarly, the silicon film deposition from trichlorosilane gas is evaluated. Because the LCM frequency decrease occurs between 570 and 600°C, the silicon film deposition is determined to occur in this temperature range.

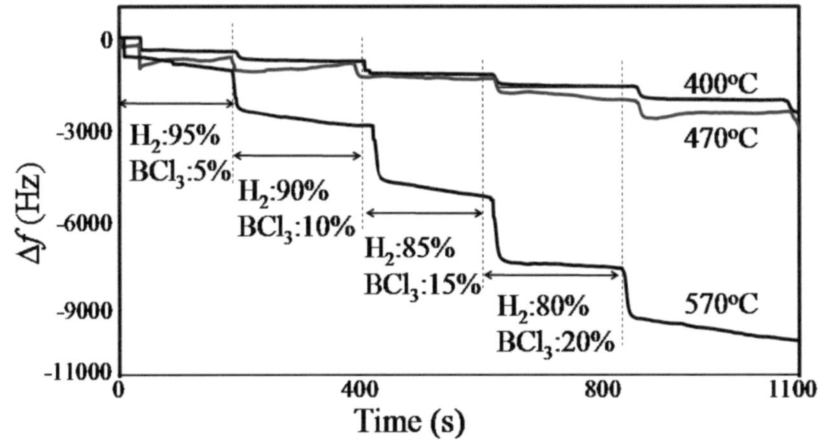

Figure 13. LCM frequency change at various boron trichloride concentrations and at 400, 470 and 570°C.

Next, the LCM frequency behaviour during the boron-doped silicon film deposition is observed at various temperatures using the trichlorosilane and boron trichloride gases, as shown in **Figure 14**. The LCM frequency shows a gradual increase and decrease in a short cycle at 330–500°C. In contrast, at 530 and 570°C, the LCM frequency continuously decreases.

Figure 15 shows the LCM frequency gradient in the temperature range between 330 and 570°C. At temperatures lower than 470°C, the frequency gradient values are near 0 Hz s^{-1}. At the temperatures higher than 500°C, the frequency gradient decreases to less than −3 Hz s^{-1}. This behaviour indicates the occurrence of a continuous film deposition. Because the LCM frequency decreases for a long period, the film deposition is determined to occur in the temperature range between 470 and 530°C, specifically higher than 530°C.

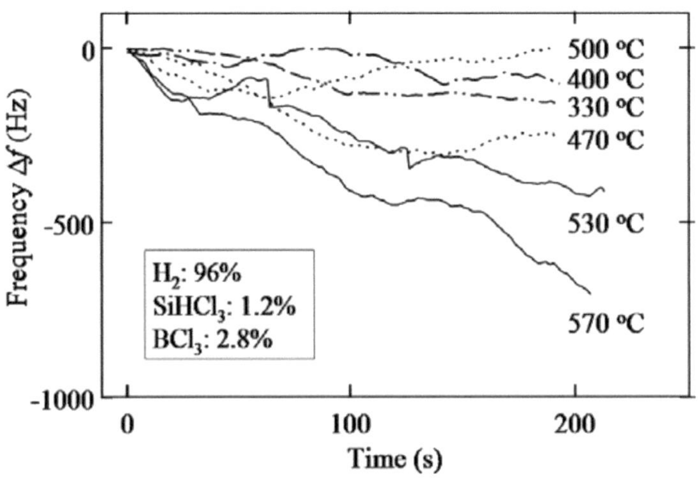

Figure 14. LCM frequency change at various trichlorosilane and boron trichloride concentrations at 330–570°C.

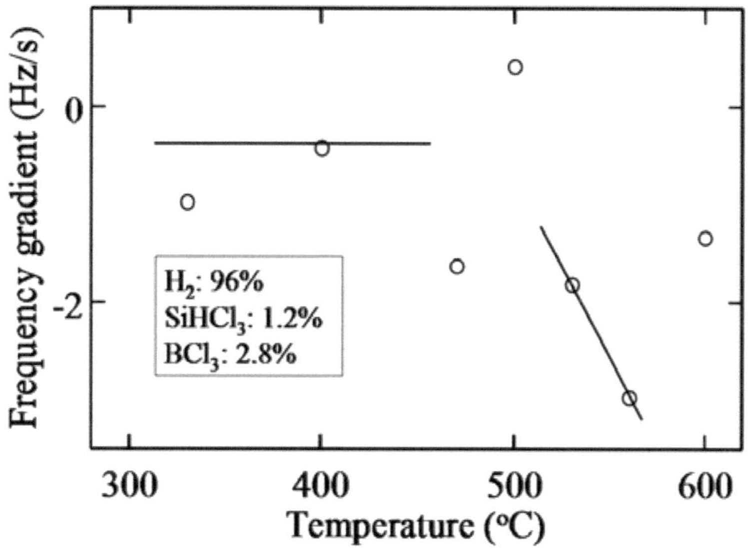

Figure 15. LCM frequency gradient in the temperature range between 330 and 570°C. The trichlorosilane and boron trichloride gases were used for the film formation.

4.2. Growth rate

The film growth rate is shown in **Figure 16**, obtained following that the frequency decrease of 1 Hz corresponds to the weight increase of 6 ng cm^{-2} [9, 13, 15]. Here, the molecular weight and the density of the boron-doped silicon film are tentatively assumed to be an average of the silicon and boron.

The boron film growth rate at 570°C is near 1.5×10^{-9} mol cm^{-2} s^{-1} at the boron trichloride gas concentrations from 1 to 4%. Additionally, the silicon growth rate shows no obvious trend versus the trichlorosilane gas concentration from 1 to 4%. Similarly, the growth rate of the boron-doped silicon film formed from both the trichlorosilane and boron trichloride gases has no obvious trend, being about 1×10^{-9} mol cm^{-2} s^{-1}. This growth rate behaviour is consistent with the saturation in the low-temperature silicon epitaxial growth process [16].

The change in the growth rate with the increasing temperature is shown in **Figure 17** as the Arrhenius plot. The boron growth rate is near 1×10^{-9} mol cm^{-2} s^{-1} and 1.5×10^{-9} mol cm^{-2} s^{-1} at 470 and 570°C, respectively. The silicon growth rate from trichlorosilane gas is about 5×10^{-10} mol cm^{-2} s^{-1} at temperatures lower than 530°C. It is too low value for determining the film deposition occurrence. Because the growth rate increases to nearly 10^{-9} mol cm^{-2} s^{-1} at 600°C, the silicon film deposition occurs at temperatures higher than 570°C.

At temperatures lower than 530°C, the film growth rate is about 5×10^{-10} mol cm^{-2} s^{-1}. Because this value is similar to that of silicon, the film deposition is negligible. In contrast, the growth rate increases at temperatures higher than 530°C. It reaches 1×10^{-9} mol cm^{-2} s^{-1} at 570°C. The boron-doped silicon film is expected to be obtained at 570°C.

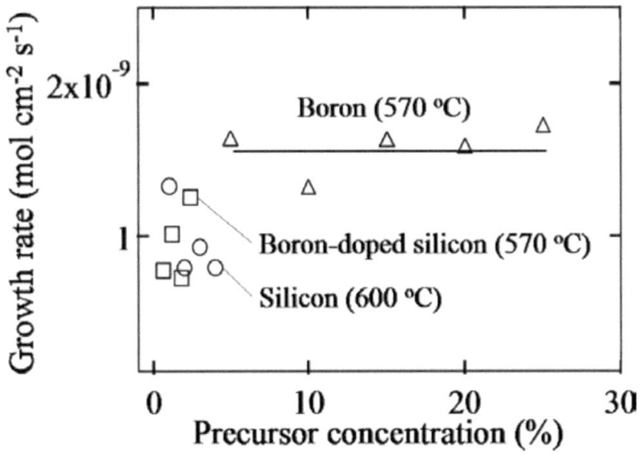

Figure 16. Growth rate at various trichlorosilane and boron trichloride gas concentrations. Silicon and boron film was grown using the combinations of trichlorosilane and boron trichloride gases of 2.4 and 1.0, 1.8 and 1.8, 1.2 and 2.8, and 0.6 and 4.0%.

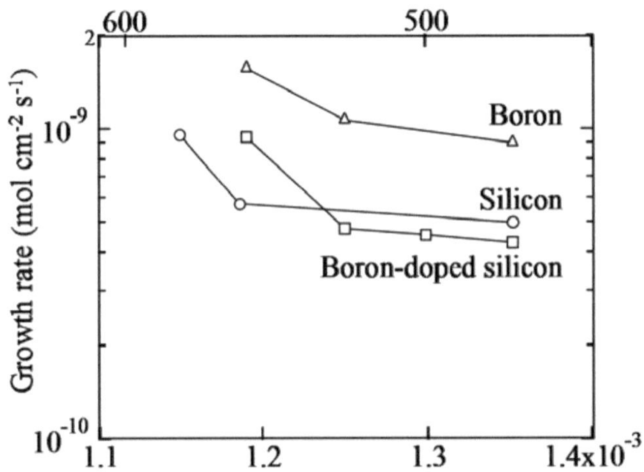

Figure 17. Film growth rates changes of silicon, boron and boron-doped silicon versus 1/*T*.

4.3. Film formation on substrate

The boron-doped silicon film is formed at 570°C using the trichlorosilane gas and boron trichloride gas at 5 and 3%, respectively. The depth profile of the boron concentration is evaluated by secondary ion mass spectrometry (SIMS), as shown in **Figure 18**. While the boron concentration in the substrate is about 5×10^{16} atoms cm^{-3}, it increases to that higher than 10^{20} atoms cm^{-3} near the film surface. Because the obtained film thickness is about 100 nm, the film growth rate is about 6–7 nm min^{-1}. This value is comparable to 4 nm min^{-1} estimated from

Figures 16 and **17**. Thus, the growth rate obtained by the LCM is consistent with that by the film growth on the substrate.

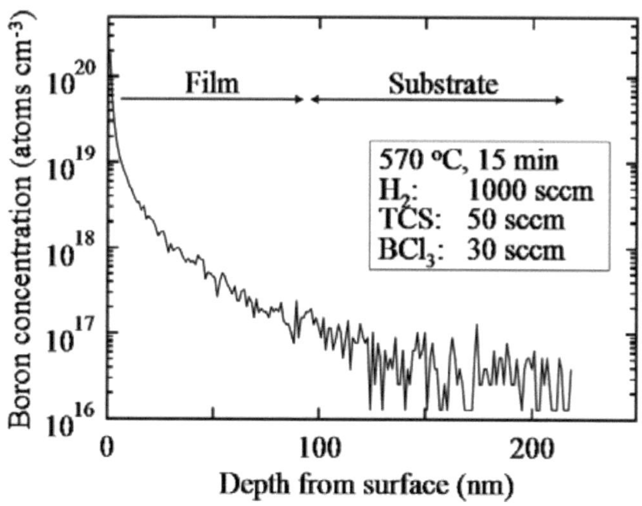

Figure 18. Depth profile of boron concentration in the obtained silicon film.

4.4. Surface process

The behaviours of film growth and doping are explained using Eqs. (9)–(12). The symbol '*' indicates the chemisorbed state. Trichlorosilane is chemisorbed at the surface to produce $*SiCl_2$ and hydrogen chloride gas; $*SiCl_2$ is decomposed by hydrogen to form silicon at the surface [15]. Similarly, $*BCl_2$ is considered to be formed at the surface; it reacts with hydrogen to produce boron.

$$SiHCl_3 \rightarrow *SiCl_2 + HCl\uparrow \tag{9}$$

$$*SiCl_2 + H_2 \rightarrow Si + 2HCl\uparrow \tag{10}$$

$$BCl_3 + \frac{1}{2}H_2 \rightarrow *BCl_2 + HCl\uparrow \tag{11}$$

$$*BCl_2 + H_2 \rightarrow B + 2HCl\uparrow \tag{12}$$

For both gases, the intermediate species are chlorides which terminate the surface. The film growth rate at low temperatures is governed by the rates of the intermediate species adsorption and the chlorine removal. Thus, the growth rate of the boron-doped silicon is influenced by the slower process, that is by the silicon film growth rate. Because the LCM detects such details of the various behaviours, it can work as a sensitive monitor for studying the film growth behaviour.

5. Temperature change related to film formation process

The surface temperature is one of the most important parameters for the CVD. In this section, the LCM is used in order to reveal and clarify the temperature behaviour related to the film formation [17]. For this purpose, the silicon film formation in a trichlorosilane–hydrogen system is used as one of the most popular systems.

5.1. LCM frequency behaviour

The LCM frequency decreases corresponding to the weight increase on the LCM surface, following the Sauerbrey equation [15].

$$\Delta f = -\frac{\Delta m\, f_0^2}{A\left(\mu_c\, \rho_c\right)^{0.5}}, \tag{13}$$

where Δf is the measured resonant frequency decrease, f_0 is the intrinsic crystal frequency, Δm is the elastic mass change, A is the electrode area, ρ_c is the density of the crystal, and μ_c is the shear modulus of the crystal.

Additionally, the LCM frequency depends on the fluid properties, as described by equation (14) [7, 8, 18].

$$\Delta f \propto \left(\rho\eta\right)^x, \tag{14}$$

where ρ and η are the density and the viscosity, respectively, of the gas mixture. The η value is obtained from the literature [11]. The x value is 1.3 [7] in this section.

These parameters produce various changes in the LCM frequency during the CVD process. The change in the LCM frequency related to the film deposition by the trichlorosilane gas is classified by Parameters (*i*)–(*viii*) schematically shown in **Figure 19**.

Parameter (*i*) in **Figure 19** is the pressure and concentration of the gas remaining in the gas system. The LCM frequency very quickly and slightly increases due to the pressure increase in the reactor when opening the trichlorosilane gas valve. Immediately after this, the LCM frequency decreases and soon recovers by the trichlorosilane gas, which is remained at a high concentration in the gas line, reaching and passing the reactor. Parameter (*ii*) is the $(\rho\eta)^{1.3}$ value of the gas mixture. Corresponding to this, the LCM frequency decreases from the initial frequency. Parameter (*iii*) is the heat capacity and the heat conduction. The increase in the heat capacity gradually causes the gas-phase temperature to decrease, resulting in an increase in the LCM frequency. Parameter (*iv*) is the reaction heat. The silicon film formation is an endothermic reaction [19]. This decreases the temperature and thus increases the LCM frequency. With the addition of the trichlorosilane gas, the film formation quickly begins and continues till termination of the trichlorosilane gas supply. Parameter (*v*) is the weight increase due to the film deposition. After a sufficient time for reaching a steady state, the weight increase by the film formation on the LCM appears as a continuous and linear decrease in the LCM frequency. Parameter (*vi*) is the $(\rho\eta)^{1.3}$ value. The decrease in this value immediately causes the LCM frequency to shift to a high value when the trichlorosilane gas supply is terminated. Parameter (*vii*) is the heat capacity and the heat conduction. Due to the heat capacity decrease because of the lack of trichlorosilane gas, the gas-phase temperature gradually increases and causes a decrease in the LCM frequency. Parameter (*viii*) is the weight increase due to the film deposition. After a sufficient period for achieving a steady state, the LCM frequency shift from that before the trichlorosilane gas supply corresponds to the increased weight of the film formed on the LCM.

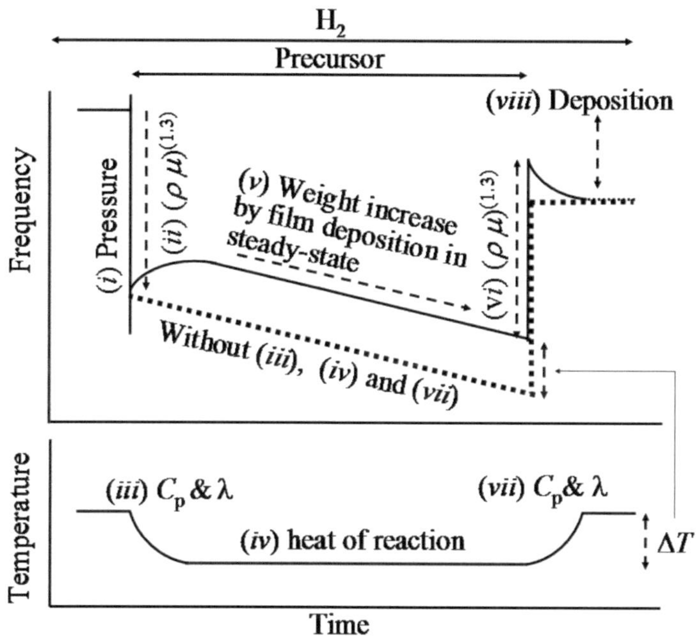

Figure 19. Schematic of changes in the frequency and the temperature of the LCM influenced by various parameters. Parameter (*i*): pressure and concentration of gas remaining in the gas system, parameter (*ii*): $(\rho\eta)^{1.3}$, parameter (*iii*): heat capacity, parameter (*iv*): reaction heat, parameter (*v*): weight increase due to the film deposition, parameter (*vi*): $(\rho\eta)^{1.3}$, parameter (*vii*): heat capacity and parameter (*viii*): weight increase due to the film deposition.

If the temperature change due to the trichlorosilane gas did not exist, the temperature during the early stage of the film formation behaves like the thick dotted line, as shown in **Figure 19**. Thus, the LCM frequency difference between the solid line and the thick dotted line is considered to be a function of the temperature shift.

5.2. Reaction heat and heat transport

By introducing the trichlorosilane gas into the reactor, the silicon film formation occurs along with the endothermic reaction heat and change of the physical properties of the gas mixture, as shown in **Figure 20**. Thus, the multiple thermal processes change the surface temperature. Here, the influence of each parameter is evaluated, following Steps 1, 2 and 3, as shown in **Figure 21**, taking into account the time constant for heat transport.

During Step1, the time constant for the surface temperature shift is evaluated without introducing a precursor in the ambient hydrogen. The quick lamp power decrease is assumed to show a significantly quick surface temperature decrease similar to that by the endothermic surface chemical reaction, as shown in **Figure 20**. After this, the gas-phase temperature of the near-surface region gradually decreases. These two processes are expected to have different time constants, such as very short and slightly long.

During Step 2, the influence of the change in the heat capacity and the heat conduction of the gas mixture are explained. The time constant for the gas-phase temperature shift induced by the precursor introduction, as shown in **Figure 20**, is evaluated at sufficiently low temperatures at which no chemical reaction occurs. The time constant for this process is expected to be longer than those for Step 1, because the gas-phase temperature shift occurs in the entire region of the reactor and not limited to the region near the surface.

During Step 3, the temperature change related to the film deposition is explained, accounting for the time constants obtained in Steps 1 and 2. During Step 3, the trichlorosilane concentration is stepwise changed from 0 to 3%. After the period corresponding to the time constants obtained in Steps 1 and 2, the LCM frequency behaviour expresses the film formation in a steady state. By extrapolation, the LCM frequency immediately after changing the precursor concentration is used for evaluating the temperature shift related to the film formation.

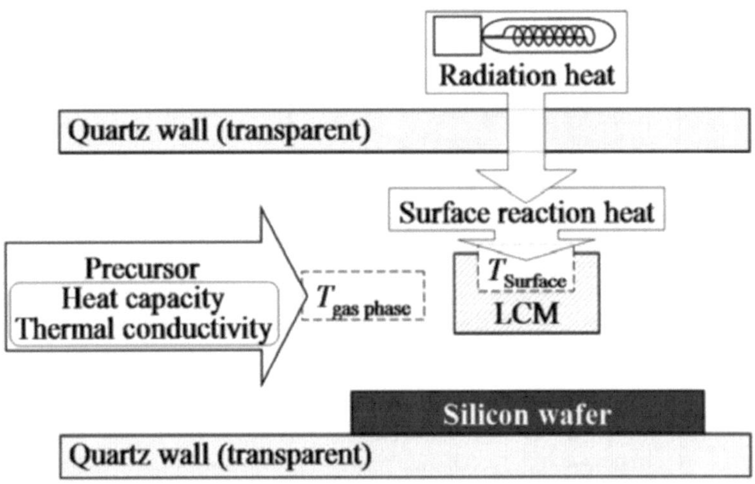

Figure 20. Parameters influencing the LCM temperature.

http://dx.doi.org/10.5772/62389

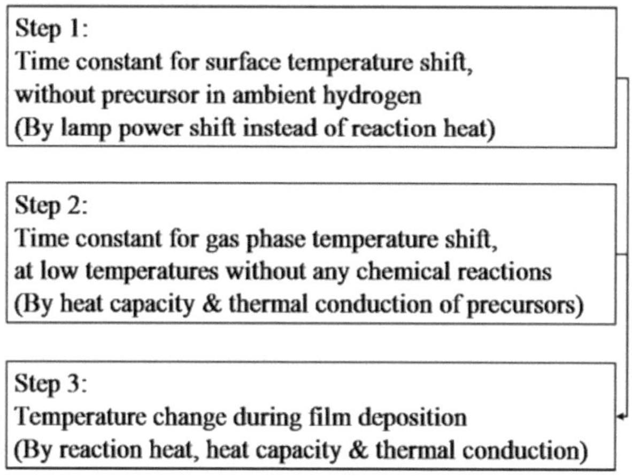

Figure 21. Three steps for evaluating temperature change during film deposition

5.3. Temperature and frequency

The entire frequency dependence on the temperature is shown in **Figure 22**. This shows the frequency difference at various temperatures from that at room temperature. With the increasing temperature, the LCM frequency decreases, as shown in **Figure 22a**. At the higher temperatures, the LCM frequency more rapidly decreases than that at the lower temperatures. The temperature gradient is shown in **Figure 22b**. The gradient decreases with the increasing temperature. The frequency change due to the temperature change is about −400 Hz/K in the temperature range between 500 and 650°C.

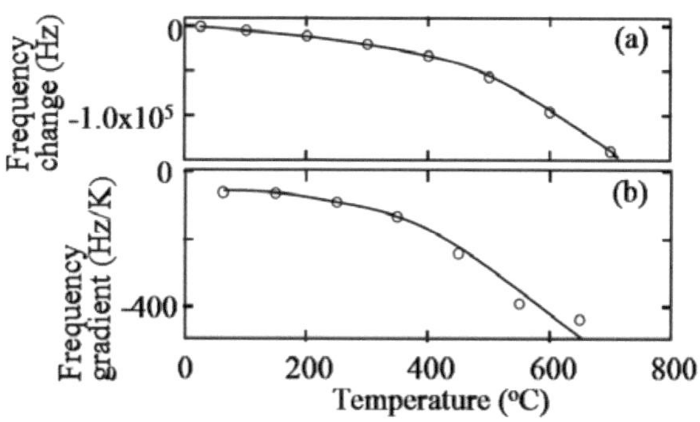

Figure 22. LCM frequency changing with the temperature. (a) frequency change from that at room temperature and (b) frequency gradient (Hz/K) at various temperatures.

5.4. Heat at surface

In order to evaluate the LCM frequency behaviour caused by the quick surface temperature decrease, such as that by the reaction heat, the LCM frequency influenced by the stepwise lamp power shift is evaluated, as shown in **Figure 23**. The lamp power corresponding to about 1 K quickly decreases at 450 and 660°C.

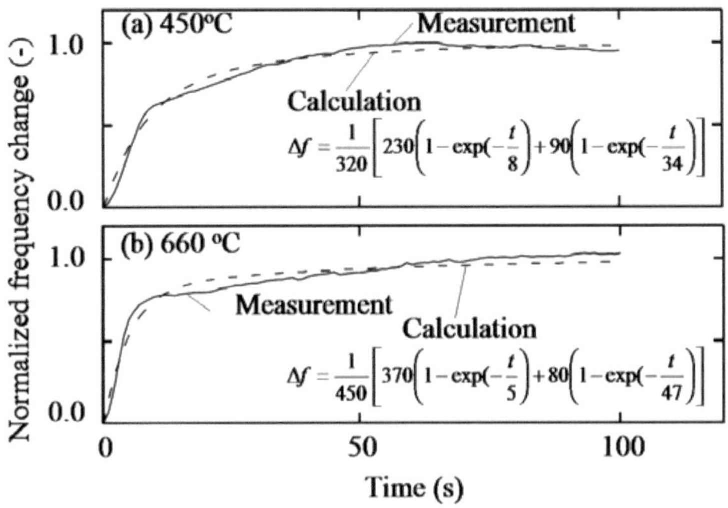

Figure 23. LCM frequency change caused by the stepwise lamp power shift corresponding to about 1 K at (a) 450°C and (b) 660°C.

Figure 23a and **23b** shows the normalized LCM frequency change caused by the stepwise lamp power shift at 450 and 660°C, respectively. In both figures, the LCM frequency very quickly increases immediately after changing the lamp power. It then moderately increases and reaches the steady state.

These behaviours are expressed as a function of time, *t* (s), assuming that the fast and slow relaxation processes.

$$\Delta f = \left[230\left(1-e^{\left(-\frac{t}{8}\right)}\right)+90\left(1-e^{\left(-\frac{t}{34}\right)}\right)\right]. \tag{15}$$

$$\Delta f = \left[370\left(1-e^{\left(-\frac{t}{5}\right)}\right)+80\left(1-e^{\left(-\frac{t}{47}\right)}\right)\right]. \tag{16}$$

In **Figure 23**, Eqs. (15) and (16) are indicated by the dotted lines. The quick temperature change, the first term, is considered to directly follow the lamp power decrease, that is the decrease in heat at the LCM surface. The slow temperature change, the second term, is due to the conduction heat transport between the surface and the gas phase very near the surface. The influence of the reaction heat and heat conduction on the LCM frequency is expected to appear within 10 and 50 s, respectively, after introducing the trichlorosilane gas.

5.5. Heat transport around surface in gas phase

The LCM frequency changes due to the thermal properties, such as the heat capacity and the heat conductivity, are evaluated at the low temperatures, which do not cause the film deposition [14]. **Figure 24** shows that the LCM frequency change is due to the stepwise concentration change of hydrogen gas and trichlorosilane gas at 380°C. At conditions C_A–C_E, the hydrogen concentration and the trichlorosilane concentration decrease and increase, respectively.

At the beginning of condition C_A, the LCM frequency quickly drops to less than −3000 Hz and recovers to −1000 Hz. The LCM frequency shift from 0 Hz to −1000 Hz corresponds to the increase in the gas density and the gas viscosity. Next, it gradually recovers to near −400 Hz. Till 500 s, the LCM frequency reaches the steady state. The gradual recovery of the LCM frequency is a result of the temperature decrease mainly due to the increase in the heat capacity of the gas mixture. Additionally, the temperature decrease at the surface is moderated by the heat balance with the gas phase *via* the heat conduction.

These heat transports overlaps and gives the gradual recovery of the LCM frequency. The LCM frequency at 380°C and at condition C_D is expressed, as shown in **Figure 25**, using the following equation.

$$\Delta f = 164\left(1 - e^{\left(-\frac{t}{120}\right)}\right). \qquad (17)$$

From the temperature shift width shown in **Figure 25**, the temperature decrease by the trichlorosilane gas concentration change at 380°C is about 1 K. The time constant of these process is about 120 s. The time constant in Eq. (17) is longer than that of the second term in Eqs. (15) and (16). The time constant is assumed to have a similar value at the higher temperatures, such as 600–700°C.

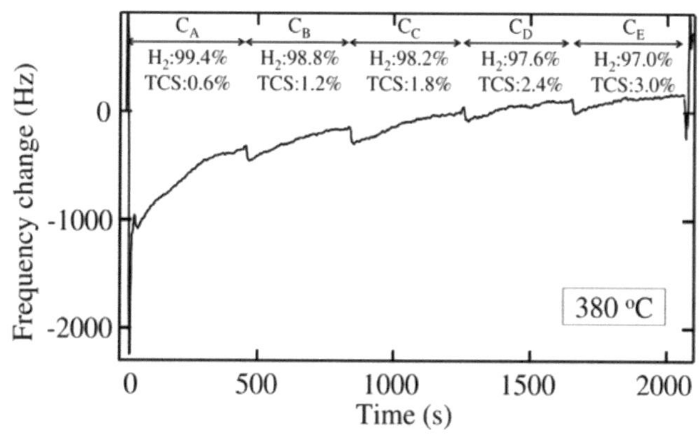

Figure 24. LCM frequency change caused by the stepwise concentration change in the hydrogen gas and trichlorosilane gas at 380°C.

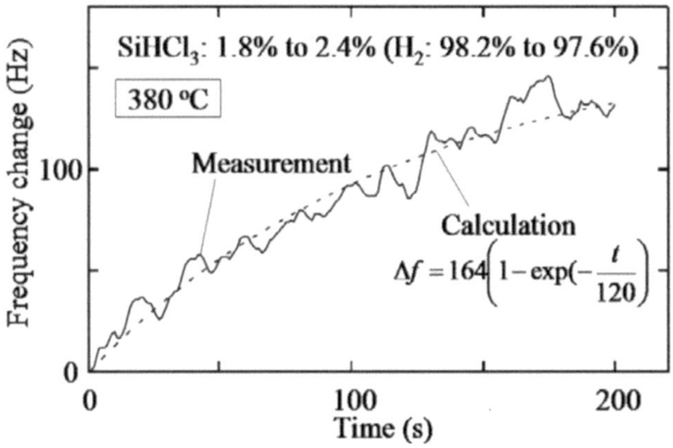

Figure 25. LCM frequency change at 380°C caused by the gas concentrations of hydrogen and trichlorosilane from condition C_C to C_D in **Figure 24**.

5.6. Film formation

The silicon film is formed at 640°C along with measuring the LCM frequency, as shown in **Figures 26** and **27**. The hydrogen gas concentration decreases from 100 to 97%, while the trichlorosilane gas concentration increases from 0 to 3%. Because the silicon film growth rate is saturated at this temperature [16], the reaction heat remains the same for conditions C_A–C_E.

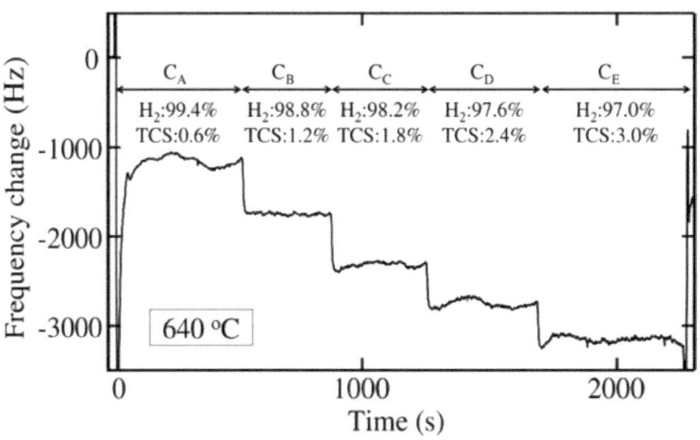

Figure 26. LCM frequency change caused by the stepwise change in the concentrations of hydrogen and trichlorosilane.

At condition C_A, the trichlorosilane gas is added to the hydrogen gas. The LCM frequency shows a significant drop and a quick recovery within several seconds. After the quick recovery, the LCM frequency gradually increases. After the peak appearance, the LCM frequency gradually decreases accompanying the fluctuation due to a temperature fluctuation. Conditions C_C, C_D and C_E show a similar LCM frequency behaviour without a significant drop, unlike that at the beginning of condition C_A.

Taking into account the time constant corresponding to the various heat processes, the LCM frequency behaviour is evaluated, as shown in **Figure 27**. This shows the LCM behaviour at condition C_D, as an example. Immediately after increasing the trichlorosilane concentration from 1.8 to 2.4%, the LCM frequency quickly decreases to about −550 Hz due to the increase in the gas density and the gas viscosity. Next, it shows a broad bottom for about 20 s. The LCM frequency gradually increases from 50 to 180 s. After showing a peak, the LCM frequency begins to decrease. This decrease is due to the film formation during the steady state.

Next, the LCM frequency gradient is evaluated. The maximum and minimum values are 1.6 Hz/s between Points A and B and 0.77 Hz/s between Points A and C, respectively. By this operation, the LCM frequency immediately after changing the trichlorosilane concentration is 245 to 410 Hz, as shown in **Figure 27**.

In addition, the flat bottom of the LCM frequency to 20 s after changing the trichlorosilane concentration may be caused by the balance among the changes in the gas density, the gas viscosity, the heat capacity and the heat conductivity. In order to obtain the possible minimum frequency value, the increasing trend in the LCM frequency between 20 and 180 s is extrapolated to that near several seconds. By this estimation, the frequency of about 50 Hz may be lower than that at the bottom. By adding these values, the LCM frequency change due to the temperature change by changing the trichlorosilane concentration from 1.8 to 2.4% is between 295 and 460 Hz. The surface temperature shift caused by changing the precursor concentration is considered to be about one degree.

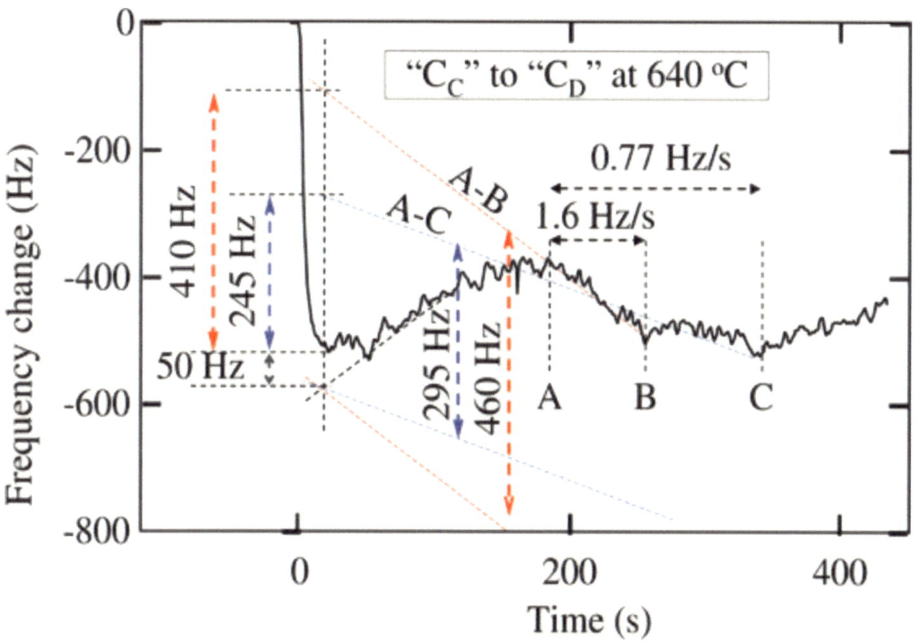

Figure 27. Evaluation of frequency change caused by the stepwise change from condition C_C to C_D in **Figure 26**.

Acknowledgements

The author would like to thank Ms. Misako Matsui, Ms. Ayumi Saito and Mr. Kento Miyazaki of Yokohama National University for their intensive work. The author would like to thank Mr. Nobuyoshi Enomoto and Mr. Hitoshi Ueno of Halloran Electronics Co., Ltd., for his technical support. A part of this study was supported by JSPS KAKENHI Grant No. 25420772.

Author details

Hitoshi Habuka*

Address all correspondence to: habuka1@ynu.ac.jp

Department of Chemical Energy Engineering, Yokohama National University, Yokohama, Japan

References

[1] S. M. Sze, Semiconductor devices: physics and technology (John-Wiley, Hoboken, NJ, 2008).

[2] A. Bouteville, "Numerical simulation applied to chemical vapor deposition process, rapid thermal CVD and spray CVD", J. Optoelectron. Adv. Mater., 7 (2005) 599–606.

[3] U. Koehler, L. Andersohn and B. Dahlheimeer, "Time-resolved observation of CVD-growth of silicon on Si (111) with STM", Appl. Phys. A, 57 (1993) 491–497.

[4] M. Schulz, J. Sauerwald, H. Seh, H. Fritze and H. L. Tuller, "Defect chemistry based design of monolithic langasite structures for high temperature sensors", Solid State Ion., 184 (2011) 78–82. doi:10.1016/j.ssi.2010.08.009

[5] H. Habuka and Y. Tanaka, "In-situ monitoring of chemical vapor deposition from trichlorosilane gas and monomethylsilane gas using langasite crystal microbalance", ECS J. Solid State Sci. Technol., 1 (2012) 62–65. doi:10.4236/jsemat.2013.31A009

[6] H. Habuka and Y. Tanaka, "Langasite crystal microbalance frequency behavior over wide gas phase conditions for chemical vapor deposition", J. Surf. Eng. Mater. Adv. Technol., 3 (2013) 61–66. doi:10.1016/j.surfcoat.2013.06.052

[7] H. Habuka and M. Matsui, "Langasite crystal microbalance frequency behavior over wide gas phase conditions for chemical vapor deposition", Surf. Coat. Technol., 230 (2013) 312–315. doi:10.1016/j.surfcoat.2013.06.052

[8] H. Fritze, "High-temperature piezoelectric crystals and devices", J. Electroceram., 26 (2011) 122–161. doi:10.1007/s10832-011-9639-6

[9] K. K. Kanazawa and J. G. Gordon II, "The oscillation frequency of a quartz resonator in contact with liquid", Anal. Chim. Acta, 175 (1985) 99–105. doi:10.1016/S0003-2670(00)82721-X

[10] R. C. Reid, J. M. Prausnitz and B. E. Poling, The properties of gases and liquids, 4 ed. (McGraw-Hill, New York, 1987).

[11] P. Pollard and J. Newman, "Silicon deposition on a rotating disk", J. Electrochem. Soc., 127 (1980) 744–752. doi:10.1149/1.2129589

[12] H. Habuka, M. Matsui and A. Saito, 17th International Conference on Crystal Growth and Epitaxy, Proc. pp. 182–186 (Warsaw, Poland, 2013).

[13] H. Habuka and K. Kote, "Langasite crystal microbalance for development of reactive surface preparation of silicon carbide film deposition from monomethylsilane gas", Jpn. J. Appl. Phys., 50 (2011) 096505-1-4. doi:10.1143/JJAP.50.096505

[14] A. Saito, K. Miyazaki, M. Matsui and H. Habuka, "In-situ observation of chemical vapor deposition using $SiHCl_3$ and BCl_3 gases", Phys. Status Solidi C, 12 (2015) 953–957. doi: 10.1002/pssc.201510002

[15] G. Sauerbrey, "Verwendung von Schwingquarzen zur Wägung dünner Schichten und zur Mikrowägung", Zeitschrift fur Physik, 155 (1959) 206–222.

[16] H. Habuka, T. Nagoya, M. Mayusumi, M. Katayama, M. Shimada and K. Okuyama, "Model on transport phenomena and epitaxial growth of silicon thin film in $SiHCl_3$–H_2 system under atmospheric pressure", J. Cryst. Growth, 169 (1996) 61. doi: 10.1016/0022-0248(96)00376-4

[17] K. Miyazaki, A. Saito and H. Habuka, "In situ measurement for evaluating temperature change related to silicon film formation in a $SiHCl_3$–H_2 system", ECS J. Solid State Sci. Technol., 5 (2016) P16–P20. doi:10.1149/2.0101602jss

[18] D. Shen, H. Zhang, Q. Kang, H. Zhang and D. Yuan, "Oscillating frequency response of a langasite crystal microbalance in liquid phases", Sens. Actuators B, 119 (2006) 99–104. doi:10.1016/j.snb.2005.12.001

[19] D. Crippa, D. L. Rode and M. Masi, Silicon epitaxy (Academic Press, New York, 2001).

Section 2

CVD: Applications in Optical, Solar Cell and Solid State Devices

Chapter 6

Low-Temperature PureB CVD Technology for CMOS Compatible Photodetectors

Vahid Mohammadi and Stoyan Nihtianov

Additional information is available at the end of the chapter

http://dx.doi.org/10.5772/63344

Abstract

In this chapter, a new technology for low-temperature (LT, 400°C) boron deposition is developed, which provides a smooth, uniform, closed LT boron layer. This technology is successfully employed to create near-ideal LT PureB (pure boron) diodes with low, deep junction-like saturation currents, allowing full integration of LT PureB photodiodes with electronic interface circuits and other sensors on a single chip. In this way, smart-sensor systems or even charge-coupled device (CCD) or complementary metal oxide semiconductor (CMOS) ultraviolet (UV) imagers can be realised.

Keywords: low-temperature boron deposition, ultrashallow p^+n junction photodiode, chemical vapour deposition, UV photodetector, CMOS imager

1. Introduction

Over the last few years, we have witnessed an increase in the demand for both high-performance ultraviolet and low-energy electron detectors. These detectors are used in high-tech applications such as optical lithography and electron microscopy, medical imaging, protein analysis and DNA sequencing, forensic analysis, disinfection and decontamination, space observation, etc.

To meet these demands, a new silicon detector technology has been proposed by TU Delft called "PureB" (pure boron) technology [1]. Conventional PureB technology is employed to fabricate high-performance photodiodes for vacuum/extreme/deep-UV (VUV/EUV/DUV) light, low-energy electrons down to 200 eV, and X-ray drift detectors [2–5]. A comparison of

PureB technology to state-of-the-art Si-based UV and electron detectors is presented in **Tables 1** and **2**, respectively.

Detector	**ETH PtSi-nSi**	**IRD AXUV**	**IRD SXUV**[a]	**IRD UVG**	**Hamamatsu (S5226)**	**PureB**[b]
Junction type	Schottky		n-on-p		p-on-n	p-on-n
EUV sensitivity (A/W) at 13.5 nm	~0.2[c]	~0.265	~0.23	Unknown	Unknown	0.267
Stability under EUV	Unknown[d]	Not stable	Good[e]	Unknown	Unknown	Δ ~ 3%[f]
VUV sensitivity (A/W) at 193 nm	~0.03	~0.1	~0.01	~0.137	~0.1	0.102
VUV sensitivity (A/W)at 157 nm	~0.02	~0.1	<0.01	Unknown	Unknown	0.123
VUV sensitivity (A/W)at 121 nm	≤0.02	Unknown	<0.01	<0.05	Unknown	0.116
Stability under DUV/VUV	Δ ~ 2%	Not stable	Unknown	Δ ~ 10%	Not stable	Δ ~ 4%
References	[6, 7]	[6, 8–10]	[9, 10]	[6, 8]	[6]	[2, 3]

[a]SXUV: Si-based n-on-p junction photodiodes with nitride metal silicide front window.
[b]Data are based on as-deposited PureB photodetectors without any post-processing thermal annealing.
[c]Value from measurement at PTB [2, 3].
[d]Stability proven only in the VUV range [6].
[e]No more information is given in the reference.
[f]Δ: reduction in responsivity based on the data reported in the mentioned references.

Table 1. Performance overview of representative commercially available Si-based UV detectors compared to the PureB UV photodetector.

Detector	**Commercial BSE**	**Commercial vCD**	**Delta doped e2V CCD 97**	**PureB BSE**
Electron signal gain[a] at 1 KeV	~46	~126	~112	213
Electron signal gain at 500 eV	~18	~52	~47	102
Electron signal gain at 200 eV	Unknown	Unknown	~13	34
Stability	Δ ~ 5%	Δ ~ 5%	Unknown	Δ ~ 4%
References	[11, 12]	[11, 12]	[13]	[4]

[a]Electron signal gain, G_{ph}, is defined as I_{ph}/I_{beam}, where I_{ph} is the output current of the photodiode and I_{beam} is the current of the incident electron beam, assuming a negligible dark current.

Table 2. Performance overview of representative commercially available, research-reported Si-based electron detectors compared to the PureB electron detector.

http://dx.doi.org/10.5772/63344

However, due to the relatively high temperature (HT) of boron deposition (500–700°C), it is hard to integrate this technology fully into a standard integrated circuit/ complementary metal oxide semiconductor (IC/CMOS) process flow.

In this chapter, a newly developed technology for low-temperature (LT, 400°C) boron chemical vapour deposition (CVD) is presented, providing a smooth, uniform, closed LT boron layer. This technology is successfully employed to create near-ideal LT PureB diodes with low, deep-junction-like saturation currents. The low-temperature deposition at 400°C makes it possible to integrate the LT PureB photodiodes fully with electronic interface circuits and other sensors on a single chip. In this way, smart-sensor systems or even charge-coupled device (CCD) or CMOS ultraviolet (UV) imagers can be realised.

What this chapter demonstrates is near ideal[1], highly sensitive, radiation-hard, low leakage current diode characteristics. Moreover, since the boron deposition is conformal and highly selective to Si, PureB technology is shown to be an attractive candidate for creating junctions on silicon nanowires and advanced CMOS transistors including a source/drain in p-type FinFETs. The impressive properties that PureB devices provide have resulted in this technology becoming commercially available very quickly for ASML EUV lithography tools and scanning electron microscopy (SEM) systems.

2. An analytical kinetic model for boron CVD

In references [14] and [15], the pattern dependency and the loading effect of CVD boron deposition are identified as sources of non-uniformity of the boron layer. Since investigating these two effects requires a thickness-monitoring technique, an end-of-line resistance measurement is introduced as a non-destructive, accurate means of monitoring the boron layer uniformity with fine resolution [16].

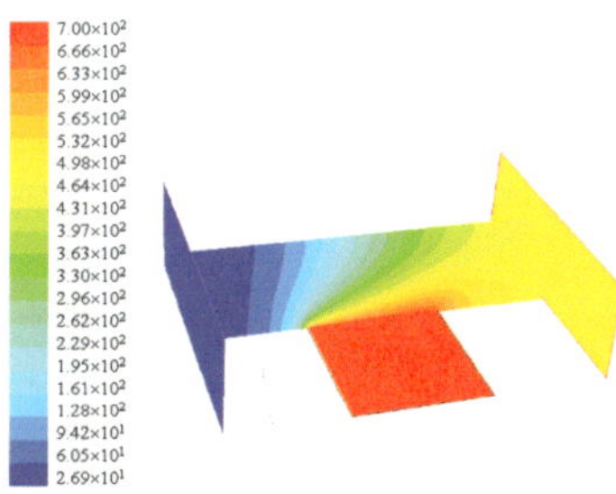

Figure 1. Temperature distribution profile simulated by commercial FLUENT© software for an ASM Epsilon One CVD reactor when the total pressure is ATM. The hydrogen gas is considered to be the main gas flowing over the susceptor, in order to simplify the simulation. The susceptor is heated up by an assembly of lamps to the deposition temperature (here 700°C) [19].

[1] Diode ideality factor, n≈1

In this section, an analytical model is established to describe the deposition kinetics and the deposition chamber characteristics that determine the deposition rate (DR) over the wafer. This pre-deposition prediction tool can be used to improve the set-up and control of the final deposition [17, 18]. It is also very useful for transferring recipes from one reactor to the other.

Some considerations such as the diffusion behaviour of the reactant (here BH_3 species) through the stationary boundary layer over the wafer with the mechanism of the diffusion, the gas-phase processes and the related surface reactions are taken into account by this model. To develop the model, the actual parabolic gas velocity and temperature-gradient profiles in the reactor are considered as well, both of which have been calculated theoretically and also simulated with FLUENT® software as shown in **Figure 1**.

The starting point for developing a mathematical model for the chemical and physical behaviour of the CVD process is the expression for the temperature and velocity profile for a fully developed flow. This can be calculated by solving the equation for energy conservation, the continuity equation and the equation for the motion of the carrier gas:

$$c_g \rho_g u_x \frac{\partial T}{\partial x} = \frac{\partial}{\partial x}\left(\mathrm{k}_0 T^{\beta} \frac{\partial T}{\partial x} \right) = 0 \tag{1}$$

$$\frac{\partial}{\partial y}\left(\mu_0 T^{\gamma} \frac{\partial u_x}{\partial y} \right) = \text{constant} \tag{2}$$

where $u_x(y)$, x, y, $\mu_0 T^{\gamma}$, and $\mathrm{k}_0 T_{\beta}$ are, respectively, the gas velocity profile, axial position in the direction of the gas flow, lateral position perpendicular to the direction of the gas flow, temperature-dependent thermal conductivity and dynamic viscosity of the carrier gas.

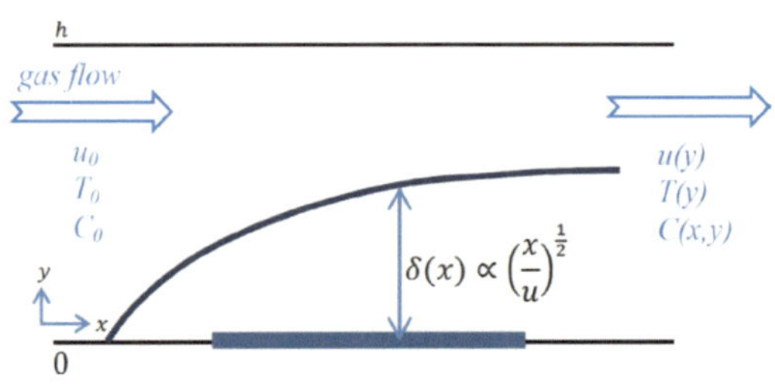

Figure 2. Schematic illustration of a classical boundary layer and reactor conditions over the susceptor.

To develop an analytical model, first we consider a system with a constant temperature in which the active component rapidly decomposes at the susceptor, $y = 0$, for all axial flow positions $x \geq 0$ (**Figure 2**). Therefore, the concentration of the reactant species at $y = 0$ is zero across the entire decomposition zone. The transport of material towards the susceptor in the y-direction goes entirely via gas-phase diffusion (i.e. laminar flow through the stationary boundary layer over the susceptor).

http://dx.doi.org/10.5772/63344

The equation for mass conservation for this case is as follows:

$$u(y)\frac{\partial C(x,y)}{\partial x} = D\frac{\partial^2 C(x,y)}{\partial y^2} \tag{3}$$

where C and D are the concentration profiles and the gas-phase diffusion coefficient of the active component in the carrier gas, respectively, and $u(y)$ is an expression for the parabolic velocity profile found by solving Eqs. (1) and (2) in the reactor chamber, which is given by

$$u(y) = 4u_0\left(\frac{y}{h} - \frac{y^2}{h^2}\right) \tag{4}$$

Equation (4) is difficult to solve for a parabolic velocity profile. Therefore, the problem is first solved for a constant flow velocity. Thereafter, the influence of a parabolic flow profile on the obtained results is evaluated.

By following the derivation procedure described in more detail in reference [18], the concentration profiles at $y = 0$ can be calculated as

$$\bar{C}(x) = 0.692C_0\exp\left[\frac{-2.52D}{h^2u_0}x\right] \tag{5}$$

This is the average concentration profile of the reactants over the susceptor with a linear velocity profile. This profile will be used to develop the final deposition rate model for boron layer deposition.

Next, the deposition rate of the boron layer deposited in a CVD system by using B_2H_6 over a bare, non-rotating silicon wafer can be calculated as [18]:

$$\mathrm{DR}^{\mathrm{B}}_{\mathrm{BnR}}(x) = 0.264\eta\gamma\beta_1\beta_2P_{\mathrm{BH}_3} \times \exp\left(-\frac{2.52D}{h^2u_0}x\right) \tag{6}$$

Where

$$\beta_1 = \frac{1}{N_0}\frac{\left(1-\theta_{\mathrm{H(B)}}\right)}{\left(m_{\mathrm{BH}_3}\mathrm{k}T\right)^{\frac{1}{2}}}$$

and

$$\beta_2 = \left(\frac{E_{\mathrm{BH_3-on-B}}}{\mathrm{k}T} + 1 \right) \exp\left(-\frac{E_{\mathrm{BH_3-on-B}}}{\mathrm{k}T} \right)$$

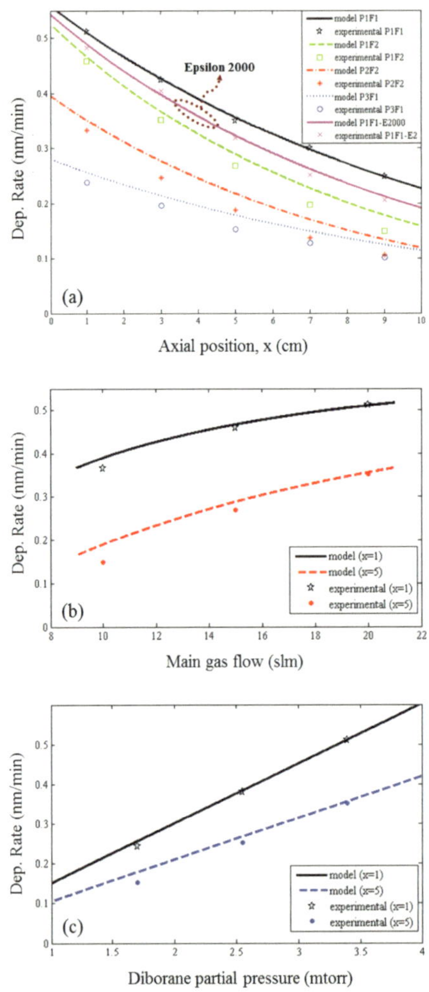

Figure 3. Boron deposition rate extracted either from the model or experimentally as a function of (a) axial position, *x*, (b) main gas flow over the susceptor and (c) diborane partial pressure. In the figure, P_1, P_2 and P_3 are the diborane partial pressure applied, that is, 3.39, 2.55 and 1.7 mtorr, respectively. F_1, F_2 and F_3 are the gas flow applied, that is, 20, 15 and 10 slm, respectively. All experiments were performed at atmospheric pressure [18].

In **Figure 3**, several experimental results are compared to model predictions. It should be noted that a parameter fitting was performed for the ASM Epsilon One. The boron deposition rates are extracted as a function of (a) the axial position, *x*, for different gas flow and diborane partial pressure conditions; (b) gas flow over the susceptor; and (c) input diborane partial pressure including curves for two different axial positions. By only adjusting the reactor/process parameters, this model was also successfully transferred from the ASM Epsilon One to the

Epsilon 2000 reactor, which has completely different reactor conditions. The experimental results and model predictions for the Epsilon 2000 are shown in **Figure 3a**. It is worth noting that this model is capable of predicting the deposition rate on any two-dimensional (2D) uniformly or non-uniformly patterned wafer such as those used for advanced device fabrication. The very small (less than 5%) deviation of the experimental results and model prediction is plausibly related to the lateral diffusion of the diborane molecules, which becomes more evident at lower gas flow and diborane partial pressure.

It can be seen that the data calculated on the basis of this model fit well with the experimental results and have been very useful not only in the development of uniform boron layers with little pattern dependence but also in the transferring of recipes from one reactor to the other.

3. Low-temperature boron deposition at 400°C

In this section, a newly developed technology for low-temperature (400°C) boron deposition is discussed. The temperature dependency of the kinetics of the boron deposition on patterned Si/SiO_2 surfaces in the temperature range of 700°C–400°C is examined. The recipe-to-deposit pure boron layer is explained. Also, selectivity issues that arise when the boron deposition temperature is reduced from 700°C to 400°C are analysed. Lastly, some provisions are recommended for minimising the undesirable boron deposition on oxide.

3.1. Temperature dependency of boron deposition

To develop a new technology for low-temperature (400°C) boron deposition on silicon, the temperature dependency of the deposition has to be studied to understand the kinetics of the boron deposition on patterned Si/SiO_2 surfaces. This study will be described in this section.

No.	Reaction[a]	Description
[R1]	$B_2H_6(g) + 2Si/B(s) \underset{T}{\Leftrightarrow} 2B\_Si/B(s) + 3H_2(g)$	B-deposition
[R2]	$H(g) + H\_Si/B(s) \underset{T}{\Leftrightarrow} \circ Si/B(s) + H_2(g)$	H desorption
[R3]	$H\_Si/B(s) + H\_Si/B(s) \underset{T}{\Leftrightarrow} Si/B\_Si/B(s) + H_2(g)$	Cross-linked
[R4]	$H_2B\_B(s) + \circ B(s) \underset{T}{\Leftrightarrow} \circ B(s) + H_2B\_B(s)$	Migration

[a]H_Si/H_B and $\circ Si/\circ B$ are the silicon/boron atoms with H-terminated dangling bonds or with free dangling bonds, respectively.

Table 3. Chemical reactions of the CVD boron deposition.

The kinetics of boron CVD is investigated in references [17] and [18]. The lateral gas-phase diffusion length of boron atoms over silicon/boron surfaces during the CVD of pure boron layers is investigated in reference [20]. The chemical reactions contributing to the pure boron CVD are listed in **Table 3**.

For high-temperature deposition (700°C), as described in reference [21], there would be enough energy to facilitate a reaction [R2]. This reaction releases hydrogen from the Si and/or B surface sites as H_2 and leaves Si and/or B sites free for deposition. In this case, the deposition starts very fast, that is, after a few seconds of the surface being exposed to diborane gas under these conditions, and a monolayer of boron appears and covers the surface [1]. **Figure 4** shows a schematic of the HT deposition mechanisms for the first few monolayers of boron layers when a Si surface is exposed to diborane. In the high-temperature deposition, the temperature-related reactions [R1]–[R4] proceed, leading to the adsorption of boron atoms, which are deposited and/or suspended, as well as migration of the boron atoms along the surface. This leads to a smooth, uniform, closed boron layer [17]. Triggering the intermediate reactions also ensures very smooth layers with a roughness that can be around 2 angstroms, as measured by HRTEM imaging and atomic-force microscopy (AFM). This is illustrated by the analysis results shown in **Figure 5** for a 2-nm-thick layer.

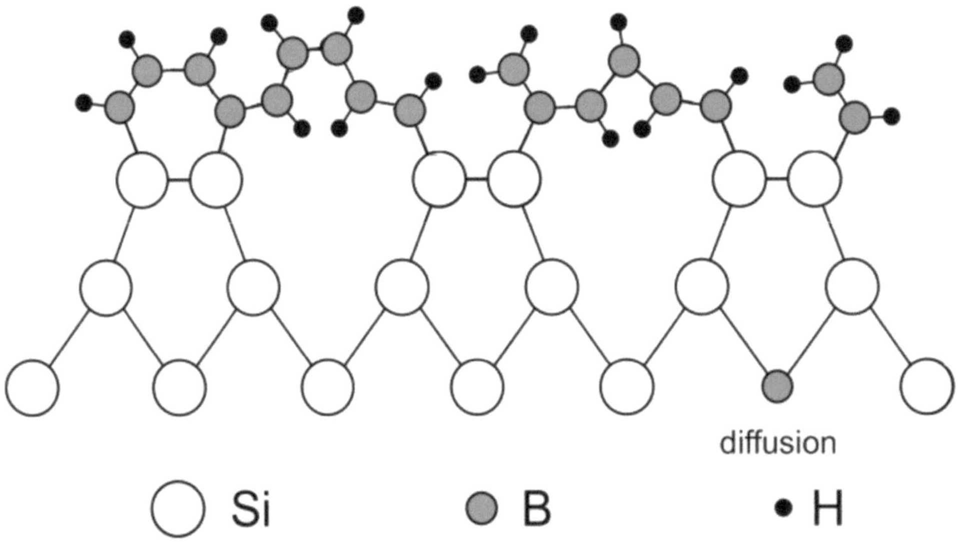

Figure 4. Schematic of the deposition of the first few monolayers of boron layers for HT, 700°C, deposition when an Si surface is exposed to B_2H_6 [22].

However, the experiments show that by lowering the boron deposition temperature from 700 to 400°C, the deposition rate drops considerably from 0.4 to less than 0.01 nm/min for deposition in an H_2 environment, with a laminar gas flow of 20 slm. In this case, even after a long deposition time at 400°C there is almost no measurable layer formed. Moreover, if such a layer is deposited, it is not closed enough to make it suitable for device applications. As a result, the performance of these devices is not comparable to the ones made by conventional 700°C boron deposition.

http://dx.doi.org/10.5772/63344

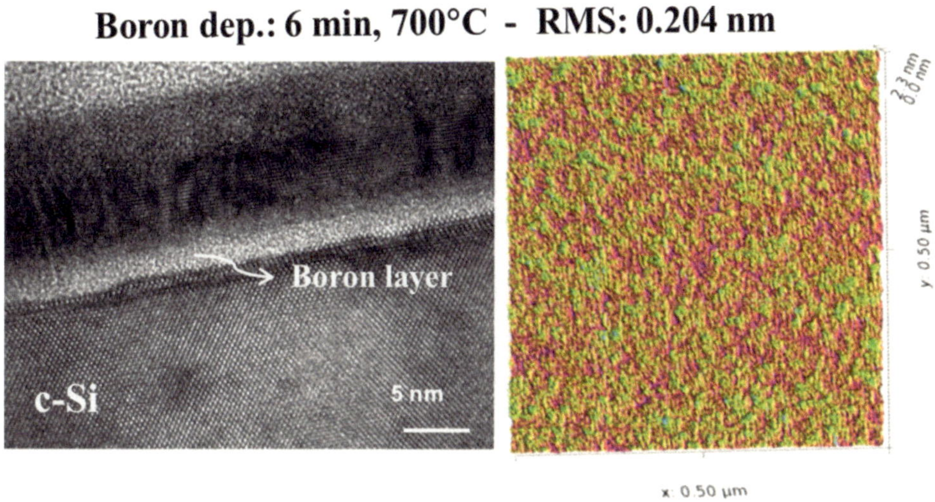

Figure 5. HRTEM image (left) and atomic-force microscopy (AFM) measurement (right) of a 6-min HT-deposited boron layer surface. The AFM image was taken in a 500×500 nm^2 scanning area. This HT layer is smooth and uniform at a thickness of about 2 nm with a root-mean-square (rms) roughness value of 0.204 nm [21].

Several factors are responsible for this negative effect. Firstly, as shown with the previously developed analytical kinetic model given in Eq. (6) [18], the deposition rate varies with the deposition temperature *T*; that is, by lowering the *T*, the DR will drop significantly. Secondly, at such a low deposition temperature, the gas-phase diffusion length is very short. Lastly, by lowering the deposition temperature, the intermediate reactions that produce adsorption at 700°C are no longer effective.

For example, due to the very low amount of energy available, reactions [R2]–[R4] are not promoted in their forward direction. In addition, the presence of hydrogen in the reactor as a carrier gas can suppress the deposition reaction of [R1] because most of the surface sites are occupied by hydrogen that will not readily desorb at 400°C. Consequently, the first monolayer coverage takes more time at this temperature than at 700°C. Also, the inhibited removal of H from the Si surface at low temperatures plays an important role in increasing the roughness. This can be seen in **Figure 6** where the schematics of the deposition of the first few monolayers of boron are shown at LT, 400°C. The HRTEM image and AFM measurement of the LT-deposited boron layer at 400°C with the newly developed recipe discussed in the next section is shown in **Figure 7**. A layer roughness of around 6–8 angstroms was found for a 5-nm-thick layer.

By switching to N_2 as a carrier gas, reactions [R1–[R3] are triggered to proceed in the forward direction, while reaction [R4] is suppressed, which is the necessary intermediate reaction for providing a smooth, uniform, closed boron layer. Thus, only lowering the temperature in the conventional 700°C PureB recipe to 400°C does not guarantee a sufficient layer deposition as a promising capping layer. Therefore, for boron deposition at 400°C, a compromise between the layer deposition rate and layer uniformity and smoothness is necessary, which demands the development of a new recipe, to be discussed in detail in the next section.

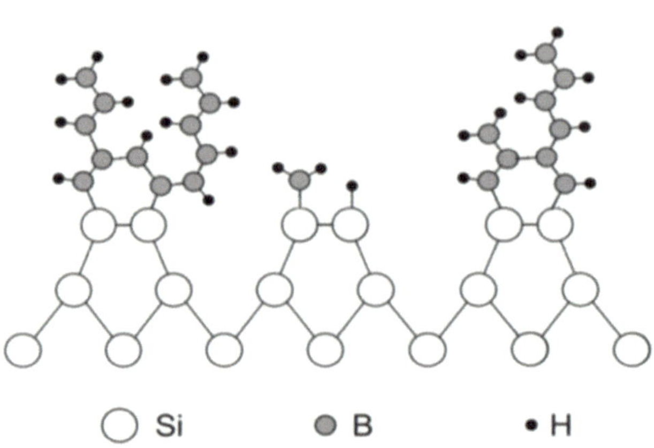

Figure 6. Schematic of the deposition of the first few monolayers of boron layers for (a) HT, 700°C, and (b) LT, 400°C, deposition when an Si surface is exposed to B_2H_6 [22].

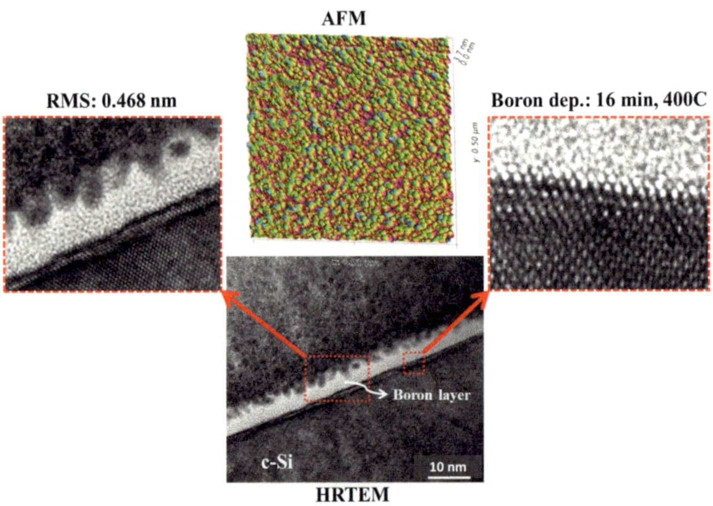

Figure 7. High-resolution TEM image (bottom) and AFM measurement (top) of a boron layer deposited at 400°C for 16 min. In both cases, the thickness of the LT boron layer was extracted to about 5 nm, with a surface rms roughness value of around 4–6 angstroms. The pictures on the left and right show a close-up of the LT boron layer surface roughness and the interface with the Si substrate, respectively [22].

The final point to discuss about the temperature dependency of boron deposition is the investigation conducted of the deposition at the intermediate temperatures 600°C, 500°C and 450°C. By lowering the deposition temperature, the gas-phase diffusion coefficient is lowered and the deposition rate based on Eq. (6) is reduced. In addition, the lower temperatures mean that intermediate reactions [R2]–[R4] are triggered less in their forward direction, resulting in a higher H-coverage of the surface. Moreover, the gas-phase diffusion lengths of boron atoms on both the Si and oxide surfaces also decrease. Altogether, this causes the deposited layer to become increasingly rough when shifting from 600°C to 500°C and 450°C. The properties of layers deposited on bare Si at temperatures between 700°C and 400°C are given in **Table 4**. What can be clearly seen is the increase in the roughness of the layers as the deposition temperature decreases.

For deposition at the lower temperatures, 450°C and 400°C, the ellipsometry measurement gave a roughness that is much greater than the average layer thickness, which may not be relevant. These parameters become very dependent on the exact deposition conditions because of both the absence of the other intermediate reactions ([R2]–[R4]) and the fact that the gas-phase diffusion lengths of the boron atoms on both the Si and oxide surfaces decrease.

Next, we will discuss the process and recipe requirements needed to make the boron deposition possible at 400°C.

Pure boron	Dep. Temp.[a]				
	700°C	600°C	500°C	450°C	400°C
Dep. time (min)	10	10	16	20	60
Thickness (nm)	3.74	3.21	3.14	1.77	1.62
Roughness (nm)	0.38	0.81	1.23	2.64	3.73

[a]The recipes only differ in time and temperature. Their structure is the same in all cases, even for 400°C, and the depositions were done only in an H_2 ambient environment.

Table 4. Thickness and roughness of deposited layers extracted from ellipsometry measurements, as a function of deposition temperature and time.

3.2. LT boron deposition at 400°C

As discussed above, the presence of hydrogen gas can suppress the forward direction of reaction [R2]. This is necessary to release the hydrogen from the surface, to decrease the H-surface coverage, and to suppress reactions [R1] and [R3]. Therefore, the deposition rate is limited by reaction [R2]. The release of hydrogen is important as it facilitates the migration of deposited boron atoms along boron/silicon surfaces via the intermediate reactions [R3] and [R4], leading to a smooth, uniform, closed boron layer deposition. However, in the case of deposition in N_2, unlike that of H_2, reactions [R1]–[R3] are not suppressed, causing hydrogen atoms to be released more easily from the surface and reducing the H-surface coverage. This leads to a higher deposition rate. However, due to the lower mobility of boron atoms in this environment, an increase in boron layer roughness was also expected. Therefore, for boron deposition at 400°C it is necessary to start the deposition in a nitrogen environment to facilitate the first monolayer deposition, and then switch to a hydrogen environment to make the surface smoother and to maintain the boron coverage over the entire silicon opening. This procedure can be repeated a few times with different cycle lengths, as shown schematically in **Figure 8**, to achieve a smooth, uniform, closed boron layer. Lowering the chamber pressure can also facilitate the release of hydrogen from the surface and decrease the H-surface coverage. Thus, deposition at a lower ambient pressure is preferable.

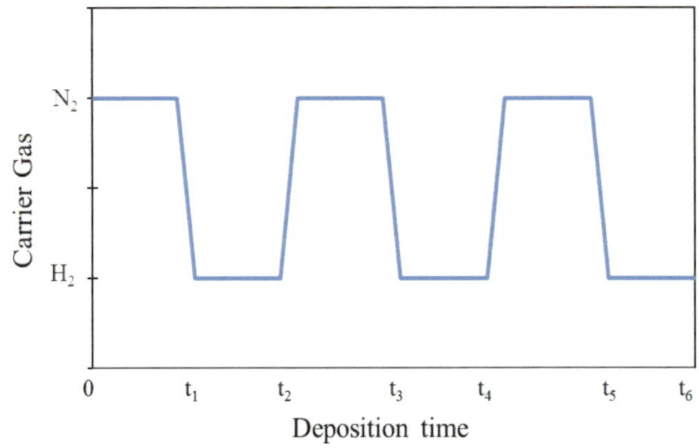

Figure 8. Schematic illustration of carrier gases switching sequences in the LT PureB recipe [22].

The HRTEM image in **Figure 9** shows an LT boron layer deposited at 400°C at a chamber pressure of 95 torr, with the newly developed recipes applied in four- and six-cycle sequences of carrier gas switching. As can be seen in these images, with more switching sequences a smoother boron layer can be created.

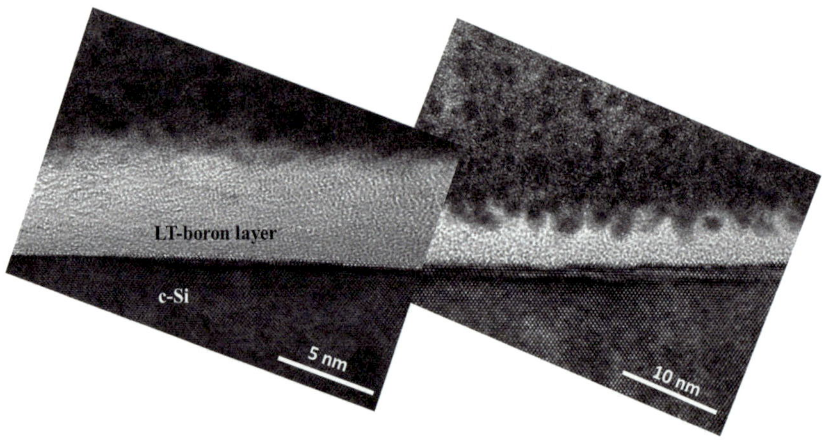

Figure 9. HRTEM image of an LT boron layer deposited after (left) the six- (right) and four-cycle recipe sequences. The deposition chamber pressure was 95 torr [22].

Besides a new recipe, some additional treatments were also needed prior to the wafers being loaded into the reactor. The first treatment was a standard ex situ-cleaning procedure prior to deposition (i.e. 10 min: HNO_3 100% at 25°C; 5 min: DI water; 10 min: HNO_3 70% at 110°C; and 5 min: DI water), immediately followed by a 4-min HF (0.55%) dip to remove native oxide and H-passivate the surface to prevent native oxide formation. This was followed by Marangoni drying, which is an effective substitute for spin-rinse drying, to prevent the formation of wet spots. These wet spots readily form on mixed hydrophilic-hydrophobic surfaces during spin-rinse drying, resulting in particle contamination from residues left behind after evaporation

http://dx.doi.org/10.5772/63344

[23]. Unlike the 700°C deposition, for the 400°C deposition there was no in situ baking step [1]. Thus, the wafers were directly loaded into the pre-prepared reactor at the 400°C deposition temperature. The lack of the in situ bake made the HF dip with Marangoni drying a very crucial step and meant that the wafers had to be loaded immediately into the N_2-purged load lock of the reactor to prevent any native oxide formation.

3.3. Issues with lowering the deposition temperature of the boron layers

During an HT (700°C) deposition, there is sufficient energy; therefore, the boron atoms that arrive above the oxide-covered surfaces are very mobile and can move around and diffuse laterally to reach the Si surface and become available for the layer deposition. However, during an LT (400°C) deposition, due to the very low amount of available energy, all those boron atoms are very limited in terms of movement and therefore they remain wherever they are. This means that the concentration of the boron atoms over oxide-covered surfaces can increase significantly so that the probability of desorption from the oxide-free sites and/or defects becomes much lower than in the 700°C case. In fact, a significant amount of boron was observed to be deposited on the oxide. This was also promoted by the longer deposition time needed to achieve a reliably closed layer because of the low deposition rate and extra thickness needed to compensate for the large amount of roughness, as can be seen in **Figure 10**.

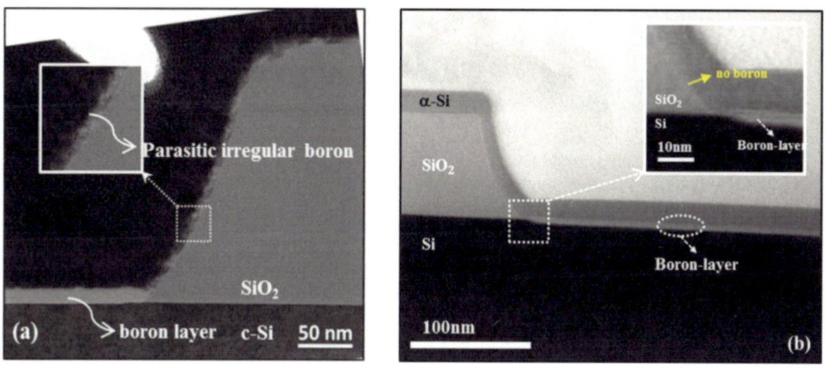

Figure 10. HRTEM images of flat/bevelled oxide surfaces of a small 40×40 μm^2Si window with (a) LT and (b) HT boron layers [21, 22].

These parasitic boron depositions on the flat/bevelled oxide are unwanted and can cause problems in the subsequent processing steps. For example, the adhesion of plasma-enhanced chemical vapour deposition (PECVD) layers of TEOS-SiO_2 was seen to degrade. This is illustrated in the SEM images of **Figure 11**. Similar adhesion issues were also witnessed in the case of physical vapour deposited (PVD) Al deposition on these surfaces. Boron depositions at 500°C can be performed without any parasitic deposition over most oxide surfaces, even at micron-sized windows. However, still for some oxide qualities this kind of unwanted parasitic deposition was observed. An example is given in **Figure 12b** and **c**, where two different dyes with the same layout are shown with and without adhesion issues for subsequent PECVD TEOS and PVD Al layer depositions.

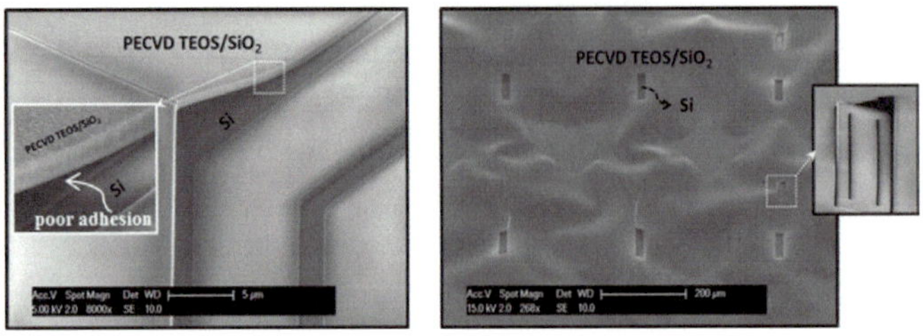

Figure 11. Examples of poor PECVD TEOS-SiO_2 adhesion as a result of parasitic boron deposition during LT deposition of boron on flat/bevelled oxide surfaces near micron-sized Si windows where the LOR is high [21].

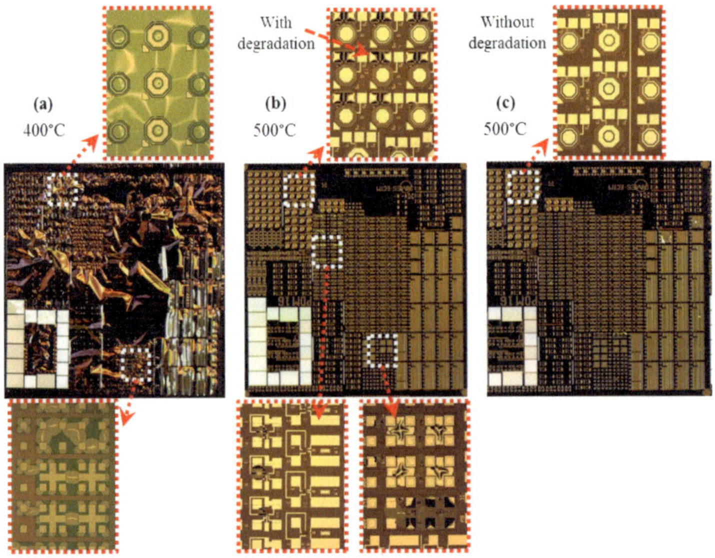

Figure 12. Examples of poor adhesion and layer delamination for the same layout. Die (a) shows PECVD TEOS deposited over 400°C boron. PVD Al was deposited after PECVD TEOS deposition over 500°C boron for two different dies on the same wafer. Die (b) displays the adhesion problems. However, die (c) does not display the same adhesion problems as die (b). The insets show the zoom-in for different structures with micron-sized openings [21].

3.4. Precautions to minimise the parasitic boron deposition at low temperatures

To minimise the undesirable parasitic boron deposition on the oxide, there are a few precautions that can be taken into account. Firstly, the chances of parasitic boron deposition can be reduced by using the higher quality oxide, which has a lower density of oxide surface-free sites and/or defects. Secondly, the global and local oxide coverage ratios can be optimised in the layout of the oxide mask to reduce the unnecessary boron atom accumulation over the oxide areas. Likewise, the deposition and process parameters, such as diborane partial pressure, gas flow and total deposition pressure, can be optimised to minimise the loading effects [15, 24]. These will otherwise also increase the accumulation of boron atoms over certain oxide areas, thus increasing the probability of parasitic boron deposition there. Also, we have found that this poor adhesion could depend on the ex/in situ treatments before the PECVD TEOS-SiO_2 or PVD Al layers are deposited. Furthermore, the adhesion of these extra layers after boron deposition is also influenced by the exact conditions on the surface. For example, a large amount of hydrogen resides on the surface just after boron deposition and is in general known to reduce adhesion. This H coverage can be reduced by carrying out the baking steps in air. An investigation of these extra treatments would have to be included in the process development when working with LT boron deposition.

By taking the aforementioned precautions into account, the LT boron deposition was successfully performed to create near-ideal LT PureB photodiodes with low, deep-junction-like saturation currents.

4. LT PureB UV photodiode fabrication

The post-metal LT boron deposition was successfully performed to create p^+n photodiodes with nm-thin, boron-only beam entrance windows and near-theoretical UV sensitivity with negligible optical and electrical degradation [25]. **Figure 13** shows the processing steps of the

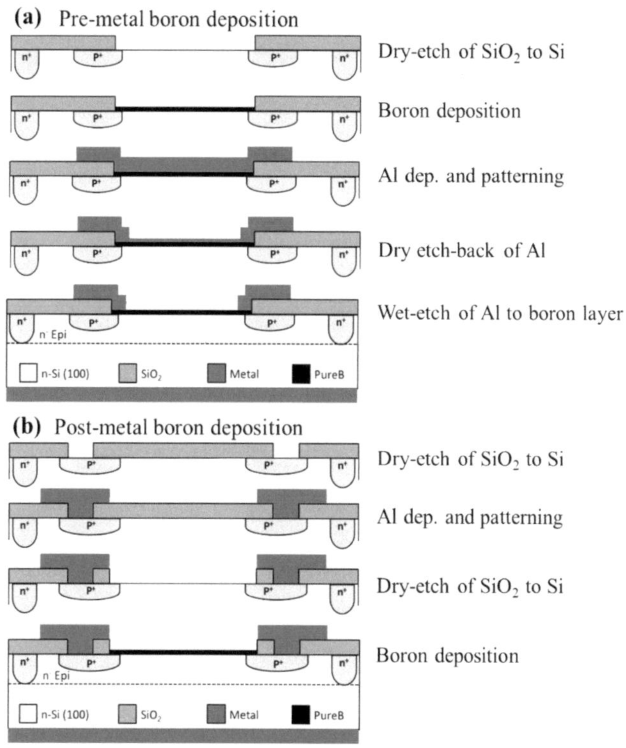

Figure 13. Processing steps of the fabricated photodiodes with boron-only beam entrance windows, for boron deposition both before (a) and after (b) metallisation [25].

fabricated photodiodes with boron-only beam entrance windows, for boron deposition both before and after metallisation.

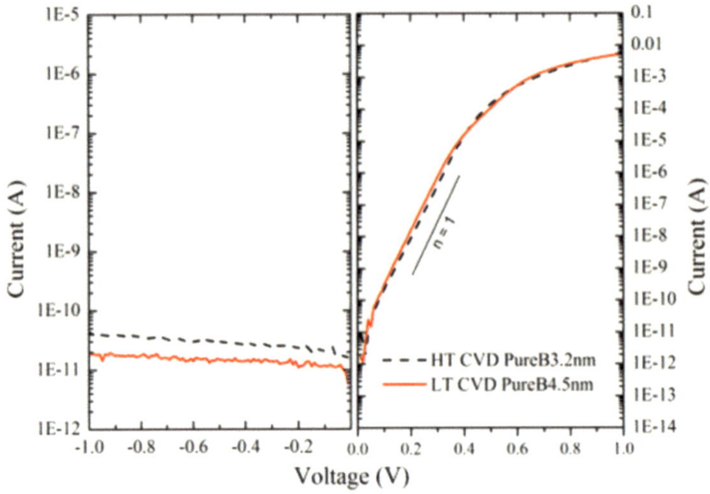

Figure 14. Measured I-V characteristics of both pre-metal HT and post-metal LT PureB photodiodes with an active area of 9.6×9.6 mm^2. Schematics of the fabricated photodiodes are shown in **Figure 13**. The boron layer thickness was measured by ellipsometry to be 3.2 and 4.5 nm for the HT and LT photodiodes, respectively [25].

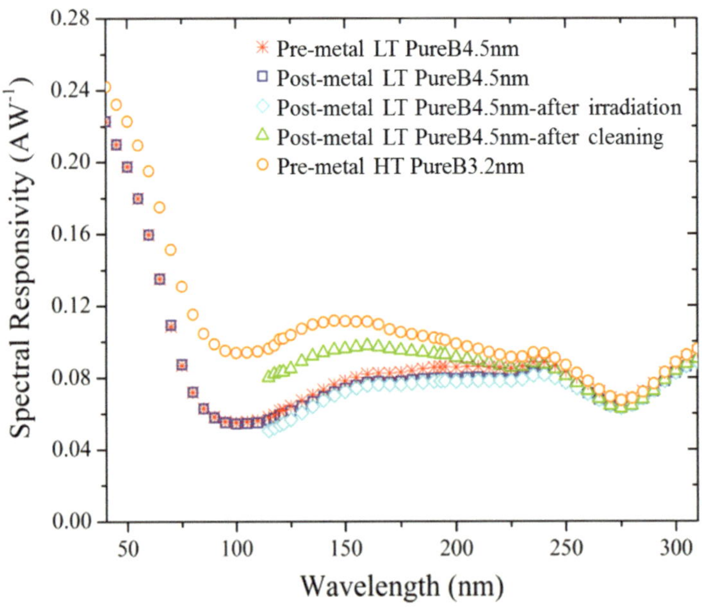

Figure 15. Measured VUV spectral responsivity as a function of wavelength for HT and LT PureB photodiodes (area: 9.6×9.6 mm^2) before and after high-level irradiation and after a final cleaning procedure [22].

http://dx.doi.org/10.5772/63344

The I-V characteristics measured for both pre-metal HT and post-metal LT PureB photodiodes with an active area of 9.6×9.6 mm² are shown in **Figure 14**. The boron layer thickness was measured by ellipsometry to be 3.2 and 4.5 nm for the HT and LT photodiodes, respectively. As can be seen, low deep-junction-like saturation currents and near-ideal diode characteristics can be provided by LT boron deposition.

Figure 15 shows the measured responsivity in the VUV spectral range (120–400 nm) for both pre-metal and post-metal LT PureB photodiodes. As can be seen in this figure, the response for pre-metal photodiodes is slightly higher than that for post-metal ones. This could be explained by a small thinning of the boron layer due to the extra processing necessary to open the light entrance window (see **Figure 13**). The HT PureB device is also shown in **Figure 15** as a reference. As will be commented on subsequently, the rough LT boron surface proves to be much less chemically resilient than the smooth HT boron layers.

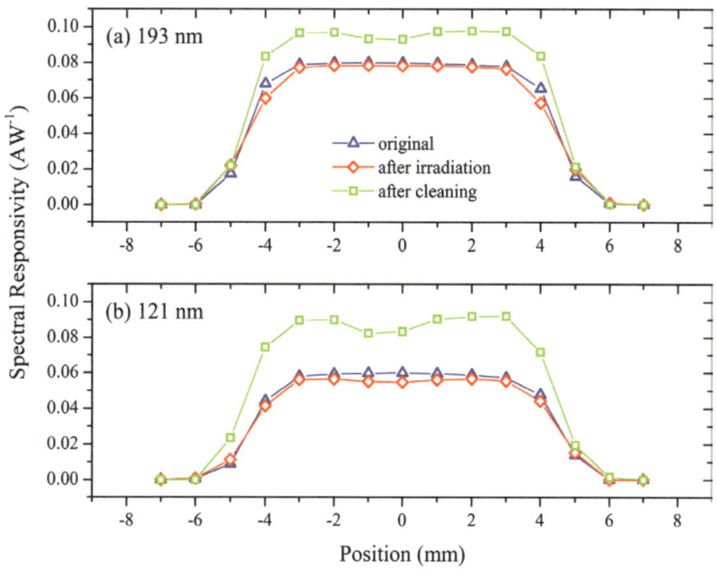

Figure 16. Measured responsivity across the surface in one direction of a 9.6×9.6 mm² LT PureB photodiode at wavelengths of 193 and 121 nm, before and after high-dose exposure, and after cleaning. The high-dose exposure was performed with a circular beam spot of ø1 mm with a radiant exposure of 37 J/cm² [22].

The optical stability of the post-metal LT PureB photodetector was measured by exposing the centre of the 10×10 mm² dies to high-dose VUV irradiation. The high-dose exposure was performed with a circular beam spot with a diameter of 1 mm, to a radiant exposure of 37 J/cm². The LT PureB photodiodes were found to be highly stable. To illustrate, the resulting responsivity at the centre of the post-metal LT PureB photodiode is included in **Figure 16**.

A slight dip in responsivity is visible after the 121-nm exposure. This is the result of the build-up of carbon contamination during exposure. To remove this layer, an ozone-cleaning procedure was performed. For HT devices, the cleaning procedure successfully removes the contamination, and the original high responsivity is regained [26]. By contrast, this treatment cannot effectively remove the carbon contamination of the LT device, probably due to the rougher surface structure as shown in **Figure 13**. As can be seen, the rougher surface makes the carbon contamination removal more difficult than the smooth surface structures. Additionally, this rougher surface structure makes the silicon surface below the boron layer more susceptible to oxidation [27]. The higher responsivity after standard ozone-cleaning suggests that the boron layer is thinned considerably, as can be seen in **Figures 15** and **16**. On closer look at these figures, a reduced thinning can be clearly seen where the carbon contamination was present for both measurement wavelengths (i.e. in the middle of the active area). Therefore, although the as-deposited LT boron layer is optically robust, it is more vulnerable to carbon contamination due to its rougher surface compared to HT boron layer, as can be seen in **Figure 9**.

5. Summary

In this chapter, an analytical model is developed to describe the deposition kinetics and deposition chamber characteristics that determine the deposition rate over the wafer, which can be used as a pre-deposition prediction tool to improve the set-up and control of the final deposition. The model has been very useful both in the development of uniform boron layers with minimal pattern dependence and in the transferring of recipes from one reactor to the other.

A new technology for low-temperature (400°C) boron deposition is presented, which provides a smooth, uniform, closed boron layer. The temperature dependency of the boron deposition on patterned Si/SiO_2 surfaces in the temperature range of 400–700°C is investigated. Some selectivity issues that arise when the boron deposition temperature is reduced from 700°C to 400°C are discussed. Some provisions are recommended to minimise the undesirable boron deposition on the oxide. For boron deposition at 400°C, it is necessary to start the deposition in a nitrogen environment to facilitate the first monolayer deposition, and then switch to a hydrogen environment to make the surface smoother and to maintain the boron coverage over the entire silicon opening.

The LT PureB technology is then successfully employed to create near-ideal LT PureB photodiodes with nm-thin, boron-only beam entrance windows and a near-theoretical sensitivity for irradiation with either VUV/DUV/EUV light down to a wavelength of 10 nm. Very low dark current of only 15 pA at -1 V bias voltage was measured for post-metal LT PureB photodiodes with an anode area of 9.6×9.6 mm^2, which we relate to the fact that no post-PureB processing was performed.

It is believed that at temperatures of 400°C and lower, no silicon doping is possible during the boron deposition. Yet, the LT boron layers provide a structure with excellent p-n junction-like I-V characteristics. Based on this observation, we make the assumption that during the damage-free LT boron deposition, which is most likely a result of a chemical reaction, a kind of effective "hole" layer is built between the boron and silicon. This interface "hole" layer creates an electric field that repels the injected minority carriers (electrons) away from the interface.

Author details

Vahid Mohammadi* and Stoyan Nihtianov

*Address all correspondence to: V.Mohammadi@tudelft.nl

Department of Microelectronics, Delft University of Technology, Delft, The Netherlands

References

[1] Sarubbi F., Nanver L.K., Scholtes T.L.M. High effective Gummel number of CVD boron layers in ultrashallow p+n diode configurations. IEEE Transactions on Electron Devices. 2010;57:1269–1278.

[2] Shi L., Nanver L.K., Laubis C., Scholze F., Nihtianov S. Electrical performance optimization of a silicon-based EUV photodiode with near-theoretical quantum efficiency. IEEE Transducers'11. 2011;art. no. 5969130:48–51.

[3] Shi L., Nihtianov S., Scholze F., Gottwald A., Nanver L.K. High-sensitivity high-stability silicon photodiodes for DUV, VUV and EUV spectral ranges. In: SPIE 8145, UV, X-Ray, and Gamma-Ray Space Instrumentation for Astronomy XVII; 21 Aug.; San Diego, CA. 2011. p. 81450N–81450N-9.

[4] Šakić A., Nanver L.K., Van Veen G., Kooijman, K., Vogelsang, P., Scholtes T.L.M., et al. Versatile silicon photodiode detector technology for scanning electron microscopy with high-efficiency sub-5 keV electron detection. In: IEEE International Electron Devices Meeting (IEDM); 6–8 Dec.; San Francisco, CA. 2010. p. 31.4.1–31.4.4.

[5] Golshani N., Beenakker C.I.M., Ishihara R. Manufacturing uniform field silicon drift detector using double boron layer. Nuclear Instruments and Methods in Physics Research, Section A. 2015;794:206–214.

[6] Kuschnerus P., Rabus H., Richter M., Scholze F., Werner L., Ulm G. Characterization of photodiodes as transfer detector standards in the 120 nm to 600 nm spectral range. Metrologia. 1998;35:355–362.

[7] Solt K., Melchior H., Kroth U., Kuschnerus P., Persch V., Rabus H., et al. PtSi–n–Si Schottky-barrier photodetectors with stable spectral responsivity in the 120–250 nm spectral range. Applied Physics Letters. 1996;69:3662–3664.

[8] Canfield L.R., Vest R.E., Korde R., Schmidtke H., Desor R. Absolute silicon photodiodes for 160 nm to 254 nm photons. Metrologia. 1998;35:329–334.

[9] Scholze F., Brandt G., Mueller P., Meyer B., Scholz F., Tümmler J., et al. High-accuracy detector calibration for EUV metrology at PTB. SPIE. 2002;4688:680–689.

[10] Scholze F., Klein R., Muller R. Characterization of detectors for extreme UV radiation. Metrologia. 2006;43:S6–S10.

[11] FEI. High-performance Microscopy Workflow Solutions [Internet]. [Updated: 2016]. Available at: http://www.fei.com/

[12] Roussel L.Y., Stokes D.J., Gestmann I., Darus M., Young R.J. Extreme high resolution scanning electron microscopy (XHR SEM) and beyond. In: SPIE 7378, Scanning Microscopy; 4 May; Monterey, CA. 2009. p. 73780W–73780W-9.

[13] Jacquot B.C., Hoenk M.E., Jones T.J., Cunningham T.J., Nikzad S. Direct detection of 100–5000 eV electrons with delta-doped silicon CMOS and electron-multiplying CCD imagers. IEEE Transactions on Electron Devices. 2012;59:1988–1992.

[14] Mohammadi V., De Boer W., Scholtes T.L., Nanver L.K. Pattern dependency of pure-boron-layer chemical-vapor depositions. ECS Transactions. 2012;45:39–48. DOI: http://dx.doi.org/10.1149/1.3700937

[15] Mohammadi V., De Boer W., Scholtes T.L.M., Nanver L.K. Local-loading effects for pure-boron-layer chemical-vapor deposition. ECS Transactions. 2013;50:333–341. DOI: http://dx.doi.org/10.1149/05004.0333ecst

[16] Mohammadi V., Van de Kruijs R.W.E., Rao P.R., Sturm J.M., Nihtianov S. Influence of the surface oxide content of a boron capping layer on UV photodetector performance. In: 9th International Conference on Sensing Technology; 8–10 Dec.; Auckland, New Zealand. 2015.

[17] Mohammadi V., De Boer W.B., Nanver L.K. Temperature dependence of chemical-vapor deposition of pure boron layers from diborane. Applied Physics Letters. 2012;101:111906.1–111906.4. DOI: http://dx.doi.org/10.1063/1.4752109

[18] Mohammadi V., De Boer W.B., Nanver L.K. An analytical kinetic model for chemical-vapor deposition of PureB layers from diborane. Journal of Applied Physics. 2012;112:113501.1–113501.10. DOI: http://dx.doi.org/10.1063/1.4767328

[19] Mohammadi V., Mohammadi S., Ramesh S., Nihtianov S. Numerical gas flow and heat transfer simulation in the ASM Epsilon 2000 CVD reactor for pure boron deposition. In: 23rd International Scientific Conference Electronics-ET; 11–13 Sept.; Sozopol, Bulgaria. 2014. p. 28–31.

[20] Mohammadi V., Nihtianov S. Lateral gas phase diffusion length of boron atoms over Si/B surfaces during CVD of pure boron layers. AIP Advances. 2016;6:025103.1–025103.8. DOI: http://dx.doi.org/10.1063/1.4941702

[21] Mohammadi V., Golshani N., Mok K.R.C., De Boer W.B., Derakhshandeh J., Nanver L.K. Temperature dependency of the kinetics of PureB CVD deposition over patterned Si/SiO2 surfaces. Microelectronic Engineering. 2014;125:45–50. DOI: http://dx.doi.org/10.1016/j.mee.2014.03.015

[22] Mohammadi V., Nihtianov S. Low temperature, 400 C, pure boron deposition: a solution for integration of high-performance Si photodetectors and CMOS circuits. In: IEEE SENSORS; 1–4 Nov.; Busan, South Korea. 2014. p. 1–4. DOI: http://dx.doi.org/10.1109/ICSENS.2015.7370555

[23] Reinhardt K.A., Reidy R.F. Handbook for cleaning for semiconductor manufacturing: Fundamentals and applications. Co-published by John Wiley & Sons, Inc. Hoboken, New Jersey, and Scrivener Publishing LLC, Salem, Massachusetts. 2011.

[24] Mohammadi V., De Boer W.B., Scholtes T.L.M., Nanver L.K. Pattern dependency and loading effect of pure-boron-layer chemical-vapor deposition. ECS Journal of Solid State Science and Technology. 2012;1:Q16–Q20. DOI: http://dx.doi.org/10.1149/2.024201jss

[25] Mohammadi V., Qi L., Golshani N., Mok K.R.C., De Boer W.B., Sammak A., et al. VUV/low-energy-electron Si photodiodes with post-metal 400°C PureB deposition. IEEE Electron Device Letters. Dec. 2013;34(12):1545–1547. DOI: http://dx.doi.org/10.1109/LED.2013.2287221

[26] Shi L., Nihtianov S., Nanver L.K., Scholze F. Stability characterization of high-sensitivity silicon-based EUV photodiodes in a detrimental environment. IEEE Sensors Journal. 2013;13:1699–1707.

[27] Mohammadi V., Shi L., Kroth U., Laubis C., Nihtianov S. Stability characterization of high-performance PureB Si-photodiodes under aggressive cleaning treatment in industrial applications. In: IEEE International Conference on Industrial Technology (ICIT); 17–19 Mar.; Seville, Spain. 2015. DOI: http://dx.doi.org/10.1109/ICIT.2015.7125599

Chapter 7

Silicon-Rich Oxide Obtained by Low-Pressure Chemical Vapor Deposition to Develop Silicon Light Sources

J. Alarcón-Salazar, R. López-Estopier,
E. Quiroga-González, A. Morales-Sánchez,
J. Pedraza-Chávez, I. E. Zaldívar-Huerta and
M. Aceves-Mijares

Additional information is available at the end of the chapter

http://dx.doi.org/10.5772/63012

Abstract

Off stoichiometric silicon oxide, also known as silicon-rich oxide (SRO), is a light-emitting material that is compatible with silicon technology; therefore, it is a good candidate to be used as a light source in all-silicon optoelectronic circuits. The SRO obtained by low-pressure chemical vapor deposition (LPCVD) has shown the best luminescent properties compared to other techniques. In spite of LPCVD being a simple technique, it is not a simple task to obtain SRO with exact silicon excess in a reliable and repetitive way. In this work, the expertise obtained in our group to obtain SRO by LPCVD with precise variation is presented. Also, the characteristics of this SRO obtained in our group are revised and discussed. It is demonstrated that LPCVD is an excellent technique to obtain single layers and multilayers of nanometric single layers with good characteristics.

Keywords: SRO, LPCVD, photoluminescence, electroluminescence, multilayer

1. Introduction

Chemical vapor deposition (CVD) is a versatile and economical technique used to deposit different materials. In the microelectronics industry, it has found a main place and it is a standard process. Currently, many efforts are being done to produce optoelectronic circuits using the mature technology of integrated circuits. A major drawback to integrate a whole silicon circuit that manages both electronic and optical signals is that silicon does not emit light efficiently. There are serious restrictions to integrate a light source in such a system [1]. Basically, two approaches have been under study to solve the problem of the light source: One of them uses a reverse-biased *pn* junction [2], the other one uses light-emitting materials that are compatible with silicon [3, 4]. In spite of the fact that both approaches have shown that it is possible to integrate a complete optoelectronic system based on Si, there is still a wide field of possibilities to improve the efficiency of the light emitters [2, 5], and in consequence, of the whole system.

Off stoichiometric silicon oxide (with empirical formula SiO_x), also known as silicon-rich oxide (SRO), is a good material to be used as a light source in all-silicon optoelectronics circuit [6]. It is obtained by different techniques, including plasma-enhanced and low-pressure chemical vapor deposition (PECVD and LPCVD) methods. The silicon excess in SRO obtained by LPCVD can be easily controlled during the deposition by the ratio of the reactive gases, in our case silane and nitrous oxide:

$$R_0 = \frac{P_{N_2O}}{P_{SiH_4}} \tag{1}$$

Dong et al. showed that for $R_0 = 10$, 20, and 30, the corresponding silicon excess is about 12, 7 and 5 at %, respectively; that is, SiO_x with $x = 1.13$, 1.50, and 1.63 [7]. However, it is worthy to mention that SiO_x is a multiphase material composed of SiO_2, elemental silicon, and SiO_y. Thus, SiO_x is an empirical formula to denote SRO (which may contain large Si excess), and SiO_y is an oxide with stoichiometry deviating a bit from SiO_2. Oxidation states obtained by X-ray photoelectron spectroscopy (XPS) of LPCVD-SRO for R_0 =10, 20, and 30 are shown in [4]. Depending on the silicon excess, SRO has different characteristics: For low silicon excess, the density of SiO_y compounds dominates, producing intense luminescence. However, for high silicon excess the density of elemental silicon increases, what reduces the emission, but increases the conductivity of the films.

In order to have intense luminescent SiO_x, high-temperature treatments are required. For SRO obtained using LPCVD, 1100°C in nitrogen is the most favorable temperature. Additionally, the emission also depends on the R_0. The emission increases as R_0 increases, being $R_0 = 30$ the top one; for R_0 higher than 30, the emission reduces again. On the other hand, the conductivity increases as the R_0 reduces, This compromise is important when electroluminescent devices are the goal. Therefore, structures that combine layers of high conductivity with layer of high emission properties are under study to obtain efficient light sources compatible with silicon.

SRO obtained using LPCVD is perhaps the most luminescent compared with SRO obtained by other methods [6]; however, in this technique, it is difficult to control the silicon excess with some precision, and to have films with controllable properties required of personal with expertise on this type of technique.

http://dx.doi.org/10.5772/63012

In this paper, details of our LPCVD deposition processes to obtain SRO single layers (SLs) and multilayers (MLs) with different R_0, including details of our LPCVD system, are described. Also, the optical and structural characteristics of our SRO layers and multilayers are reviewed. Electroluminescence (EL), cathodoluminescence (CL), and photoluminescence (PL) results will be presented and discussed.

2. Our system

We have two homemade LPCVD reactors, one for two- and another for four-inch wafers. Both reactors have the same layout, thus we will describe only one of them in a general way. Our laboratory is a teaching and research facility, therefore every day different materials have to be deposited and the equipment has to be very versatile. For this reason, we found that controlling it manually produces better results than using automatic parts, then the control of the gas flux using rotameter (ABB model 10A6131NB1B1X00) give us enough functionality. High throughput is not required and our main concern is to have good films with repetitive characteristics. Normally, polysilicon, silicon nitride, silicon oxynitride, and SROs are deposited in the reactor; however, we are not limited to only those materials. Perhaps, obtaining SROs with good characteristics is the most demanding, that is because small differences in silicon excess produce big changes in its characteristics. In the following paragraphs, we will concentrate on describing the details to obtain SRO in a controllable manner.

As shown in **Figure 1**, our LPCVD system is hot wall type that allows having a uniform temperature in the whole deposition chamber area. The heating element is a three-zone furnace, and a flat zone of ±2°C can be obtained. In the past, an analysis using multivariated experiment was carried out to study different parameters involved in the deposition process [8]. Based on that, we decide to maintain the wafer horizontally on a flat quartz wafer holder. The working temperature profile was chosen with an increasing slop to obtain lesser thickness variation, as shown in **Figure 2**. The increasing temperature compensates the changes of the boundary layer and produces a more uniform deposition through the flat wafer holder [9]. The deposition temperature allows to deposit SRO from $R_0 = 5$ to $R_0 = 100$.

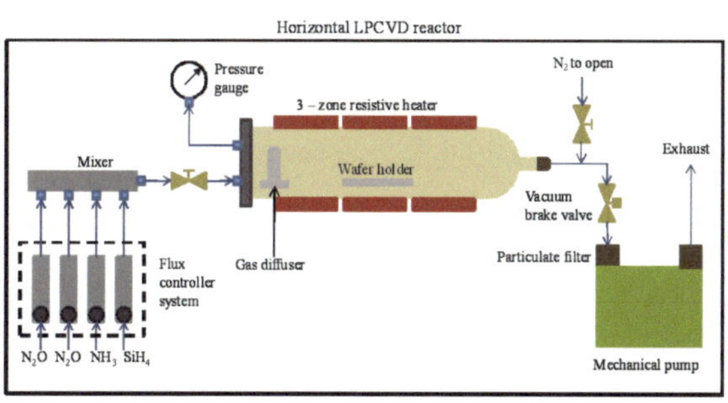

Figure 1. Schematic of the LPCVD system. A three zones furnace and mechanical pump are used.

To obtain SRO, the reactive gases are N_2O and SiH_4 at 5%, the silane is diluted in N_2. The high dilution of silane is a restriction of the system in order to increase the versatility of the reactor. Thereby, we have no possibilities to vary the chamber pressure varying a gas carrier. **Figure 3** shows a calibration graph of the pressure of silane and nitrous oxide as a function of the gas flux. The flux of the N_2O is controlled by two rotameters as shown in the schematic of **Figure 1**. Double control of nitrous oxide allows for an efficient way to produce nanometric layers in multilayers with different R_0, that is, layers of different silicon excess. For the two previous figures, the vacuum valve was 75% open and the base vacuum was at least 6 mTorr.

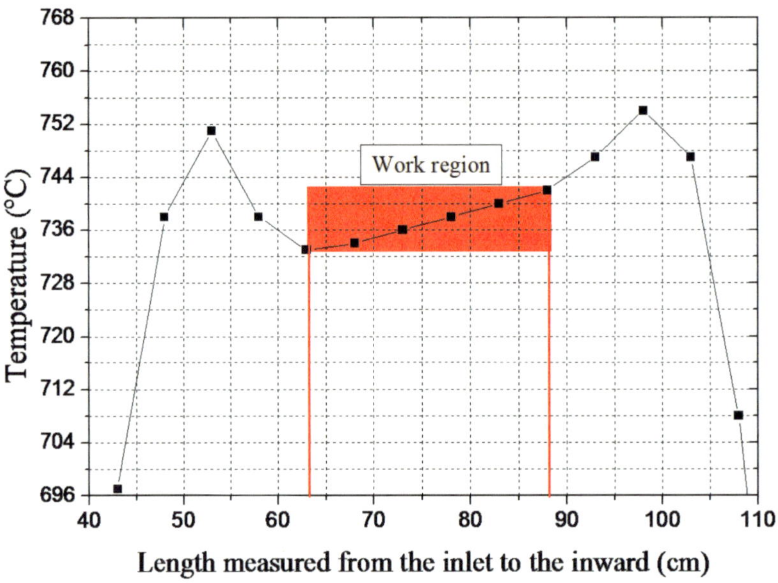

Figure 2. Working temperature profile to deposit SRO with different R_0's. The profile increases along the deposit area to compensate for the boundary layer.

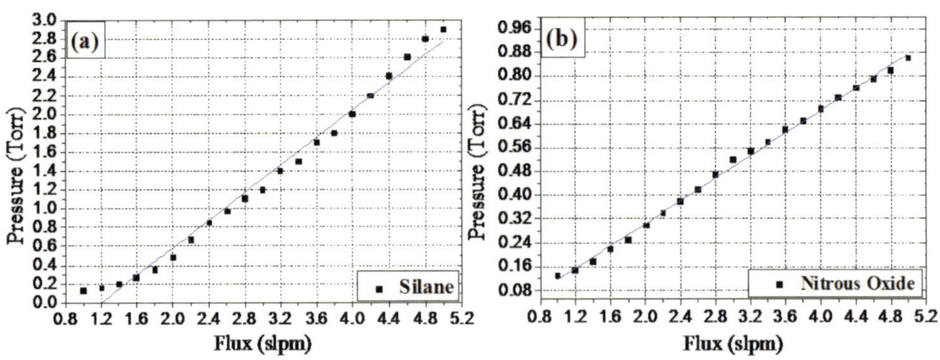

Figure 3. Calibration graph of pressure as function of the gas flux of (a) silane and (b) nitrous oxide. In case of N_2O, the flux is controlled by two rotameters.

http://dx.doi.org/10.5772/63012

The characteristics of the SRO depend strongly on the silicon excess; the flux ratio R_0 is used to control such excess. In the LPCVD technique, it is easy controlling the R_0 by the ratio of partial pressure of each gas, defined in Eq. (1). In our case, the dilution of silane has to be taken into account. Therefore, Eq. (1) has to be written adding a multiplying factor (F) as

$$R_0 = \frac{P * P_{N_2O}}{P_{SiH_4}} \tag{2}$$

Considering that only 0.05 parts of gas corresponds to silane, F takes a value equal to 20. However, due to the natural variation of the gas provided by the supplier, when a new tank is used, a new calibration is carried out in such a way that the refractive index (n) and PL are kept within the known values. Then, normally, the multiplicative factor is different than 20. This procedure is also done periodically to assure the gases aging do not alter the characteristics of the SRO films.

In the everyday procedure to obtain always the same conditions, we fixed the P_{SiH4} and varied the P_{N2O} as required by the R_0; **Table 1** shows the set values used for different R_0. To deposit single layers, the flux of silane is varied until the partial pressure is obtained, then the silane valve is closed and the nitrous oxide flux is set until the partial pressure is obtained. Next, both valves are shutoff and wafers are loaded. When the base vacuum is established, silane and nitrous oxide valves are simultaneously opened. The reactive gases mix and react producing variations in pump extraction and a reduction of the total pressure is observed. That is, the total pressure is different from that of the sum of the partial pressures. Also, a variation of the gases flux is observed. To obtain a repetitive process, it is necessary to adjust the fluxes to the values they had before the gases are mixed. This procedure assures to obtain repeatable characteristics. Also in **Table 1**, the refractive index, the rate of deposition, and the thickness variation from beginning to end of the wafer holder are presented.

R_0	Silane (SiH_4)		Nitrous oxide (N_2O)		Deposit rate (nm/min)	Refractive index	Thickness variation (%)
	P (Torr)	Flux (slpm)	*P* (Torr)	Flux (slpm)			
5	0.97	2.6	0.20	1.5	7.10	2.742	–
10			0.30	2.0	5.20	1.74 ± 0.05	13
20			0.60	3.4	3.80	1.64 ± 0.03	3
25			0.74	4.5	3.10	1.57 ± 0.01	2
30			0.88	5.2	3.05	1.52 ± 0.01	2
50			1.5	9.8	6.50	1.44 ± 0.01	12

Table 1. Deposit conditions in LPCVD system for different R_0's, also the refractive index, the deposit rate and the thickness variation is presented.

There are many partial pressures combinations that fulfill Eq. (1). However, depending on each laboratory conditions, it is recommendable to set a linear relationship between the flux ratio and the partial pressure ratio of each R_0. In our knowledge, it is very important to maintain a linear relationship of the ratios of pressure and flux (**Figure 4**) between the different R_0's. Nevertheless, different deposit conditions such as pump valve aperture, system cleanliness, dilution of silane, and aging produce different relations. In our experiments, all the mentioned factors were taken into account varying the multiplicative factor (*F*) of Eq. (2). **Figure 4** shows different relationship varying the *F* due to different conditions of the system. In our system, using the vacuum valve open at 75% of the maximum aperture and purging with nitrogen the gas lines a couple of hour before the deposit are enough to maintain a linear relationship, and with it we obtain repetitive results.

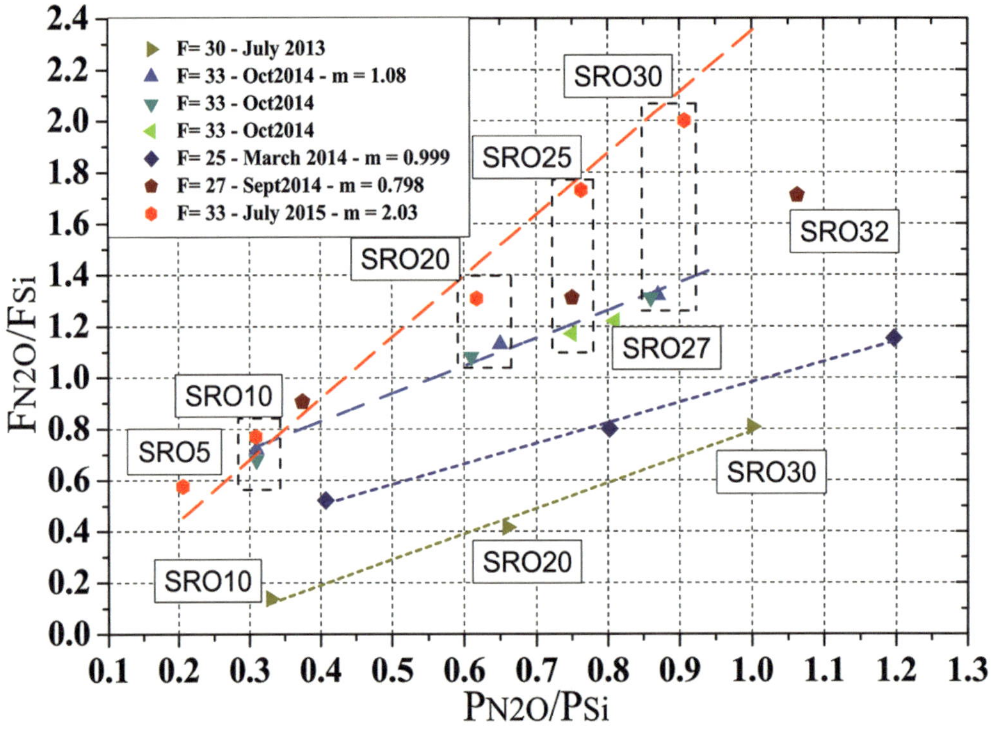

Figure 4. Linear relationship between flux ratio and partial pressure ratio of each R_0.

To deposit a multilayer with different R_0's, the procedure is similar to that of single layers. However, in this case, the flux of N_2O is set using the two flux controls. Using the gas that goes through one of them, one R_0 is fixed (Ro1). The other R_0 (Ro2) is fixed by using the second rotameter or both simultaneously, depending on the deposit conditions. The nitrous oxide valves are open during the time needed to obtain the Ro1 or Ro2 successively. The time required for each layer is established to have the nominal thickness, and it is calculated from the rate of deposit. If the layer being deposited is thin, then the time can be some seconds.

As it is expected in a low-pressure system, deposition in our system works under surface reaction kinetics limited [9, 10]. It implies that the deposit rate is low and good step coverage is obtained, and also by-products are trapped in the film, and that is why off stoichiometric silicon oxide is obtained. The LPCVD also has shown good step coverage, and in order to corroborate our step coverage results, the sticking factor and the gas arrival dates are estimated.

The sticking coefficient (S) is defined as

$$S = \frac{R_r}{A_r} \tag{3}$$

where R_r is the reaction rate and A_r is the arrival rate, respectively. The reaction rate is related with deposit rate; meanwhile the arrival rate is the velocity of the total flux of precursor gases that enters into the chamber. In our case, A_r is estimated as

$$A_r = \frac{F_T}{A} \tag{4}$$

where F_T is the sum of the fluxes of precursor gasses and A is the tubing area (our system uses ¼ inches diameter). Using Eqs. (3) and (4) and conditions from **Table 1**, the sticking coefficient is determined for each R_0. **Table 2** shows results of A_r, R_r, and S for each R_0 from 5 to 50. The sticking coefficient has values in the order to 10^{-11}. This value is rather too low; however, independent of the numerical value, it agrees with what is expected from a low-pressure system; see, for example, [9–11].

R_0	A_r (m/min)	R_r (m/min)	S
5	32.36	7.10E-09	5.48E-11
10	36.31	5.20E-09	3.58E-11
20	47.36	3.80E-09	2.01E-11
25	56.04	3.10E-09	1.38E-11
30	61.57	3.05E-09	1.24E-11
50	97.88	6.50E-09	1.66E-11

Table 2. Arrival and reaction rates and sticking coefficient for SRO-LPCVD deposited at 736°C.

3. Experimental procedure

SRO films were deposited on <100> and low resistivity (5–10 Ω cm) silicon substrates by LPCVD at 736°C. The ratio between reactive gases nitrous oxide (N_2O) and silane (SiH_4) was varied to obtain films with different silicon excess. Single layers with R_0 values of 5, 10, 20, 25, 30, 35, and 50 were deposited, and will be labeled as SRO_5, SRO_{10}, SRO_{20}, SRO_{25}, SRO_{30}, SRO_{35}, and SRO_{50}, respectively, for clarity. Also, two multilayer structures were fabricated. The multilayer is a stack of seven layers. One structure intercalates three SRO_{25} layers with four SRO_5 layers, and the second one with four SRO_{10} layers. After deposition, all samples were thermally annealed at 1100°C for 3 h in nitrogen ambience to induce the silicon agglomeration.

Thickness and refractive index of all samples, including multilayer structure, were determined using a null ellipsometer Gaertner L117 with a laser He–Ne of 632.8 nm wavelength. The PL emission spectra were obtained with a Fluoromax-3 spectrometer; all the films were excited with UV radiation (300 nm) and the luminescence was measured from 370 to 1000 nm with a resolution of 1 nm. Optical filters were used in order to guarantee the wavelength of excitation beam. CL measurements were performed using a luminoscope equipment model ELM2-144, 0.3-mA current and 5 kV were used. The luminescence spectra (PL and CL) were measured at room temperature.

For electrical and electroluminescent studies, Metal-Insulator-Semiconductor (MIS) devices were fabricated, and we refer to them as light emitting capacitor (LEC). A ~250-nm-thick semitransparent n^+ polycrystalline silicon (Poly) gate was deposited onto the SRO film surface by LPCVD. After a photolithography process step, square-shaped gates of 4-mm^2 area were defined. The backside contact was formed with 1-μm thick aluminum layer by evaporation. Finally, the devices were thermally annealed at 480°C in forming gas.

A source meter Keithley model 2400 was used to obtain current versus voltage (I-V) curves. EL spectra were obtained by biasing the device with a constant DC voltage. The light emitted was collected with an optical fiber located facing the Poly gate and connected to the Fluoromax 3 spectrometer.

4. Composition of SRO by LPCVD

SRO is a multiphase material composed of silicon oxides of different stoichiometry and Si nanocrystals (Si-ncs). In XPS spectra of this material, each Si 2p core level band is composed of bands originated in Si at different oxidation states (Si^0, Si^{1+}, Si^{2+}, Si^{3+}, Si^{4+}), which manifest themselves at different energies. The position of the peaks corresponding to Si^0 and Si^{4+} (SiO_2) is well known and is easily distinguishable [12], but the peaks related with silicon suboxides cannot be distinguished unequivocally in a complex spectrum composed of different Si oxide species; they have been usually studied at Si/SiO_2 interfaces [13–16]. In this way, a quantitative analysis of such highly convoluted spectra is not straightforward. The material can be conveniently considered as composed of SiO_2, elemental Si and SiO_y with $0 < y < 2$. **Figure 5** shows an XPS spectrum of SRO with $R_0 = 1$ as example. The three phases can be clearly distinguished.

http://dx.doi.org/10.5772/63012

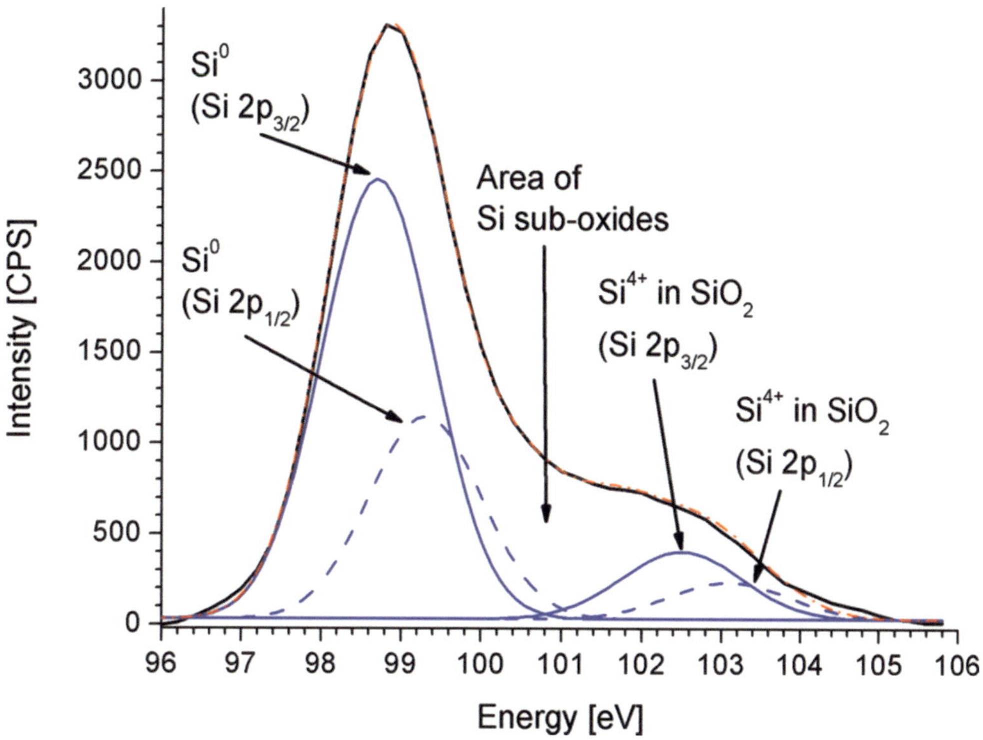

Figure 5. XPS spectrum of SRO with R_0 =1 [17].

Table 3 presents the compositions of SRO with different R_0's [17, 18]. As can be observed, the amount of elemental Si and the amount of SiO_2 monotonically increase and decrease, respectively, when decreasing the R_0. SiO_y oscillates around 43%, for R_0's higher than 3. The oscillating amount could be understood in the sense that SiO_y is in reality a combination of different stoichiometries, which vary in proportion depending on the R_0.

R_0	30	20	10	3	1
% Si	2	3	11	20	72
% SiO_y	43	47	40	46	16
% SiO_2	55	50	49	34	12

Table 3. Composition of SRO with different R_0s.

From **Table 3**, it is possible to make a fit of the monotonically varying data (Si and SiO_2). For the fit is considered that R_0 = 0 means 100% elemental Si, and that 0% elemental Si is obtained by R_0 = 100 (no Si-ncs are observed from R_0 = 30 [17]). From the fitting curves, it is possible to calculate the curve for SiO_y as 100 - %Si - %SiO_2. All calculated curves are shown in **Figure 6**.

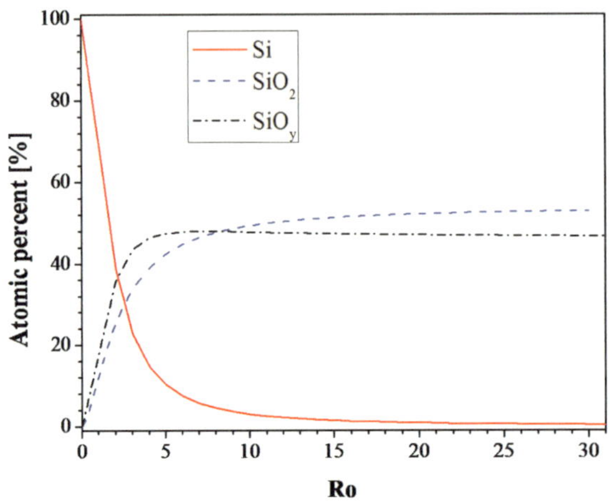

Figure 6. Calculated curves for %Si, %SiO_2, and %SiO_y, obtained by fitting measured data.

For R_0's below 8, the amount of SiO_y is larger than the amount of SiO_2. This is caused by the high Si excess. At larger R_0's, the amount of SiO_2 is larger than the one of SiO_y, nevertheless, the amounts are almost constant (vary linearly with R_0, with a small slope) at R_0's over 20, when the amount of elemental Si is close to zero. Following this tendency, the amount of SiO_y is not zero even at $R_0 = 100$ (the proportion is 44% SiO_y by 56% SiO_2). This result implies that the oxides obtained by LPCVD are in a large percentage non-stoichiometric, even at large R_0's. This nature of the oxides may mean a large number of defects, many of them being luminescent, as will be made clear in Section 6.

For our CVD system, one can write the chemical reaction as

$$SiH_4 + (2+x) * N_2O \rightarrow SiO_x + 2H_2O + (2+x) * N_2 \quad (5)$$

where

$$SiO_x = a * Si + b * SiO_y + c * SiO_2 \quad (6)$$

with a, b, and c being the atomic proportions of the different phases ($a + b + c = 1$).

It is worthy to mention that certain amount of nitrogen is incorporated in SiO_x during the deposition, but it may be of maximum 1%. Lower amounts of nitrogen are presented in samples of smaller R_0's [19]. These amounts do not change the material structurally, but may enhance its luminescence [20, 21].

It is also important to know the form how elemental Si is present in the samples. Through transmission electron microscopy (TEM) studies, it has been possible to evidence Si-ncs in samples with R_0's below 20. By larger R_0's, Si is in amorphous state or dispersed in the oxide matrix. **Figure 7** shows a plot of the sizes of the Si-ncs versus R_0 [17, 22, 23]. As can be observed,

http://dx.doi.org/10.5772/63012

the Si-nc size decreases almost linearly with the increasing of R_0, for R_0's above 3. SRO with R_0 = 1 is closer to semi-insulating polysilicon (SIPOS), presenting much bigger grains.

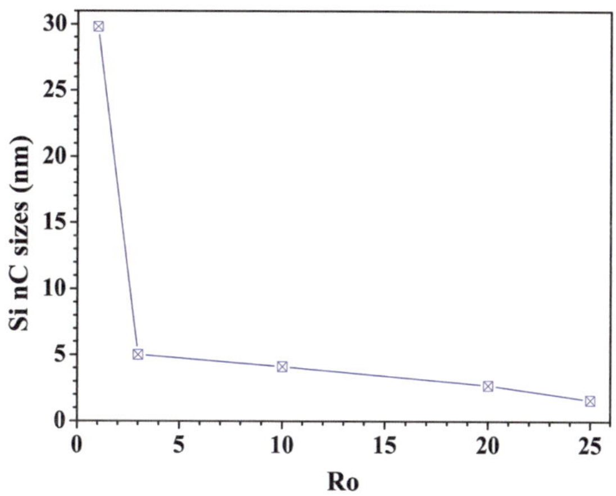

Figure 7. Si-nC sizes depending on R_0.

5. Electrical characteristics

Figure 8 shows the current density (*J*) as a function of the electric field (*E*), which it is defined as the ratio of applied voltage (*V*) and the thickness of the SRO film (t_{SRO}). This J-E behavior corresponds to LEC with a single layer of SRO. All LECs are forwardly biased (accumulation mode) considering the substrate as reference.

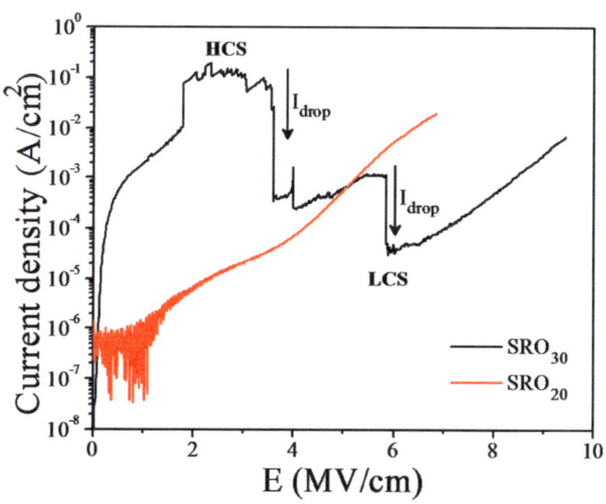

Figure 8. J-E curve of SRO_{30} and SRO_{20}-based LECs. SRO films thermally annealed at 1100°C.

As we can see, the presence of defects including the Si-nps, either crystalline or amorphous, and their density in the SRO films affect clearly the current transport when they are used in MIS devices. LEC with SRO_{30} films show a high current state (HCS) at low electric fields, and then after the applied voltage increases, the current is switched to a low conduction state (LCS), as shown in **Figure 8**. The switching from the HCS to LCS shown by SRO_{30}-based devices was observed by our group before and for both forward and reverse bias [24–28]. That effect was related to the annihilation of conductive paths created by adjacent stable Si-nps and unstable silicon nanoclusters (Si-ncls) through structural changes and by the possible creation of defects (breaking off Si-Si bonds) [24, 25, 27]. Recent studies regarding the same electrical switching in SRO films was observed and related with a conductive filament [29–32]. The conductance switching behavior observed in that SRO films was explained also by structural changes through an electroforming process. In fact, the structural changes in the conductive filament was analyzed by in situ imaging TEM analysis, showing that the conductance switch is related with a crystallization and an amorphization process of Si-nps that creates the conductive filament [30]. These observations are in agreement with our asseverations about the conductance switching observed in our SRO_{30}-based LECs [24].

In the HCS regime, current jumps and drops, which are observed independently of the temperature of annealing, have been related to the creation and annihilation of the preferential conductive paths and with the appearing or disappearing of electroluminescent spots (EL dots) on the LEC surface [24, 25, 27, 28]. A clear correlation between current jumps/drops and EL dots appearing/disappearing was observed [27]. Once the current fluctuations disappear, through an electrical annealing, the current behavior stabilizes, as reported in [24, 28].

On the other hand, the electrical behavior of most of LECs with SRO_{20} films does not show current fluctuations. This effect has been related with the presence of well-separated and crystalline silicon nanoparticles (or Si-ncs) and mainly on the density of Si-nps [28]. The Si-nps density estimated from energy-filtered transmission electron microscopy (EFTEM) images of SRO_{20} films thermally annealed at 1100°C is ~2.46×10^{12} cm^{-2}, about twice the Si-nps density in SRO_{30} with ~1.1×10^{12} cm^{-2} [27]. Therefore, a uniform network of conductive paths becomes possible as the Si-nps density increases, allowing a uniform charge flow through the whole capacitor area. Meanwhile, as the Si-nps density decreases (SRO_{30} films), the distance between them increases reducing the amount of available paths, with a resulting set of discrete and preferential conductive paths within the oxide.

Basically, there are four main mechanisms known to contribute in the carrier transport through a Si-rich oxide layer, including the direct tunneling, Fowler Nordheim tunneling (F-N), Poole-Frenkel (P-F) and the trap-assisted tunneling (TAT) [33–37]. It has been found that the TAT conduction mechanism predominates in our SRO_{30}-based LECs, where the trap energy (ϕ_t) was estimated at about 1.99 eV [28]. This implies that traps, which are placed at ~2 eV below the conduction band, would be assisting the tunneling event. On the other hand, the P-F tunneling was found as the charge transport in the SRO_{20}-based LECs. A relative permittivity (ε_r) value of 9.16 was obtained from the P-F fit, closer to the relative permittivity of silicon (ε_{Si} = 11.9), similar to other reports [36]. Relatively high permittivity values are a good indication of the large amount of silicon present as Si-nps within our SRO_{20} films. Moreover, it was found that the Si-nc size obtained through a relation between size and permittivity (obtained by P-F fit) is very close to that measured by high-resolution transmission electron microscopy (HRTEM) [28].

6. Electro-optical characteristics

6.1. Single layer

6.1.1. Photoluminescence

PL spectra of annealed films from $R_0 = 5$ to $R_0 = 50$ are depicted in **Figure 9**. After annealing, all SRO samples present a main emission from ~600 to 850 nm and a negligible emission for some samples from ~380 to 500 nm. The emission intensity increases when the silicon excess decrease until a maximum of $R_0 = 30$ and then the PL intensity decreases until it practically disappear.

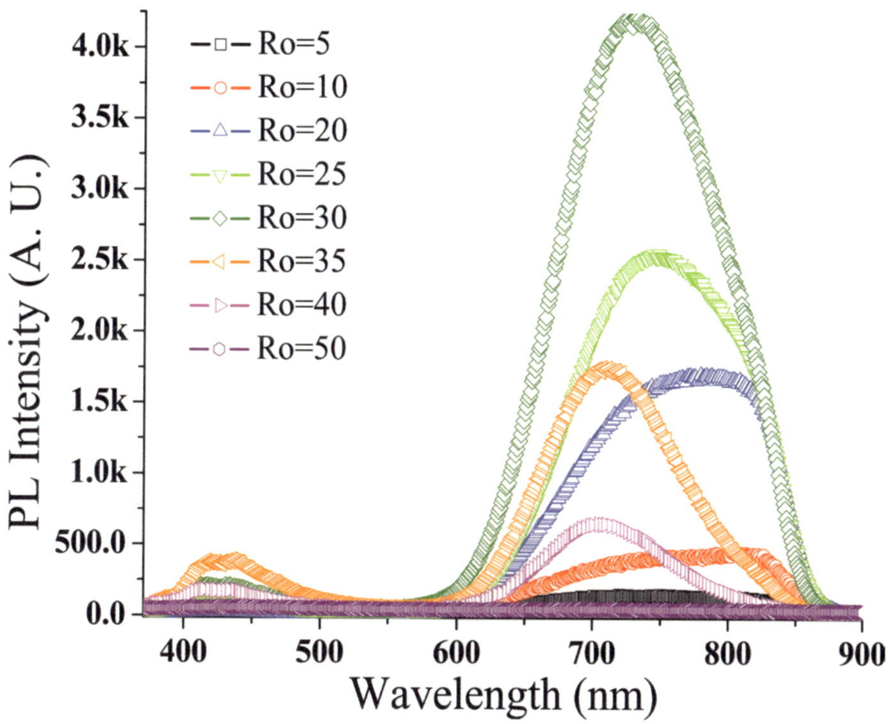

Figure 9. Photoluminescence of annealed SRO films with R_0 from 5 to 50.

As can be observed, the PL emission exhibits a shape dependence on the silicon excess, which could indicate different emission mechanism. Because of this, the multi-Gaussian deconvolution of PL spectra was performed for some annealed samples, and the set of band positions have been determined (**Figure 10**). Each spectrum can be well fitted to a superposition of three Gaussian distributions: a main band (1) and two shoulders (2 and 3). Fit peaks are centered at (1) 710–730, (2) 780–790, and (3) 820 nm with FWHM of (1) 50–60, (2) 20–29, and (3) 18 nm, respectively.

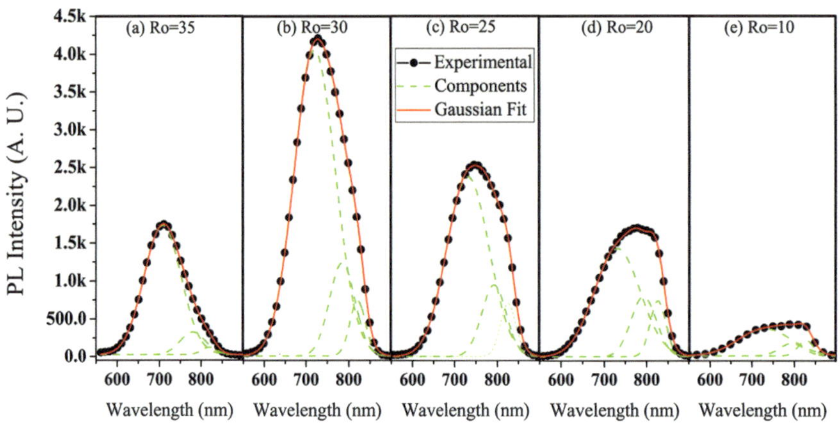

Figure 10. PL spectra and fits from SRO films with different silicon excess. Symbols are experimental data, lines are the Gaussian fits and dash lines are distributions.

Peak position and intensity vary according to the silicon excess, as shown in **Figure 11**. There is a blue wavelength shift for all components when the silicon excess decreases (except for R_0 = 5), see **Figure 11(a)**. The main contribution of the luminescence is the peak 1 that increases rapidly as R_0 increases until R_0 = 30 and then decreases, as can be seen in **Figure 11(b)**. While peaks 2 and 3 slightly increase when R_0 increases, due to this, samples with $R_0 < 25$ shows a shoulder in the near infrared region (NIR). Since there are different components that change with silicon excess, it can be assumed that PL emission is related with at least three different types of emission centers (or emission mechanism).

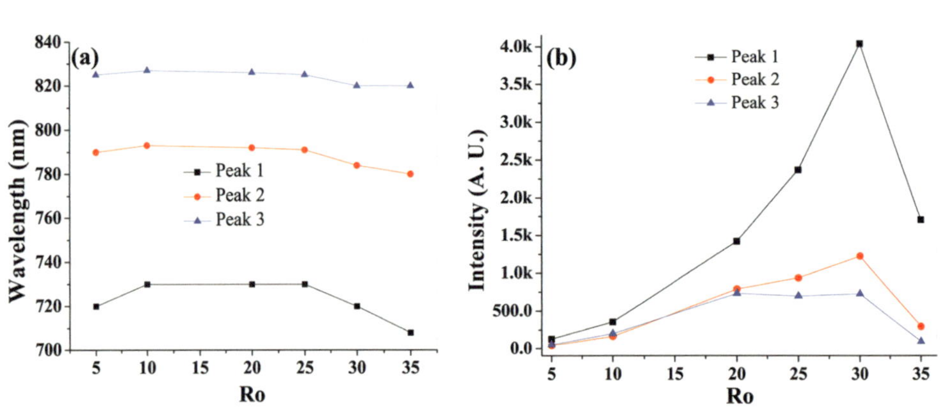

Figure 11. (a) Position and (b) intensity of the fit distributions for different silicon excess.

6.1.2. *Cathodoluminescence*

CL spectra from SRO films with different silicon excess are depicted in **Figure 12(a)**. The CL spectra of SRO with thermal treatment consist of a broad emission in the visible and NIR from ~400 to 850 nm. After annealing, intensity of the blue band at ~460 nm increases with increasing the R_0. On the other hand, the intensity of the red-NIR CL band seems to have a maximum for $R_0 = 20$.

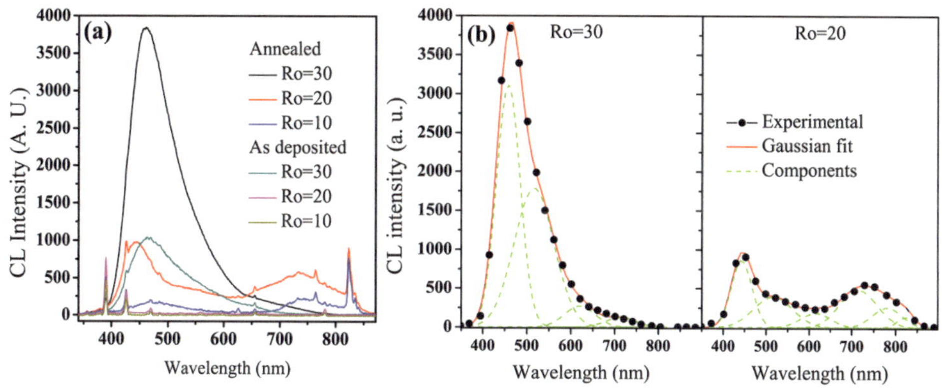

Figure 12. (a) Cathodoluminescence spectra from SRO films with different silicon excess. (b) Gaussian fit of CL experimental spectrum, the best fit requires four distributions for $R_0 = 30$ and six distributions for $R_0 = 20$.

As CL emission has asymmetrical shape for all SRO samples, it can be assumed that CL emission is also due to different luminescent centers. Hence, multi-Gaussian deconvolution of CL spectra was also obtained, shown in **Figure 12(b)**. The best fit of CL spectra requires four and six components for $R_0 = 30$ and 20, respectively. Peaks (or distributions) were obtained at about 460, 522, 643, and 714 nm for $R_0 = 30$ and 447, 541, 645, 714, 780, and 823 nm for $R_0 = 20$. Peaks obtained at 714, 780, and 823 nm in CL are centered in the same position than distributions obtained from PL spectrum in $R_0 = 20$. Furthermore, peak at 714 nm was obtained for Gaussian fit, in PL and CL in $R_0 = 30$. Then, the red emission of the CL emission can be ascribed to the same PL emissive centers. PL distributions in higher wavelength are not observed in CL due to either destruction of the emissive centers or inefficient emission from low-energy emissive centers [38]. The latter one could occur increasingly because cathode-excited electrons acquire so high energy that they arrive to the higher emissive positions where they emit in the blue region (higher energy); however, almost none of the excited electrons reach that with lower energy; then, the red emission is not likely to occur in CL. Therefore, there could be several different kinds of emission traps located at different energy levels in SRO.

Depending on the emission wavelength, multiple luminescence centers have been reported in SiO_2 films. Luminescent emission at 460 nm (2.7 eV), 520 nm (2.4 eV) and 650 nm (1.9 eV) nm are mainly related to defects such as oxygen deficiency-related centers (ODC) or oxygen vacancies [39–41], *E'*δ defect or peroxide radical [42] and non-bridging oxygen hole centers (NBOHC) [40, 41, 43], respectively. Since CL and PL measurements have shown luminescent peaks (or distributions) close to those wavelengths, such defects could be inside the SRO films.

6.1.3. *Electroluminescence*

Figure 13 shows the electroluminescence spectra from the SRO-based LECs. Blue electroluminescence is observed in the SRO_{30} film, as observed in **Figure 13(a)**. Nevertheless, this blue EL in whole area of LECs is obtained only after the current drop. The main EL peak remains at 468 nm even for different thermal annealing temperatures [28]. A long spectral shift, blue shift, of almost ~227 nm has been observed between the EL and PL band of the SRO_{30} films. Devices with SRO_{20} films emit a broad EL spectrum in the red region (713 nm), as observed in **Figure 13(b)**. An additional EL peak of low intensity is also observed at 468 nm. There exists also a blue shift of the EL spectra with respect to PL spectrum in SRO_{20} films. Nevertheless, both EL and PL spectra in SRO_{20} films appear in the red region, which could indicate that the same luminescent centers are involved. Images of the blue and red LECs are shown in the insets of the **Figure 13**. As we can see, intense EL is emitted in the whole area of the LEC devices.

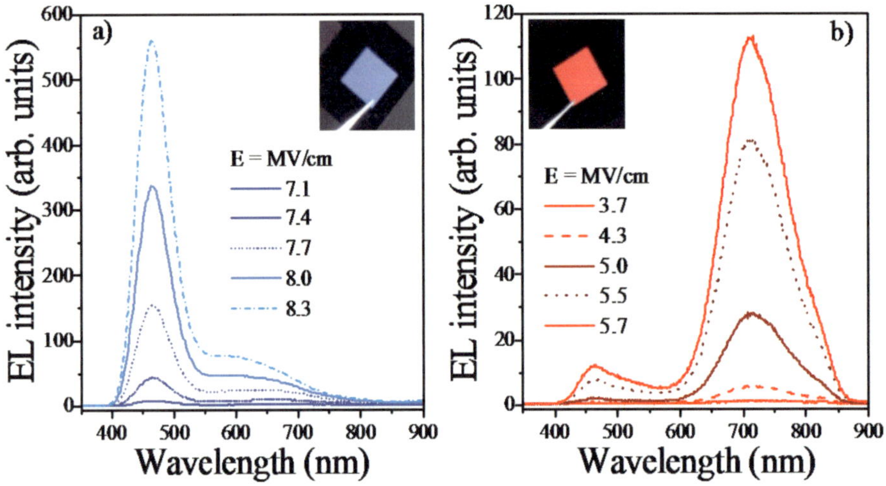

Figure 13. EL spectra of (a) SRO_{30} and (b) SRO_{20} based LECs biased at different electric fields. Insets show images of each SRO-based LECs.

The spectral shift between PL and EL has been reported before and it has been explained according to three different mechanisms [35, 44, 45]. Our experimental results have suggested that the red EL observed in SRO_{20} films can be related with the combination of some surface defect on the Si-ncs, while the blue EL in SRO_{30} devices is consistent with the defect emission which could be intrinsically present or generated by electric field within the SRO matrix [28].

6.2. Multilayer

6.2.1. *Photoluminescence*

Multilayer structures were fabricated in order to improve the optical properties of the SRO films. Two samples were obtained, one of those is a combination of low silicon excess (R_0 = 25) and high silicon excess (R_0 = 5) and the second one is a combination of R_0 = 25 and R_0 = 10.

http://dx.doi.org/10.5772/63012

PL spectra of annealed multilayers and single layer are shown in **Figure 14(a)**. As can be seen the intensity emission is improved in the multilayer samples, where multilayer SRO_{10}/SRO_{25} (ML-10/25) is the most intense. In order to obtain the components of every layer, the multi-Gaussian deconvolution of PL spectra was also obtained. **Figure 14(b)** and **(c)** shows the position and the intensity of the three peaks obtained from the Gaussian deconvolution. There is a blue-shift wavelength for the three peaks that can be due to the participation of high silicon excess (R_0 = 5 or 10) on the PL. The intensity improvement could be due to the three components together, peak 1 is from R_0 = 25 and the improvement of peaks 2 and 3 comes from high silicon excess (R_0 = 5 or 10).

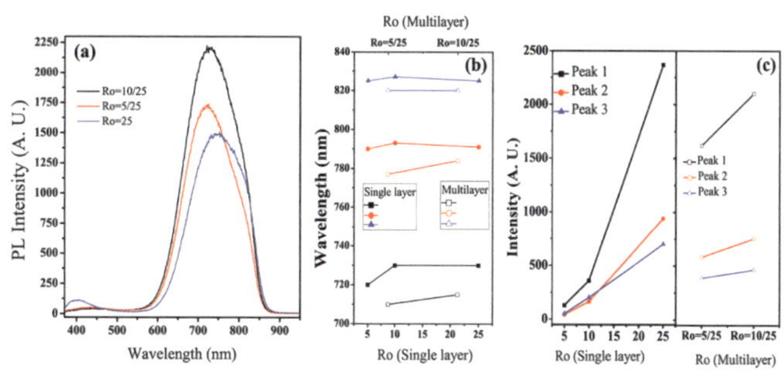

Figure 14. (a) Photoluminescence of annealed multilayer SRO films (R_0 = 10/25 and R_0 = 5/25) and a single layer of R_0 = 25. (b) Position and (c) intensity of the fit distribution for single layer (R_0 = 5, 10, and 25) and multilayers.

6.2.2. Electroluminescence

Figure 15 shows a scheme of the multilayered structure fabricated with its dimensional characteristics (left side), and a TEM image of the structure, which exposes the layers composing the SRO multilayer (right side). The goal of this structure is to improve the electro-optical properties of the LECs. In this multilayer, the luminescent properties of three layers with low silicon excess (SRO_{25}) are combined with four conducting layers (SRO_5).

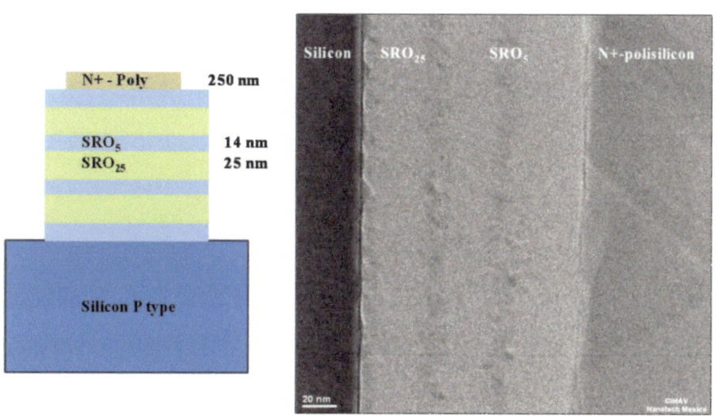

Figure 15. Scheme of a light emitter capacitor with a multilayered SRO films (left side) and a TEM image of the SRO nanometric multilayer (right side).

The electroluminescence in this ML-SRO-based LEC is observed at forwardly bias considering the substrate as reference. A broad band with the main peak at about 600 nm is observed with E = 3.7 MV/cm, as we can see in **Figure 16**. As the voltage increases, the spectrum is divided in two. One peak at 680 nm and other at 450 nm, the blue peak increases to higher intensity than the red one. Apparently, SRO_5 layers increase the conductivity across the structure and SRO_{25} layers produces the emission. The spectrum behavior shown in our samples have been observed in other reports [26].

However, the electric field needed to turn on the emission on an ML-SRO-based LEC is lower than an LEC of single layer (see **Figures 13** and **16**). This proves that the electro-optical properties of a ML-SRO-based LEC are improved, thereby the conductivity of the structure is increased by layers of high silicon excess, and luminescence response is conserved using layers of high R_0.

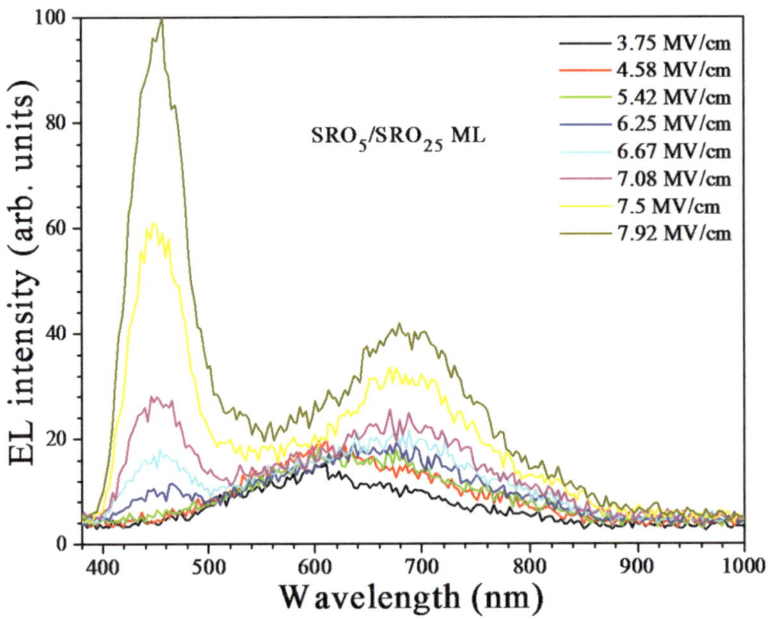

Figure 16. EL spectra of a multilayered SRO_5/SRO_{25}-based LEC under different electric fields.

7. Conclusion

In this chapter, details of a homemade hot-wall LPCVD system were presented. Also, important aspects of how to obtain SRO in a reliable and repetitive form were addressed. We show that in our system, it is possible to obtain single layers with variable silicon excess, and also good quality multilayered structures of nanometric layers. The structural, electrical, and luminescent characteristics of single- and multilayered structures were reviewed and discussed.

Acknowledgements

Authors recognize the financial support of the CONACYT, particularly, J. Alarcón-Salazar for his grant with number 353251. Also, authors thank the microelectronic laboratory technicians Pablo Alarcon, Victor Aca, and Armando Hernández for their help during the fabrication process.

Author details

J. Alarcón-Salazar[1*], R. López-Estopier[2], E. Quiroga-González[3], A. Morales-Sánchez[4], J. Pedraza-Chávez[1], I. E. Zaldívar-Huerta[1] and M. Aceves-Mijares[1]

*Address all correspondence to: j.alarcon.sal@gmail.com

1 Electronics Department, National Institute of Astrophysics Optics and Electronics (INAOE), Luis Enrique Erro 1, Tonantzintla, Puebla, México

2 Professor CONACYT – IICO-UASLP, Karakorum, San Luis Potosí, México

3 Institute of Physics, Autonomous University of Puebla, México

4 Advanced Materials Research Center (CIMAV), Monterrey, Apodaca, Nuevo León, México

References

[1] Guangzhao Ran, Hongqiang Li, and Chong Wang. On-chip silicon light source: from photonics to plasmonics. Front. Optoelectron. 2012;5(1):3–6. DOI:10.1007/s12200-012-0221-x

[2] Kingsley A. Ogudo, Diethelm Schmieder, Daniel Foty, and Lukas W. Snyman. Optical propagation and refraction in silicon complementary metal–oxide–semiconductor structures at 750 nm: toward on-chip optical links and microphotonic systems. J. Micro/Nanolith. MEMS MOEMS. 2013;12(1):013015. DOI:10.1117/1.JMM.12.1.013015

[3] A. Anopchenko, A. Marconi, F. Sgrignuoli, L. Cattoni, A. Tengattini, G. Pucker, Y. Jestin, and L. Pavesi. Electroluminescent devices based on nanosilicon multilayer structures. Phys. Status Solidi A. 2013;210(8);1525–1531. DOI:10.1002/pssa.201200957

[4] M. Aceves-Mijares, A. A. González-Fernández, R. López-Estopier, A. Luna-López, D. Berman-Mendoza, A. Morales, C. Falcony, C. Domínguez, and R. Murphy-Arteaga. On

the origin of light emission in silicon rich oxide obtained by low-pressure chemical vapor deposition. J. Nanomater. 2012;2012:890701. DOI:10.1155/2012/890701

[5] A. A. González-Fernández, J. Juvert, M. Aceves-Mijares, C. Domínguez. Monolithic integration of a silicon-based photonic transceiver in a CMOS process. IEEE Photonics J. 2016;8(1):7900213. DOI:10.1109/JPHOT.2015.2505144

[6] A. Morales, J. Barreto, C. Domínguez, M. Riera, M. Aceves, and J. Carrillo. Comparative study between silicon-rich oxide films obtained by LPCVD and PECVD. Phys. E Low Dimens. Syst. Nanostruct. 2007;38(1–2):54–58. DOI:10.1016/j.physe.2006.12.056

[7] D. Dong, E. A. Irene, and D. R. Young. Preparation and some properties of chemically vapor-deposited Si-rich SiO and SigN, films. J. Electrochem. Soc. 1978;125(5):819–823. DOI:10.1149/1.2131555

[8] M. Aceves, L. A. Hernandez, and R. Murphy. Applying statistics to find the causes of variability of aluminum deposition: a case study. IEEE Trans. Semicond. Manuf. 1992;5(2):165–167. DOI:10.1109/66.136280

[9] H. O. Pierson. Handbook of chemical vapor deposition (CVD): Principles, technology, and applications. 2nd ed. New Jersey, USA: Noyes Publications; 1999.

[10] Hwaiyu Geng. Chemical vapor deposition. In: Hwaiyu Geng, editor. Semiconductor manufacturing handbook. 1st ed. New York, USA: McGraw-Hill Handbooks; 2005.

[11] Y.-P. Zhao, J. T. Drotar, G.-C. Wang, and T.-M. Lu. Morphology transition during low-pressure chemical vapor deposition. Phys. Rev. Lett. 2001;87(13)1–4. DOI:10.1103/PhysRevLett.87.136102

[12] F. Verpoort, P. Persoon, L. Fiermans, G. Dedoncker, and L. Verdonck. SiO2/Si(100)model support with AES and XPS in combination with MLCFA. J. Chem. Soc., Faraday Trans. 1997;93(19):3555. DOI:10.1039/A702279D

[13] G. Hollinger and F. J. Himpsel. Probing the transition layer at the SiO2-Si interface using core level photoemission. Appl. Phys. Lett. 1984;44(1):93–95. DOI:10.1063/1.94565

[14] F. J. Grundthaner, P. J. Grundthaner, R. P. Vasquez, B. F. Lewis, and J. Maserjian. High-resolution X-ray photoelectron spectroscopy as a probe of local atomic structure: Application to amorphous SiO_2 and the Si-SiO_2 interface. Phys. Rev. Lett. 1979;43(22): 1683. DOI:10.1103/PhysRevLett.43.1683

[15] P. J. Grundthaner, M. H. Hecht, F. J. Grundthaner, and N. M. Johnson. The localization and crystallographic dependence of Si suboxide species at the SiO2/Si interface. J. Appl. Phys. 1987;61(2):629. DOI:10.1063/1.338215

[16] K. Hirose, H. Nohira, K. Azuma, and T. Hattori. Photoelectron spectroscopy studies of SiO2/Si interfaces. Prog. Surf. Sci. 2007;82(1):3. DOI:10.1016/j.progsurf.2006.10.001

[17] E. Quiroga-González, W. Bensch, M. Aceves-Mijares, Z. Yu, R. López-Estopier, and K. Monfil-Leyva. On the photoluminescence of multilayer arrays of silicon rich oxide with

high silicon content prepared by low pressure chemical vapor deposition. Thin Solid Films. 2011;519:8030–8036. DOI:10.1016/j.tsf.2011.06.020

[18] R. López-Estopier, M. Aceves-Mijares, and C. Falcony. Cathodo- and photo-luminescence of silicon rich oxide films obtained by LPCVD, cathodoluminescence. In: Naoki Yamamoto, editor. Chathodoluminescence. InTech; Shanghai China; 2012. DOI: 10.5772/34888

[19] E. Quiroga, W. Bensch, Z. Yu, M. Aceves, R. A. De Souza, M. Martin, V. Zaporojtchenko, and F. Faupel. Structural characteristics of a multilayer of silicon rich oxide (SRO) with high Si content prepared by LPCVD. Phys. Stat. Sol. A. 2009;206(2):263–269. DOI: 10.1002/pssa.200824365

[20] R. López-Estopier, M. Aceves-Mijares, and C. Falcony. Photoluminescence of silicon rich oxide films with different silicon excess and nitrogen content. In: 3rd International Conference on Electrical and Electronics Engineering (ICEEE); Mexico City. IEEE; 2006. DOI:10.1109/ICEEE.2006.251868

[21] R. López-Estopier, M. Aceves-Mijares, J. Carrillo, Z. Yu, and C. Falcony. Effect of nitrogen in the photoluminescence of silicon rich oxide films prepared by LPCVD. In: 2nd International. Conference on Electrical and Electronics Engineering (ICEEE) and XI Conference on Electrical Engineering (CIE); Mexico City. IEEE; 2005. p. 227–230. DOI:10.1109/ICEEE.2005.1529614

[22] E. Quiroga, W. Bensch, M. Aceves, Z. Yu, J. P. Savy, M. Haeckel, and A. Lechner. Silicon rich oxide with controlled mean size of silicon nanocrystals by deposition in multilayers. In: 10th International Conference on Ultimate Integration of Silicon; March 18–20; Aachen, Germany. IEEE; 2009. p. 349–352. DOI:10.1109/ULIS.2009.4897607

[23] J. A. Luna-López, M. Aceves-Mijares, J. Carrillo-López, and A. Morales-Sánchez. Photoconduction in silicon rich oxide films obtained by low pressure chemical vapor Deposition. J. Vac. Sci. Technol. A. 2010;28(2):170–174. DOI:10.1116/1.3276781

[24] A. Morales-Sánchez, J. Barreto, C. Domínguez, M. Aceves, and J. A. Luna-López. The mechanism of electrical annihilation of conductive paths and charge trapping in silico-rich oxides. Nanotechnology. 2009;20(4):045201. DOI:10.1088/0957-4484/20/4/045201

[25] A. Morales-Sánchez, J. Barreto, C. Domínguez, M. Aceves-Mijares, J. A. Luna-López, M. Perálvarez, and B. Garrido. DC and AC electroluminescence in silicon nanoparticles embedded in silicon-rich oxide films. Nanotechnology. 2010;21(8):085710. DOI: 10.1088/0957-4484/21/8/085710

[26] A. A. González Fernández, M. Aceves Mijares, A. Morales Sánchez, and K. M. Leyva. Intense whole area electroluminescence from low pressure chemical vapor deposition-silicon-rich oxide based light emitting capacitors. J. Appl. Phys. 2010;108(4):043105. DOI:10.1063/1.3465335

[27] A. Morales-Sánchez, M. Aceves-Mijares, K. Monfil-Leyva, A. A. González, J. A. Luna-López, J. Carrillo, C. Domínguez, J. Barreto, and F. J. Flores-Gracia. Strong blue and red

luminescence in silicon nanoparticles based light emitting capacitors. Appl. Phys. Lett. 2011;99(17):1711102. DOI:10.1063/1.3655997

[28] L. Palacios Huerta, S. A. Cabañas Tay, J.-A. Luna López, M. Aceves, A. Coyopol, and A. Morales-Sánchez. Effect of the structure on luminescent characteristics of SRO-based light emitting capacitors. Nanotechnology. 2015;26(39):395202. DOI: 10.1088/0957-4484/26/39/395202

[29] J. Yao, Z. Sun, L. Zhong, D. Natelson, and J. M. Tour. Resistive switches and memories from silicon oxide. Nano Lett. 2010;10(10):4105. DOI:10.1021/nl102255r

[30] J. Yao, L. Zhong, D. Natelson, and J. M. Tour. In situ imaging of the conducting filament in a silicon oxide resistive switch. Sci. Rep. 2012;2(242):1-5. DOI:10.1038/srep00242

[31] A. Mehonic, A. Vrajitoarea, S. Cueff, S. Hudziak, H. Howe, C. Labbé, R. Rizk, M. Pepper, and A. J. Kenyon. Quantum conductance in silicon oxide resistive memory devices. Sci. Rep. 2013;3:2708. DOI:10.1038/srep02708

[32] A. Mehonic, S. Cueff, M. Wojdak, S. Hudziak, C. Labbé, R. Rizk, and A. Kenyon. Electrically tailored resistance switching in silicon oxide. Nanotechnology. 2012;23(45): 455201. DOI:10.1088/0957-4484/23/45/455201

[33] B. H. Lai, C. H. Cheng, and G. R. Lin. Multicolor ITO/SiOx/P-Si/Al light emitting diodes with improved emission efficiency by small Si quantum dots. IEEE J. Quantum Electron. 2011;47(5):698. DOI:10.1109/JQE.2011.2109699

[34] S. Cueff, S. Labbé, O. Jambois, Y. Berencén, A. J. Kenyon, B. Garrido, and R. Rizk. Structural factors impacting carrier transport and electroluminescence from Si nano-cluster-sensitized Er ions. Opt. Express. 2012;20(20):22490. DOI:10.1364/OE.20.022490

[35] B. H. Lai, C. H. Cheng, and G. R. Lin. Electroluminescent wavelength shift of Si-rich SiOx based blue and green MOSLEDs induced by O/Si composition Si-QD size variations. Opt. Mater. Express. 2013;3(2):166. DOI:10.1364/OME.3.000166

[36] J. M. Ramírez, Y. Berencén, L. López-Conesa, J. M. Rebled, F. Peiró, and B. Garrido. Carrier transport and electroluminescence efficiency of erbium-doped silicon nancrystals superlattices. Appl. Phys. Lett. 2013;103:081102. DOI:10.1063/1.4818758

[37] R. Perera, A. Ikeda, R. Hattori, and Y. Kuroki. Trap assisted leakage current conduction in thin silicon oxynitride films grown by rapid thermal oxidation combined microwave excited plasma nitridation. Microelectron. Eng. 2003;65(4):357–370. DOI:10.1016/S0167-9317(02)01025-0

[38] A. N. Trukhin., H. J. Fitting, T. Barfels, and V. Czarnowski. Cathodoluminescence and IR absorption of oxygen deficient silica – influence of hydrogen treatment. J. Non-Cryst. Solids. 1999;260(1–2):132–140. DOI:10.1016/S0022-3093(99)00558-X

[39] M. Cervera, M. J. Hernández, P. Rodríguez, J. Piqueras, M. Avella, M. A. González, and J. Jimenez. Blue-cathodoluminescent layers synthesis by high-dose N+, C+ and B+ SiO_2 implantation. J. Lumin. 2006;117(1):95–100. DOI:10.1016/j.jlumin.2005.03.015

[40] H. J. Fitting. Can we make silica luminescent? Opt. Mater. 2009;31(12):1891–1893. DOI: 10.1016/j.optmat.2008.07.015

[41] V. A. Gritsenko, Y. G. Shavalgin, P. A. Pundur, H. Wong, and W. M. Lau. Cathodoluminescence and photoluminescence of amorphous silicon oxynitride. Microelectron. Reliab. 1999;39(5):715–718. DOI:10.1109/ICMEL.2000.840567

[42] M. Goldberg, H. J. Fitting, and A. Trukhin. Cathodoluminescence and cathodoelectroluminescence of amorphous SiO_2 films. J. Non-Cryst. Solids. 1997;220(1):69–77. DOI: 10.1016/S0022-3093(97)00225-1

[43] H. J. Fitting, T. Barfels, and A. N. Trukhin. Cathodoluminescence of crystalline and amorphous SiO_2 and GeO_2. J. Non-Cryst. Solids. 2001;279(1):51–59. DOI:10.1016/S0022-3093(00)00348-3

[44] D. C. Wang, J. R. Chen, J. Zhu, C.-T. Lu, and M. Lu. On the spectral difference between electroluminescence and photoluminescence of Si nanocrystals: A mechanism study of electroluminescence. J. Nanopart. Res. 2013;15(11):2063. DOI:10.1007/s11051-013-2063-x

[45] D. Chen, Z. Q. Xie, Q. Wu, Y. Y. Zhao, and M. Lu. Electroluminescence of Si nanocrystals-doped SiO_2. Chin. Phys. Lett. 2007;24(8):2390. DOI:10.1088/0256-307X/24/8/064

Chapter 8

High-Density Plasma-Enhanced Chemical Vapor Deposition of Si-Based Materials for Solar Cell Applications

H. P. Zhou, S. Xu and S. Q. Xiao

Additional information is available at the end of the chapter

http://dx.doi.org/10.5772/63529

Abstract

High-quality and low-cost fabrication of Si-based materials, in which many fundamental and technology problems still remain, have attracted tremendous interests due to their wide applications in solar cell area. Low-frequency inductively coupled plasma (LFICP) provides a new and competitive solution, thanks to its inherent advantages of high-density plasma, low sheath potential, and low electron temperature, etc. The plasma characteristic-dependent microstructures, optical and electronic properties of the LFICP CVD-based hydrogenated amorphous/microcrystalline silicon and silicon oxides are systematically studied. Remote-LFICP combing the high-density plasma nature of ICP and mild ion bombardment on growing surface in remote plasma allows the deposition of high-quality Si-based materials providing excellent c-Si surface passivation. The mechanism of surface passivation by LFICP CVD Si-based materials, interaction between plasma species and growing surface are analyzed in terms of the plasma properties. These results pave the way for LFICP CVD utilization in Si-based high-efficiency and low-cost solar cell fabrication.

Keywords: LFICP CVD, remote-plasma, Si-based materials, solar cells, passivation

1. Introduction

Plasma-enhanced chemical vapor deposition (PECVD) is an excellent alternative for depositing thin films at lower temperature than those utilized in a conventional CVD process. Plasma generated by a glow discharge absorbs the electrical energy, and hence the neutral gases were dissociated into reactive radicals, ion, neutral atoms and molecules, and other highly excited species through the electron impact instead of conventional thermal energy. These atomic and molecular fragments interact with a substrate and, depending on the nature of these interactions, either etching or deposition processes occur at the substrate. Therefore, the substrate can be maintained at a lower temperature keeping the reactions for the thin-film deposition. These salient advantages of PECVD, i.e., low temperature nature, together with the high quality of the deposited films make PECVD be an important route for the fabrication of Si-based materials, which have been the dominant material system for the present solar cell research and commercial market. PECVD-based silicon nitride (SiN_x:H) with the feedstock gases of SiH_4 and NH_3 act as the passivation and antireflection layer of the bulk Si solar cells. Fabrication of Si thin-film solar cells consisting of the stack of p-, i-, and n-type hydrogenated amorphous/microcrystalline Si (a-/μc-Si:H) films also relies on PECVD technology. The interface quality and energy conversion efficiency of HIT (heterojunction with intrinsic thin layer) solar cell with the record efficiency of 25.6% [1] closely depends on the PECVD process because the symmetric intrinsic passivation layers and doped layers (front emitter and back surface field layers) are also prepared by PECVD technology [2].

Although PECVD-associated Si-based materials are dominant over other materials in present solar cell market, there are a variety of challenges due to the nature of the utilized PECVD method. The common PECVD uses the capacitively coupled plasma (CCP) between parallel plates driven by a 13.56 MHz RF power generator. The ionization degree of the neutral gas is at a low level of $\sim 10^{-6}$–10^{-4} because of the low electron density of $\sim 10^{9}$–10^{10}/cm^3 in a CCP discharge mode. In principle, this means a low deposition rate of the targeted materials. As a consequence, the deposition rate of PECVD-based a-/μc-Si:H is about 0.5 nm/s [3], which is not desirable for cost-efficient, thin-film solar cell productions. Furthermore, the intensive coupling of the plasma density and ion energy in CCP complicates the deposition of high-quality thin films and the achievement of high-efficiency solar cells because strong ion bombardment would result in excessive defect formation in the films and on the growing surface/interface, and hence significant deterioration of the energy conversion efficiency of the solar cells [4]. These disadvantages of PECVD can be partially remedied by using very high-frequency (VHF) PECVD, which can effectively decouple the plasma density and electron density due to the increasing electron input power fraction of the total input power with frequency [5, 6] and enhance the deposition rate. However, there is not any report on the higher efficiency of the VHF-PECVD-based solar cells [7].

High-density plasma represented by inductively-coupled plasma (ICP) has increasingly attracted more interests due to its wide application areas such as microelectronics, display, and solar cells, etc. The prominent advantage of high density of ICP makes it very appropriate for the deposition of Si-based materials (a-/μc-Si:H, silicon carbide, silicon oxide, and silicon nitride), which are widely used in solar cells. Although there are numerous reports on the ICP-based deposition of these materials, the as-deposited Si-based materials with low defect density and excellent interface features are few, not to mention the involved high-efficiency solar cells. We have been making many efforts on low-frequency (460 kHz) ICP (LFICP) excitation, characterization, and application in nano-fabrication [8] and a wide range of thin films' synthesis [9, 10]. A discharging photograph of the LFICP facility and its schematic structure are shown in **Figure 1(a)** and **1(b)**, respectively. The power of the low-frequency power generator is coupled into the reactive chamber sealed by a quartz window via a matching/tuning unit to sustain the reactive gases discharge.

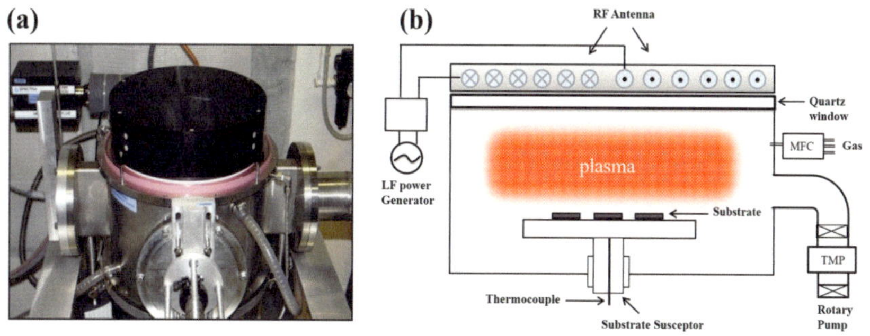

Figure 1. Discharging photograph of the LFICP facility (a), and its schematic structure (b).

In comparison with the conventional CCP source, the LFICP source has several prominent merits as follows [11, 12]: (i) high densities of the plasma species; in particular, the electron number density n_0 can reach 10^{13} cm^{-3} in the pressure range of a few pascal; (ii) low plasma sheath potentials of several or tens of volts, which are beneficial for the reduction in ion bombardment on the deposited films; (iii) low electron temperatures of a few electron volts in a broad range of discharge conditions; and (iv) excellent uniformity of the plasma parameters in the radical and axial directions. By using this novel plasma generator, Si-based materials have been fabricated with a high-deposition rate at low temperature. Hydrogenated, amorphous silicon carbide [13] (a-SiC_x:H, presently used as transparent doped layers in thin film and heterojunction solar cells) with a wide carbon component x (0.09–0.71) has been attained by adjusting the feedstock gas ratio of CH_4/SiH_4 dissociated by LFICP generator. Furthermore, low-temperature (500 °C) deposition of nanocrystalline SiC (nanocrystal size of about 6 nm) has been developed by using LFICP [14]. Low-temperature (100–200 °C) and high-rate (>1 nm/S) deposition of nc-Si in the direct plasma region of LFICP have been realized [15]. In our previous works [16, 17], it was also found that the severe etching on silicon surface directly exposed to the high-density hydrogen containing plasma drastically changes the morphology and microstructures of the surface. As a consequence, it is very easy to attain μc-/nc-Si:H with a high deposition rate. However, this high-density ion region is not suitable for solar cell fabrications where the defect density in the silicon films and on the interface need to be elaborately minimized in terms of the plasma characteristics. Apparently, the severe ion bombardment in a direct ICP would introduce additional defects reducing the photovoltaic performance of the solar cells. Remote-LFICP [18] is a feasible solution to avoid the ion bombardment damage on the surface through separating the plasma generation region and deposition region. High-density plasma of hydrogen gas is generated and mainly confined in the region near the antenna, i.e., the plasma generation region. The energetic species transport into the growth region and react with the injected SiH_4 neutral gas forming a variety of precursor radicals for the deposition of high-quality thin film at low temperatures.

High-quality Si-based materials have been attained by means of remote-LFICP:

i. As-deposited a-Si:H exhibits competitive surface passivation ability on crystalline silicon, thanks to the minimized inner and interface defect density [19];

ii. Silicon nitride with controlled bandgap due to the excellent dissociation ability of LFICP facility was realized [20]. Meanwhile, the remote-LFICP-deposited silicon-rich a-SiN_x:H displays excellent Si surface passivation (surface recombination velocity of 36 cm/s for low resistivity n-type Si wafer) [21].

iii. Remote-LFICP-based a-SiO_x:H provides effective surface passivation of low resistivity Si wafer (surface recombination of 70 cm/s obtained) [22].

The aim of this work is to systematically report the high-density plasma-enhanced chemical vapor deposition of device-grade Si-based materials by means of LFICP and its solar cell applications. The as-deposited Si-based materials will be studied in combination with the LFICP characteristics. Their soar cell applications mainly focus on the Si surface passivation using remote-LFICP-CVD-based Si-based materials. This article is organized as follows. Section 2 makes a brief introduction on a-/μc-Si:H and then focuses on the processing parameters dependent microstructures and properties of the LFICP-CVD intrinsic and doped a-/μc-Si:H thin films. Section 3 focuses on the applications of LFICP-CVD a-/μc-Si:H for c-Si surface passivation. The influence of the plasma characteristics on the properties of LFICP-based materials and growing surface properties are intensively investigated.

2. a-/μc-Si:H thin films

2.1. a-Si:H

Amorphous silicon (a-Si) is a solid-state material made of silicon atoms which are arranged in a short-range order. With the absence of the constraints of periodicity in elemental amorphous silicon, it is difficult for each silicon atom to be linked up with four others. Compared with crystalline silicon (c-Si), the average bond angles between nearest-neighboring atoms are distorted. Some bonds are even broken and result in so-called "dangling bonds". These two main defects of the lattice of a-Si:H, i.e., bond distortion and dangling bonds, create large densities of allowed states within the forbidden band gap in such a way that the material cannot be effectively doped N- or P-type and is not electronically useful. However, if the substance is hydrogenated, the hydrogen atoms saturate the dangling bonds and reduce the density of states within the forbidden gap. The bond distortion results in band tails near the valence and conduction bands. In these band tails, the electrons (or holes) are localized in space and do not directly participate in the electronic transport. The deep states near the middle of the band gap results from the non-passivated dangling bonds. These defect centers are "amphoteric"—there are three charge states, i.e., +e, 0, and –e charges, leading to two levels (transitions between the 0/+ and −/0 charge states). The dangling bonds act as recombination centers for free electrons and holes.

2.2. μc-Si:H

μc-Si:H is known to be a complex material consisting of crystalline and amorphous silicon phase plus grain boundaries [23]. **Figure 2(a)** shows a typical transmission electron microscopy (TEM) of μc-Si:H. Apparent columnar grains are observed. The crystalline growth starts at nucleation centers near the interface between film and substrate. During the competitive growth between the nucleation centers, the diameter of the remaining crystalline increases leading to the cone-shaped crystallites near the substrate surface [24]. **Figure 2(b)** displays the microstructure evolution from highly crystal μc-Si:H to amorphous a-Si:H. The highly crystal and amorphous structure are described in the left and right margin, respectively. From left to right, the column size decreases and more amorphous phase is introduced between the crystalline grains. This interrupts the columnar crystalline structure in the lateral direction.

Practically, μc-Si:H is closely related to a-Si:H, as the deposition technique for both materials is the same. However, the deposition parameters, such as discharge power density, working pressure, substrate temperature, etc., in the RF discharge deposition method, collectively determine the ultimate microstructures of the obtained silicon thin films. CCP-PECVD is the most widely studied technique for the deposition of a-/μc-Si:H films. In this technique, hydrogen-diluted silane is dissociated by electron impact from the RF field between two parallel plate electrodes to form SiH_x radicals and atomic hydrogen. The generated species diffuse/drift to the growing surface. The species and their reaction products are either incorporated into the growing film or are re-emitted from the surface into the gas phase. The radicals then undergo various secondary reactions during their transport to the substrate.

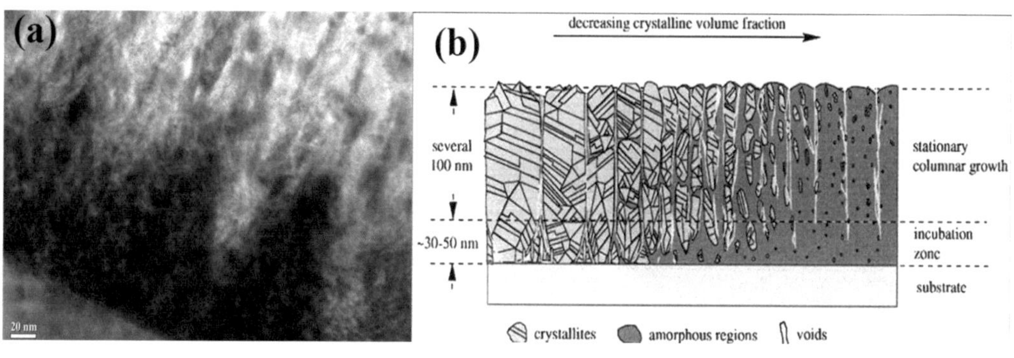

Figure 2. TEM micrograph of a μc-Si:H film grown on silicon substrate by means of LFICP (a), and a schematic drawing to show the microstructure evolution from μc-Si:H (left) to a-Si:H (right) (Reproduced from Ref. [24]).

2.3. a-/μc-Si:H by LFICP CVD

Although method of ICP has the advantages over the traditional CCP as demonstrated above, the reports on its applications in silicon thin-film deposition are not abundant, not to mention the solar cell fabrication. For reference, some main progresses are summarized as follows. Li et al. [25] used the built-in one-turn coil inductance antenna driven by 13.56 MHz RF power to fabricate μc-Si:H with tunable preferred orientation through controlling the hydrogen dilution. By the similar method with external conductance coil power by the same frequency generator, Shen et al. [26, 27] obtained low defect density ((3 x 10^{15}/cm^3) a-Si:H films, from which a single-junction a-Si:H solar cell with energy conversion efficiency of 9.6% was successfully fabricated. The researchers from Hiroshima University realized the high-rate deposition (up to 4 nm/s) of μc-Si:H with a low-level defect density in the year of 2001 [28, 29]. They systematically studied the films' optoelectronic properties such as the optical band gap, dark and photo conductivity, defect density, crystalline fraction ratio, etc. It seems that as-deposited silicon films are device grade and fulfill the common requirements for solar cell fabrications. Unfortunately, they have not reported their subsequent achievements in their research. The reports on the application of ICP-grown silicon thin films in heterojunction solar cells are not available. The only related work on the surface passivation by ICP-grown a-Si:H was reported in Ref. [30], where minority carrier lifetime of 53 μs was obtained on a P-type silicon substrate (resistivity 1–20 Ωcm) with a 15-nm-thick a-Si:H passivation layer.

For ICP reactor, there are two types of antenna configurations, namely cylindrical and planar. The present LFICP driven by a low frequency of 460 kHz power generator belongs to a planar coil configuration. As mentioned above, high-density ion region in ICP is not suitable for a-Si:H thin-film deposition for silicon surface passivation due to the severe ion bombardment. In order to decrease the ion bombardment, the distance *d* between the antenna and the substrate stage is increased by extending the deposition chamber height. In the a-/μc-Si:H deposition, two distance values of 33 cm or 53 cm were adopted for comparison purpose. The dependence of a-/μc-Si:H film microstructure and properties on the deposition parameters were studied. The n-doped and p-doped μc-Si:H layer were obtained with the introduction of dopant gases of phosphine (PH_3) and diborane (B_2H_6), respectively. The electrical properties of the doped films were analyzed.

2.4. Discharge power effect

The a-Si:H/μc-Si:H thin films were deposited on the double-side polished silicon and glass substrates cleaned by standard methods. The sample holder was heated to a temperature of 140 °C monitored by an internally equipped thermal couple. The flux of silane (H_2) was kept constant at 5 sccm (20 sccm). The inductive RF discharge power was varied between 1.0–2.5 kW. The thicknesses of all the deposited thin films were measured directly from the cross-sectional scanning electron microscopy (SEM) image by using a JEOL JSM-6700F filed emission scanning electron microscope. The deposition rate can be calculated from the thickness divided by deposition duration. FTIR measurements were performed on a Perkin-Elmer FTIR 1725X spectrometer in the mid-infrared range from 400 to 4000 cm^{-1} with an increment of 1 cm^{-1}. The Cary 510 Bio spectrometer (300–1100nm) was used to measure the UV-VIS transmission spectra.

2.4.1. Optical band gap and deposition rate

The UV-VIS transmission spectra (shown in **Figure 3(a)**) of the thin films deposited on glass at different RF powers were measured to extract the parameter of optical band gap E_g. Evident interference fringes due to multi-layer interface reflections were observed in the spectra range longer than the absorption edge. The absorption edge occurs a redshift with deposition power, indicating shrinkage of the optical band gap. The transmission T was transformed into absorption α by the relation of $\alpha\infty$-ln(T) using the thickness of the films directly read from the cross-sectional SEM images. The optical band gap E_g of a-/μc-Si:H (Si an indirect band gap semiconductor) can be deduced from the Tauc equation

$$h\gamma\alpha(h\gamma)=B(h\gamma - E_g)^2 \qquad (1)$$

where B is the edge width parameter related to the width of band tails in the film [31], and hv the photon energy. In general, B factor is also taken as a measure of the overall structural disorder and a higher value for B would indicate a lower degree of structure disorder [32]. In our calculation, the obtained value of B (>800 $cm^{-1/2}.eV^{-1/2}$) is larger than that of literatures, indicating less disorder and defects in the thin films.

The obtained E_g as a function of applied RF power together with the average deposition rate are both plotted in **Figure 3(b)**. A transition point of E_g was observed at power of 1.8 kW. Below 1.8 kW, E_g encounters a decrease from 2.30 eV for 1.0 kW to 2.10 eV for 1.8 kW. On the contrary, E_g increases from 2.10 eV for 1.8 kW to 2.14 eV for 2.5 kW. This changing trend of E_g is related to the crystallinity of the thin films and the hydrogen content, which will be discussed below based on the experiments of Raman scattering and FTIR. The increasing RF power results in the steady increase in the average deposition rate from 7 to 21 nm/min. These values obtained at antenna-substrate distance d of 33 cm are about half of those corresponding to distance of 11 cm [33].

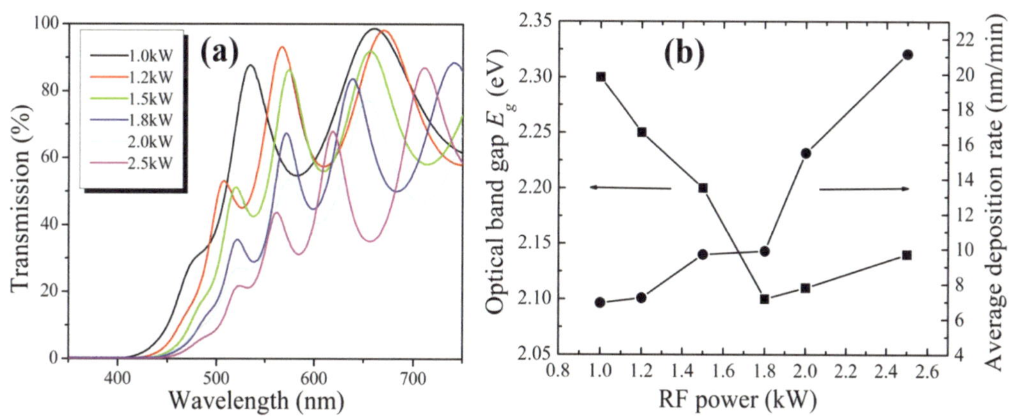

Figure 3. The UV transmission spectra of the thin films deposited at different RF powers on glass substrate (a), and deduced E_g (filled squares) and average deposition rate (filled circles) (b).

Another feature of the present films is the higher optical band gap E_g than that reported in Refs. [34–36] where E_g was obtained by the methods other than Tauc method (spectroscopic ellipsometry and surface photovoltage spectroscopy).

Although it was recognized that the Tauc method will lead to a larger value for E_g, it will give a correct variation trend of the value with respect to an independent parameter like RF power, substrate temperature, and so on. Herein, the values of E_g are above 2 eV, larger than the values (<1.9 eV) [33] obtained in earlier works performed on the direct-plasma ICP system with other similar deposition parameters. This should be correlated to the increased hydrogen content (following FTIR results) in the thin films.

2.4.2. Raman scattering and crystallinity of the thin films

The micro-Raman scattering experiments were performed to estimate the crystal phase information in the thin films. The obtained spectra are displayed in **Figure 4(a)**. It was observed that the dominant peak gradually shifts from (480 to (520 cm^{-1} through a mixed phase state at the power of 1.8 kW. This change reflects phase evolution from completely amorphous to highly crystal with an applied RF power in the films. In fact, the crystallinity value can be estimated by using different methods described in Ref. [37]. In this work, the method based on Gaussian line fitting [38] is used, where I_a

$$fc = (I_c + I_i) / (I_a + I_c + I_i) \quad (2)$$

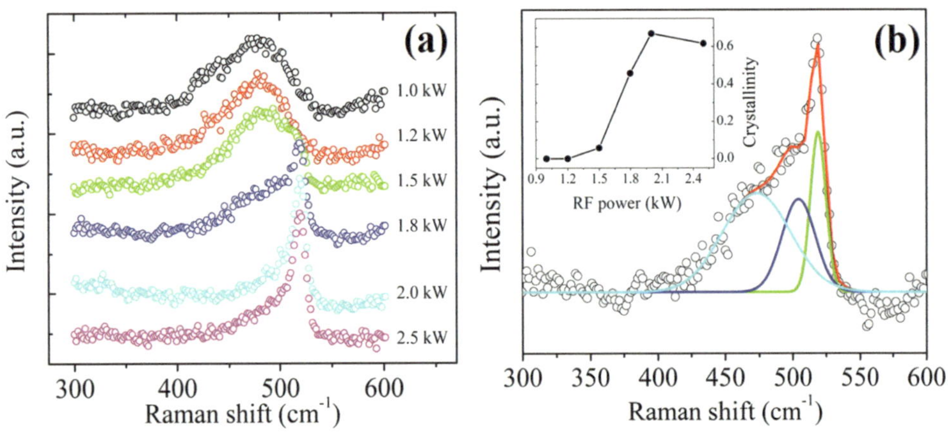

Figure 4. RF power-dependent Raman scattering spectra (a), and the Gaussian-shaped curve decomposition of the Raman spectrum in the thin film deposited at RF powers of 1.8 kW. The inset of (b) shows the RF power-dependent crystallinity of the thin films.

I_i, and I_c are the integrated area of each phase, i.e., the amorphous ((480 cm^{-1}), intermediate ((510cm^{-1}), and crystalline phase ((520 cm^{-1}), respectively, in the Raman spectra. A typical deconvolution result is shown in **Figure 4(b)**, where the cyan, blue, and green curves (from left to right) denote the amorphous, intermediate, and crystal component, respectively. The experimental (open circle) and overall fitting result agrees well, indicating the validity of the deconvolution procedure. The calculated crystallinity as a function of RF power is included in the inset of **Figure 4(b)**. As expected, the crystallinity increases monotonously in the power range of 1.0–2.0 kW because of the production of denser and more energetic ion/radicals in the plasma and area near the surface of the substrate. Higher plasma density means more frequent impacts between the electron and the precursor gases, and thus a larger deposition rate. More energetic ion/radicals imply that the adsorbed radicals on the surface will have a greater diffusion length. This assists the adsorbed radicals to reach the most energetically favorable sites and gives rise to better crystallization of the synthesized thin films [39]. The crystallinity decreases from 67.0 to 61.6% when the RF power is further increased from 2.0 to 2.5 kW. This should be related to the rapid increase of deposition rate in the range of 2.0–2.5 kW shown in **Figure 3(b)**. One possible interpretation may be that: a higher (>2.0 kW) RF power excites more and more precursor ions like SiH, SiH_2, and SH_3 to improve the growth rate; on the other hand, the overly rapid growth minimizes the hydrogen etching effect on the existing weak and strained Si bonds. It is generally accepted that hydrogen etching is also an important mechanism to improve the crystallinity of Si thin films [40].

2.4.3. FTIR and hydrogen content in the thin films

The FTIR spectra of the thin films deposited with various RF power are shown in **Figure 5(a)**. Within a wavenumber range of 500–2300 cm^{-1}, three Si-H bond-related bands were observed. The first one, located at around 640 cm^{-1} was unambiguously attributed to Si-H rocking/wagging mode. Its exact location gradually changed from 644 to 630 cm^{-1} with increasing RF power as displayed in **Figure 5(b)** (the filled circles). This is also correlated to the phase evolution from amorphous to crystal silicon [41]. The second one comprising of two sub-peaks located at 845 and 890 cm^{-1} were identified as the bending mode of dihydrides SiH_2 [41]. The third one at the region of (2100 cm^{-1} is recognized as the SiH_x stretching mode. One can see that these hydrogen-related signals become weak with increasing RF power, implying the decreasing bonded hydrogen content in the thin films. Actually, the bonded hydrogen content in the thin films can be estimated from the Si-H bond rocking/wagging mode using the following equation:

$$C_H = A_\omega \int \frac{\alpha(\omega)}{\omega} d\omega = A_\omega I_\omega \tag{3}$$

where the oscillator strength A_ω is of value of 1.6 × 10^{19} cm^{-2}. The calculated hydrogen content at.% (defined as density ratio of hydrogen to that of hydrogen plus silicon) versus applied RF power is plotted as filled square in **Figure 5(b)**. The hydrogen content is in the range of 7.9–-17.4% and exhibits a general decreasing trend with RF power. Meanwhile, much lower values (<8%) were obtained using the same method in the case of direct plasma (i.e., the film deposition region located in the plasma generation region) [15, 33]. A local fluctuation of hydrogen content was also observed at a power of 1.8 kW, where the thin film is of transition state from amorphous to crystal phase. Further increase in RF power from 2.0 to 2.5 kW leads to a little improvement of hydrogen content, which is consistent with the decreasing crystallinity in this power range shown in **Figure 4(b)**.

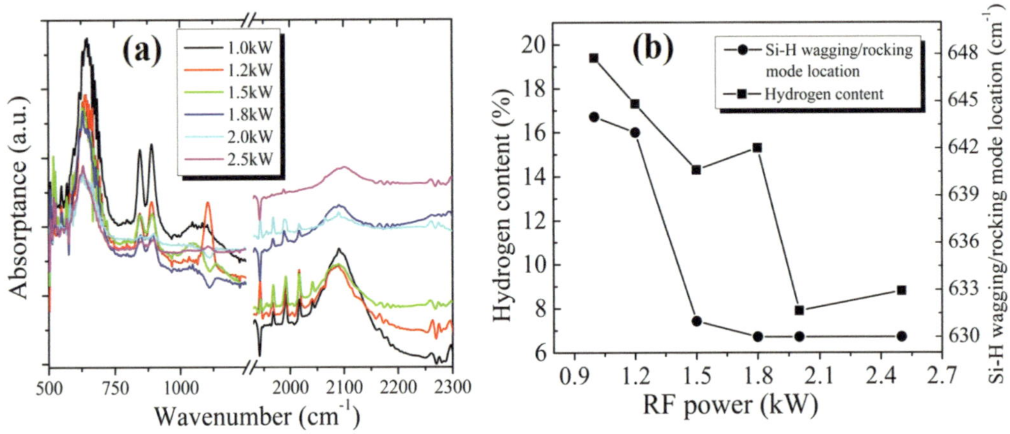

Figure 5. FTIR absorption spectra of the thin films deposited at different RF powers (a), and the RF power-dependent hydrogen content (filled squares) and Si-H wagging/rocking mode location (filled circles) (b).

2.5. Hydrogen dilution effect

Hydrogen in the deposited films originates from the diluent hydrogen gas and hydrogen atom bonded in silane molecules. From the point view of the deposition process, hydrogen facilitates the dissociation of silane through increasing the ion/electron impact, influences the surface energy of the growing surface via terminating the surface silicon atom, and etches the growing surface. Therefore, the hydrogen dilution should have a substantial effect on the microstructures of the ultimate silicon thin films. In this set, the RF discharge power and substrate are fixed at 1.5 kW and 140 °C, respectively. The total flow rate of SiH_4 and H_2 is kept at 20 sccm. The respective flow rate of SiH_4 and H_2 are changed with varying dilution ratio $[H_2]/[SiH_4]$ = 0, 0.25, 0.5, 1, 2, 3, and 4. The working pressure in the deposition chamber is kept at 7.2 Pa.

Figure 6 shows the average deposition rate and E_g deduced from the Tauc method described above. As expected, the deposition rate monotonically decreases with the decreasing silence concentration (i.e., increasing hydrogen dilution ratios). When the dilution ratio changes from 0 (pure silane) to 4, the deposition rate drops sharply from 42 nm/min to 10 nm/min. However, the case of E_g is more complicated with increasing dilution ratios. Within the range of ratio 0–2, E_g increases from 2.17 to 2.32 eV. Further increase in dilution ratio results in decreasing E_g. At a dilution ratio of 4, E_g is 2.21 eV. It has been mentioned above that E_g of a-/μc-Si:H is related to the crystallinity and hydrogen content in the films. The corresponding Raman spectra shows that all the films deposited under present conditions are of amorphous structure without clear crystalline fractions. As pointed out above, the wagging/ rocking mode location in FTIR spectra is also related to the phase configuration in the films.

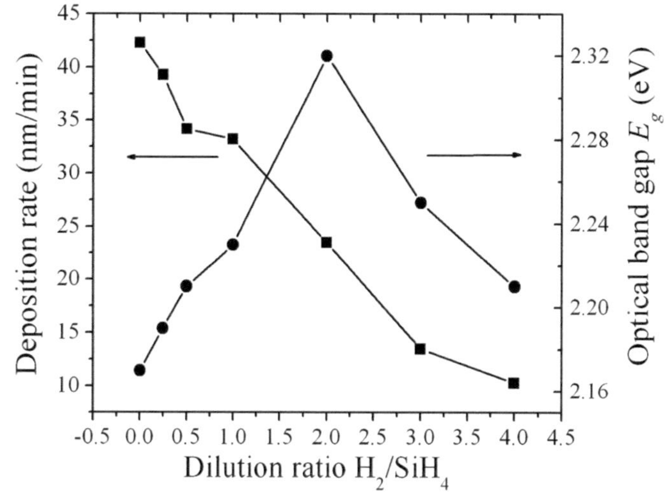

Figure 6. Dilution ratio-dependent optical band gap and average deposition rate.

Figure 7 shows the extract waging/rocking vibration mode. With a small amount of hydrogen introduction (H_2/SiH_4 = 0.25), the hydrogen content C_H increases from 26.3 to 27.5% and the wagging/rocking mode location shifts from 649 to 644 cm^{-1}. C_H drastically drops to 24.4% at a dilution ratio of 0.5. The values of C_H and the location of wagging/rocking mode are comparably steady in the dilution range of 0.5–2. Beyond 2, C_H decreases drastically to 12.2% and the mode location shifts to 631 cm^{-1} at a ratio of 4. The location shift to the small wavenumber side with increasing dilution ratios implies the phase evolution to the crystalline orientation.

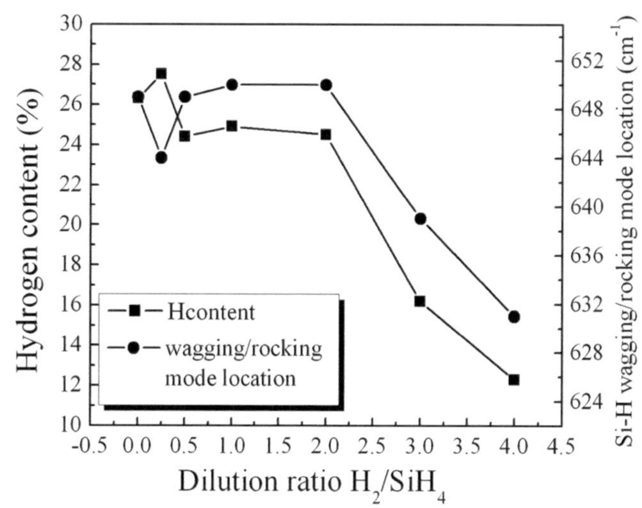

Figure 7. Hydrogen content and the wagging/rocking mode location in the a-Si:H films as functions of the dilution ration H_2/SiH_4.

In view of the above experimental results, hydrogen dilution in LFICP method is also a key parameter influencing the hydrogen content C_H in the films. However, its effect on the crystallinity of the films is less pronounced than that of the discharge power.

2.6. Working pressure effect

One of the inherent features of LFICP is low working pressure up to several tens of Pa. In this section, the working pressure is altered in the range (1–19Pa) enabling a stable discharge. The discharged power was kept at 2.5 kW. No external heating was applied to the substrate. The flow rate of SiH_4 and H_2 are 12.5 sccm. The Raman scattering experiments revealed the amorphous structure of these films. **Figure 8** displays the working pressure-dependent E_g and average deposition rate. The deposition rate monotonically increases with working pressure from the 68 nm/min (1 Pa) to 142 nm/min (19 Pa). On the contrary, E_g becomes narrower with working pressure: 2.34 eV for 1 Pa, and 1.92 eV for 19 Pa.

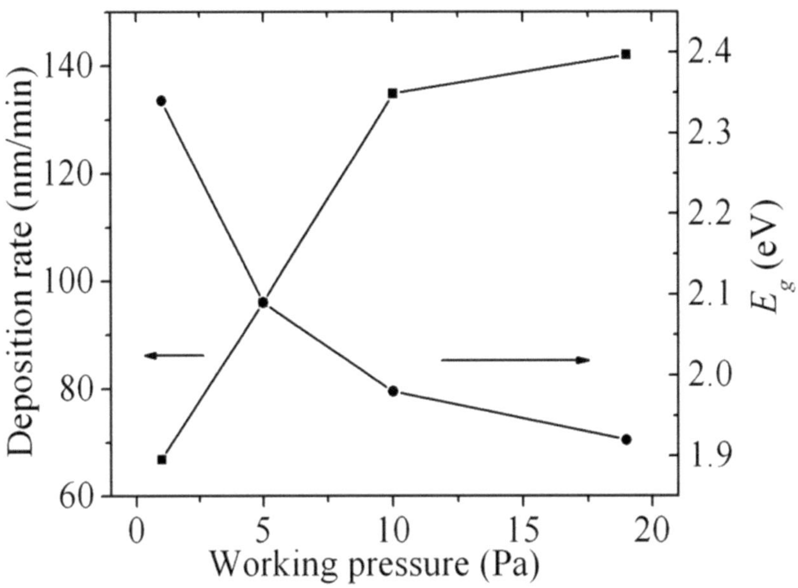

Figure 8. The deposition rate and E_g as functions of working pressure.

Figure 9(a) shows the FTIR absorbance spectra of these films. The intensity of the waging/rocking mode becomes weaker with increasing work pressure, indicating the effusion of the bonded hydrogen with increase in the working pressure. The detailed hydrogen content as a function of working pressure is shown in **Figure 9(b)**. The hydrogen content was reduced about 20% when the working pressure is increased from 1 to 19 Pa. The location of this mode experiences a slight shift to higher wavenumbers with working pressure, which implies an amorphous phase change. It is well known that an increase in the gas pressure enhances the collisions in the plasma and reduces the electron and ion energies, and hence decreases the crystallinity.

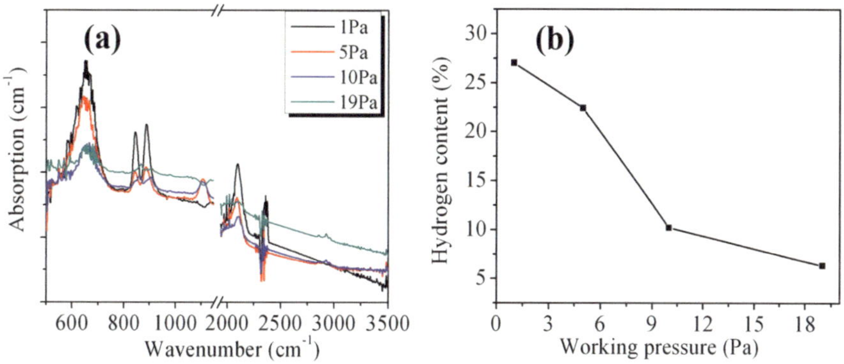

Figure 9. FTIR absorptance spectra of the a-Si:H thin films deposited at different pressures (a), and deduced hydrogen content C_H in the films (b).

It is worth noting that the stretching mode also decreases considerably with increasing working pressure. Assuming the stretching mode corresponds to SiH-associated peak at 2000 cm^{-1} and the SiH_2-associated peak at 2100 cm^{-1}, we will note the severe decrease of SiH_2 fraction in this mode. This should be responsible for the reduced hydrogen content in the a-Si:H films.

In summary, the working pressure substantially influences the optical band gap E_g and hydrogen content of the films. In present LFICP deposition, the high E_g values larger than 2 eV is mainly correlated to the high hydrogen content C_H which corresponds to the ultra-low working pressure.

2.7. Effect of substrate temperature

The substrate temperature substantially affects the plasma species diffusion on the surface and nucleation process, and hence the properties of the resulting a-/μc-Si:H thin films. In this set, the feed gas flow rate ratio is SiH_4/H_2 =5:20 (in sccm) and the discharge power is 2.0 kW. The substrate temperature is varied in the range of 27–150 °C. **Figure 10** shows the temperature-

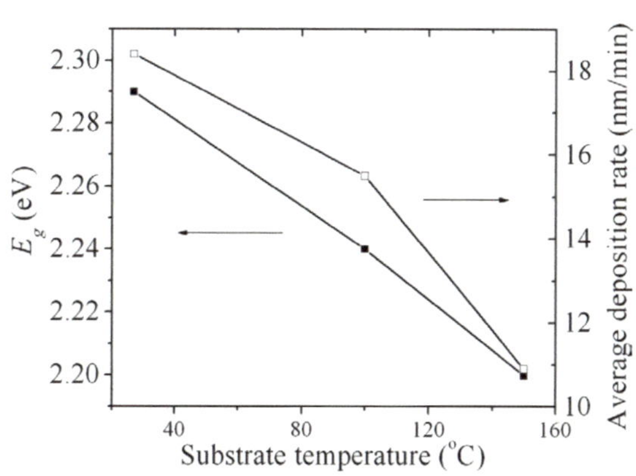

Figure 10. The substrate temperature-dependent deposition rate and E_g.

dependent deposition rate and E_g. The increasing substrate temperature leads to the decrease in both of these two parameters. In the studied temperature range, deposition rate and E_g decreases by about 7.5 nm/min and 0.9 eV, respectively.

Figure 11 shows the substrate temperature-dependent Raman spectra. The obtained films are highly crystalline under the present deposition conditions. The elevated substrate temperature results in the enhanced intensity of the 520 cm^{-1} peak, and thus more order structure in the films.

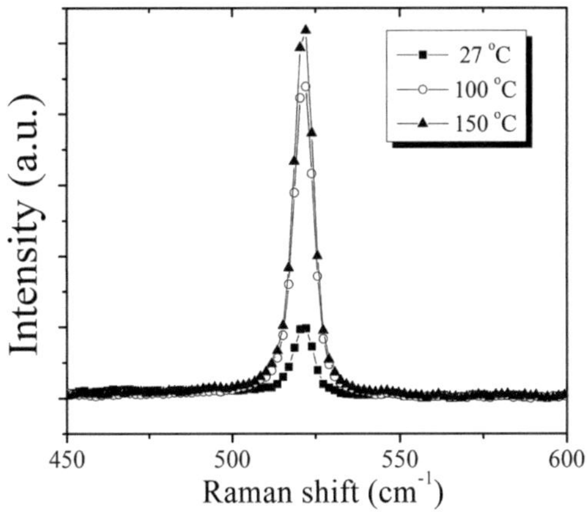

Figure 11. Substrate temperature-dependent Raman spectra.

The wagging/rocking mode in FTIR transmission spectra provides the bonded hydrogen numbers information which is displayed in **Figure 12**. The bonded hydrogen content C_H is very sensitive to the substrate temperature, decreasing by about 21% with substrate temperature from room temperature to 150 °C.

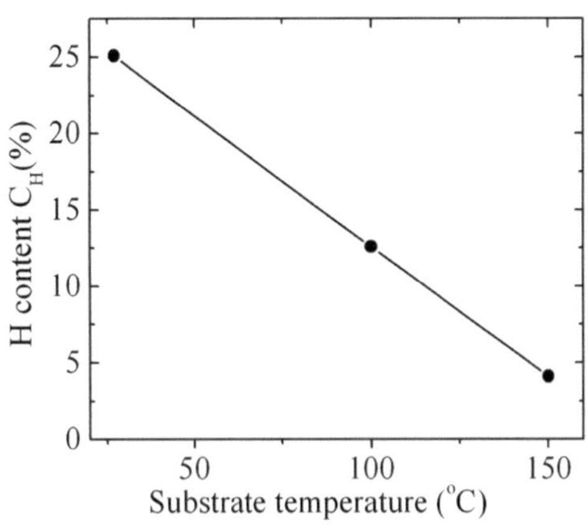

Figure 12. Substrate temperature-dependent hydrogen content in the films.

http://dx.doi.org/10.5772/63529

2.8. Effect of antenna-substrate distance

In the axial direction, the plasma properties should be dependent on the antenna-substrate distance *d* between the RF antenna and the depositing substrate. Although the detailed plasma characterizations in different distances have not been conducted, the different plasma densities, electron temperatures, and sheath potentials, etc. are expected. In principle, the neutral particles will become more and more important among species contributing to the film deposition due to the enhanced ion recombination with increasing distance *d*. It was recognized that the severe etching and ion bombardment produces additional defects on growing surface/substrate directly contacting with the high-density plasma [16]. Therefore, the deposition region should avoid direct contact with the high-density plasma during deposition. **Figure 13** shows the Raman spectra of the films deposited at different *d* values. The film is completely crystal in the case of direct plasma at $d = 11$ cm. In comparison, microcrystalline silicon is obtained when the distance is widened to 33 cm (namely, a semi-remote plasma configuration). The film is completely amorphous at the distance of $d = 53$ cm (a remote plasma configuration). Meanwhile, the atom network evolution from crystalline to amorphous is accompanied with the deposition rate reduction shown in **Figure 14**. The drastic reduction of deposition rate with distance *d* reflects the decreasing number of precursors contributing to Si:H deposition.

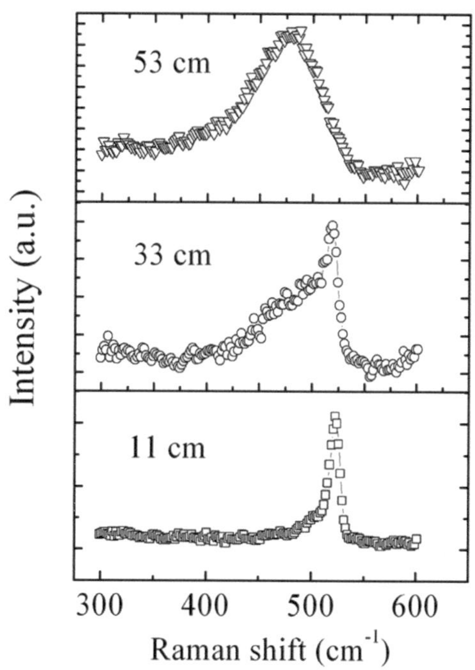

Figure 13. Raman scattering spectra of the silicon thin films deposited at different antenna-substrate distance *d* values. The other deposition parameters are as follows: discharge power 1.8 kW, substrate temperature 100 °C, and gas pressure of 7.2 Pa.

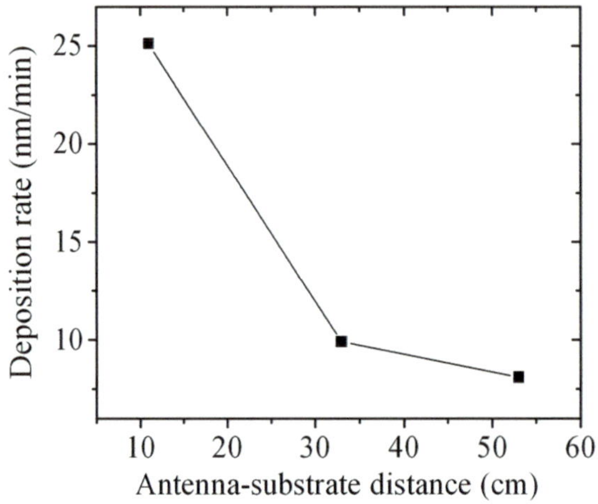

Figure 14. The antenna-substrate distance-dependent deposition rates.

Figure 15 shows the evolution of FTIR absorbance spectra with respect to the distance *d*. All the H-associated signals are enhanced with distance *d*. The hydrogen content C_H deduced from the wagging/rocking mode is represented in the inset. C_H increases sharply with *d* possibly due to the reduced plasma heating effect.

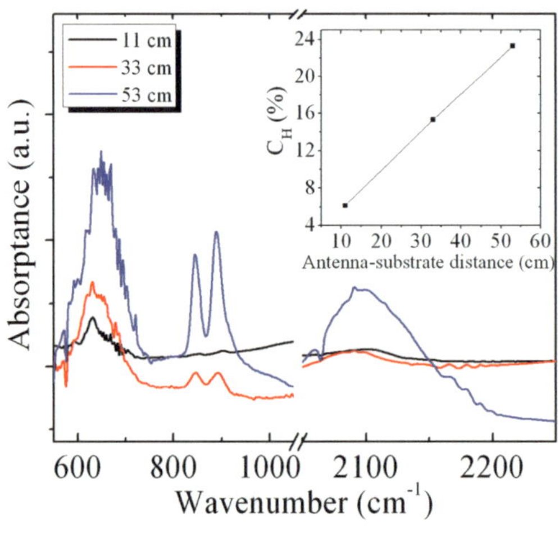

Figure 15. Comparison of FTIR absorbance spectra of the films deposited at different antenna-substrate distance values.

2.9. Doped a-/μc-Si:H films by LFICP CVD

In this section, the process parameters-dependent microstructure and electrical properties of the n- and p-doped a-/μc-Si:H films are studied to evaluate the doping ability of LFICP. Doping of a-Si:H by PECVD was first demonstrated by Spear and Le Comber in Ref. [43]. The disordered structure of a-Si:H was supposed to allow the incorporation of dopant into the amorphous network, thus impeding the doping of the material. Subsequently, the underlying doping mechanism of a-Si:H was further clarified by Street [44]. Nevertheless, the doping efficiency of a-Si:H (<1%) as a disordered material is much lower than in crystalline silicon [45].

Here, the feed gas silane is pure without hydrogen pre-dilution. The n-type dopant resource gas phosphine (PH_3) and p-type dopant resource gas diborane (B_2H_6) both were H_2-diluted to a 15% concentration. In the following discussions, the gas consumptions are directly denoted as the nominal flow rate. It was found that the antenna-substrate distance *d* also significantly affects the doping capability of LFICP. As an example, **Figure 16** shows the hole carrier concentration evolution with respect to the parameter of *d* under the same other parameters: SiH_4:H_2:B_2H_6 = 5:20:0.4 (in sccm), discharge power 2.0 kW and low substrate temperatures of 150 or 100 °C. When *d* is increased from 11 to 53 cm, the hole concentration drops sharply spanning a wide range of 10^{19}–10^{14} cm^{-3}. The result at 100 °C is also included in the case of 33 cm. Recalling the Raman results at different *d* values (**Figure 13**), one will find that carrier density is inversely proportional to the crystallinity of the films. This is consistent with the fact that amorphous network is more difficult to dope. In the next section, it will be revealed that larger-distance-deposited a-Si:H films provide more effective surface passivation effect on silicon. Hence, doping and passivation ability for LFICP-grown a/-μc-Si seem to be a contradiction.

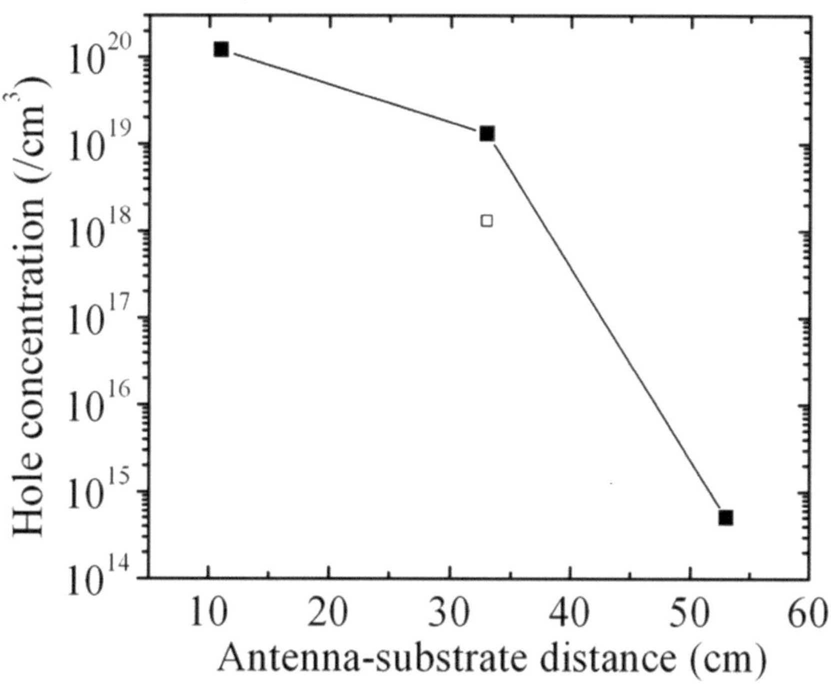

Figure 16. The hole carrier density as a function of antenna-substrate distance *d* in the p-doped films deposited at substrate temperature 150 °C (solid square) and 100 °C (open square).

2.9.1. p-doping

Different-flux phosphine gas was introduced into the feed gases of SiH_4:H_2 = 6:60 (in sccm) discharged by RF power of 3 kW with substrate temperature of 100 °C. **Figure 17(a)** shows the electrical parameters (from Hall effect measurement using the Ecopia HMS-3000 system) of as-deposited p-doped a-/μc-Si:H films. The hole density p reaches its maximum value of 1.2×10^{19} cm^3 at an intermediate B_2H_6 flux of 0.4 sccm. The ultimate conductivity σ displays the same trend with carrier density. The highest σ value is 0.58/Ωcm obtained at B_2H_6 flux of 0.4 sccm. The corresponding Raman scattering spectra of the films are shown in **Figure 17(b)**. A systematical shift from microcrystal to amorphous structure occurs with the gradual addition of dopant gas B_2H_6. The flux of 0.4 sccm is an optimum one, where the optimized amount of the dopant is introduced into the host μc-Si:H structures.

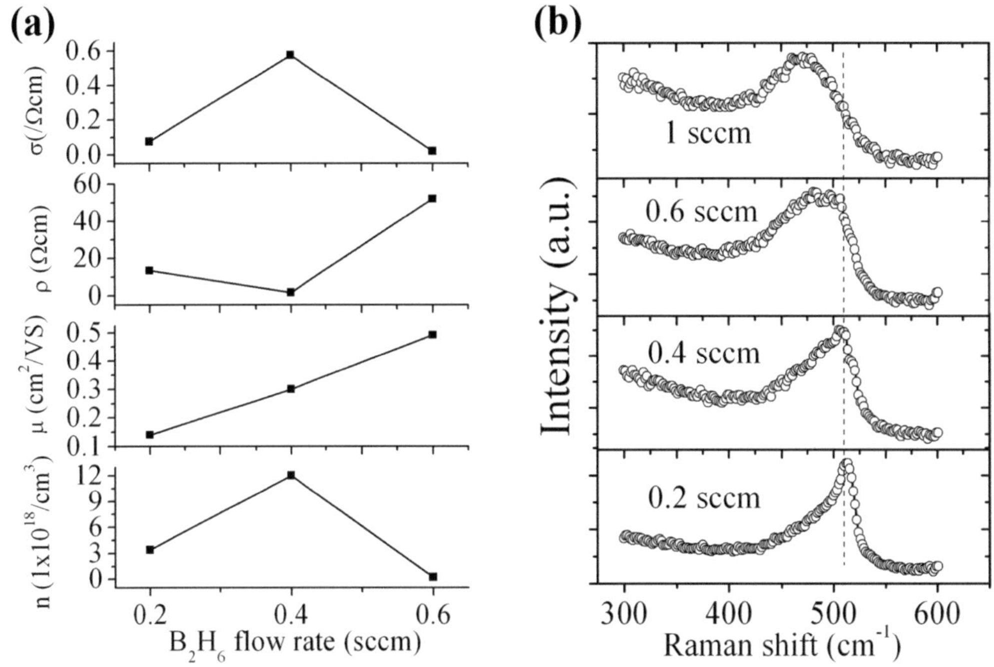

Figure 17. B_2H_6 flux-dependent electrical properties including the hole carrier density, mobility, resistivity, and conductivity (a), and corresponding Raman spectra of the a-/μc-Si:H films (b).

It is found that the doping is strongly dependent on the deposition temperature. Increasing substrate temperature improves the conductivity of the deposited films. **Figure 18(a)** shows the electrical properties of the doped a-/μc-Si:H films deposited in the substrate temperature range of 100–200 °C. The flux of B_2H_6 is kept at 0.4 sccm. The increased temperature facilitates the activation of the dopant in the film leading to the improvement in the hole density and degraded mobility. The resulting conductivity σ almost increases linearly with substrate

http://dx.doi.org/10.5772/63529

temperature and σ reaches 2.1/Ωcm. **Figure 18(b)** shows the corresponding Raman scattering spectra. The peak gradually shifts from 505 to 510 cm^{-1} indicating the microstructure shift to the microcrystalline side. This ordered network resulted from the higher substrate temperature is beneficial to doping behavior.

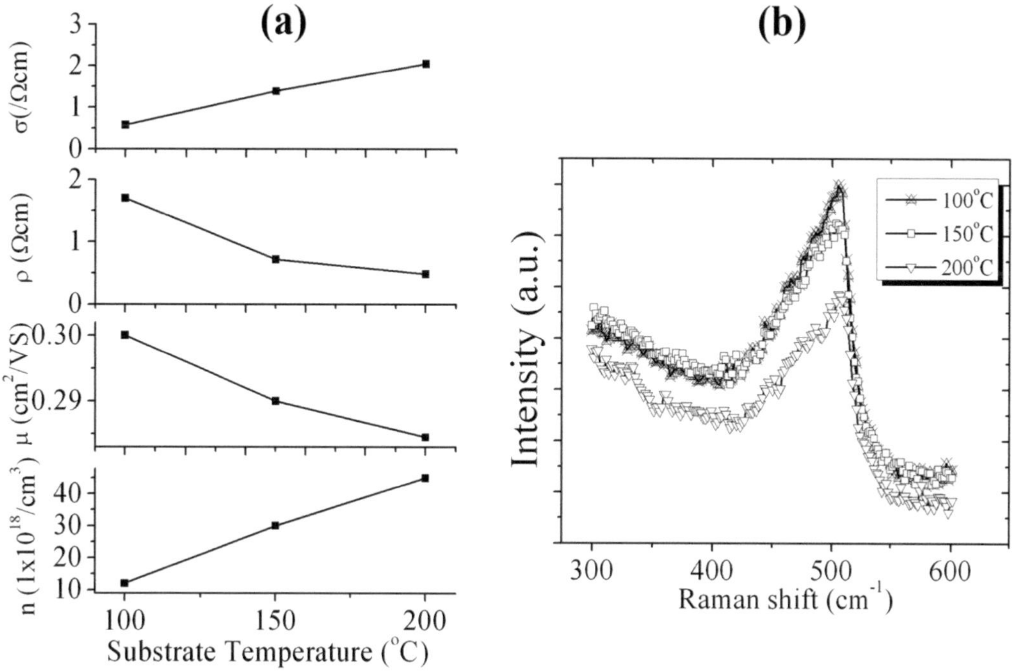

Figure 18. Electrical properties of the doped films with B_2H_6 = 0.4 sccm at different substrate temperatures (a), and the corresponding Raman scattering spectra (b).

2.9.2. *n-doping*

It seems that the n-doping is more difficult than p-doping in the case of higher antenna-substrate distance. Herein, the substrate temperature ranges from 100–200 °C, PH_3 flux covers the range of 0.5–1.0 sccm, and the feed gas ratio SiH_4:H_2 is changed from 5:40 to 5:60 (in sccm). The discharge power is set at 2.75 kW. **Figure 19** shows the Raman scattering spectra of the a-/μc-Si:H films n-doped by different PH_3 flux. The microstructure change with the PH_3 introduction is not as monotonic as shown in **Figure 17** associated with the case of B_2H_6. **Table 1** presents the deposition parameters and associated electrical properties of the obtained n-doped a-/μc-Si films. At PH_3 flux of 0.5 and 0.62 sccm (100 °C), the electron density n is below 10^{14}/cm^3. The former flux 0.5 sccm is not adequate for effective doping and the latter 0.62 sccm leads to an amorphous-oriented network according to **Figure 19**. The optimum set is 0.75 sccm PH_3 and 200 °C, where the highest conductivity is 3.45/Ωcm.

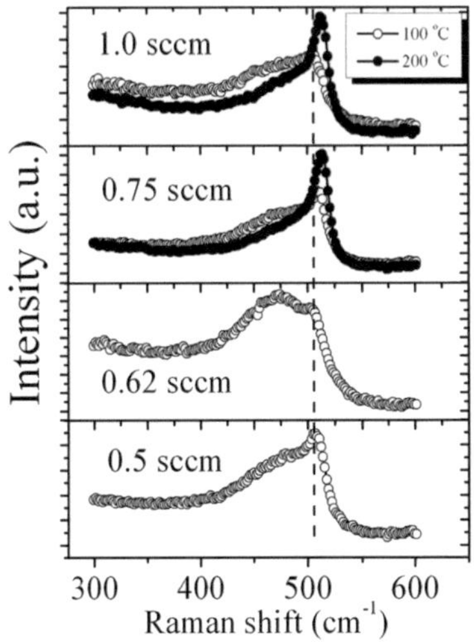

Figure 19. Raman scattering spectra of a-/μc-Si:H n-doped at different PH_3 fluxes.

Power	SiH_4	H_2	PH_3	Pressure	*T*	n	*μ*	*ρ*	*σ*
(kW)	**(sccm)**	**(sccm)**	**(sccm)**	**(Pa)**	**(°C)**	**(/cm³)**	**(cm²/VS)**	**(Ωcm)**	**(/Ωcm)**
2.75	5	60	0.5	12	100	4.11E+14	17.85	859	0.001
2.75	5	60	0.75	12	100	4.95E+15	6.34	199	0.005
2.75	5	60	0.75	12	200	1.03E+20	0.21	0.29	3.45
2.75	5	60	1	12	200	7.56E+19	0.031	2.68	0.38

Table 1. Deposition parameters and properties of n-doped a-/μc-Si:H thin films.

Figure 20 shows the n-doped a-/μc-Si:H films deposited with different hydrogen dilution ratio. Up to the ratio of SiH_4/H_2 = 5:50, the film is a completely amorphous network impeding the

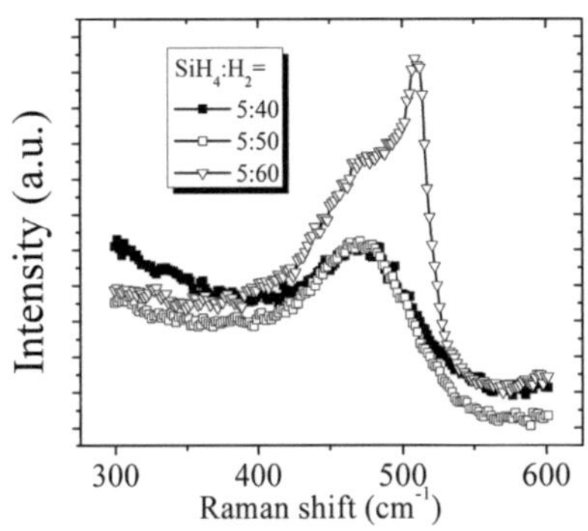

Figure 20. Raman scattering spectra of the n-doped a-/μc-Si:H films deposited at different hydrogen dilution ratio.

http://dx.doi.org/10.5772/63529

doping behavior. The structure turns into microcrystalline at the ratio of SiH_4/H_2 = 5:60, where the better conductivity can be obtained in present experiments.

Figure 21 shows a typical top view **(a)** and cross-sectional **(b)** SEM image of a n-doped μc-Si:H film deposited at a temperature of 200 °C and dilution ratio of H_2:SiH_4 = 60:5. Nanostructured Si particles are uniformly distributed on the surface in **Figure 21(a)**. There are no any observable voids or apparent defects. The apparent column growth is observed in **Figure 21(b)**. The interface between the film and the Si substrate is clear and free of incubation layer.

Figure 21. Top view (a) and cross-sectional SEM image of the n-doped μc-Si:H.

3. c-Si surface passivation by LFICP CVD a-/μc-Si:H

In Sections 1–2, the intrinsic and doped a-/μc-Si:H films based on LFICP-CVD method are studied in terms of the plasma parameters. The optical band gap, bond configuration, microstructure, and hydrogen content in the films are dependent on the deposition parameters of RF power, gas ratio, working pressure, and substrate temperature. In particular, the properties of the LFICP-grown a-/μc-Si:H films are sensitive to the antenna-substrate distance *d*. A high *d* value is inclined to a-Si:H over μc-Si:H growth due to avoiding direct immersion in high-density plasma. However, increasing *d* makes it difficult to dope the a-/μc-Si:H films. Improving substrate temperature and hydrogen dilution can effectively compensate the doping obstacle from increased *d*. The conductivity of 2.10 and 3.45/Ωcm are achieved in p- and n-doped μc-Si:H at a low substrate temperature of 200 °C.

In this section, the crystalline silicon surface passivation by LFICP-CVD a-/μc-Si:H is investigated. In terms of the antenna-substrate distance *d*, a-/μc-Si:H films display different passivation level evaluated by the carrier lifetime measurements. The passivation effect of the film, which is strongly dependent on a variety of processing conditions, e.g., the pre-deposition treatment of c-Si, the film deposition parameters, and post-deposition annealing, is physically analyzed. The carrier recombination mechanisms will be firstly reviewed.

3.1. Recombination and passivation in Si-based solar cells

Usually, the loss in photovoltaic solar cell comprises surface light loss, the transmitted light loss, thermalization loss (thermal electron loss). In a practical heterojunction (HJ) silicon solar cell, the a-/μc-Si:H layer is of several nanometers thick, which marginally absorbs sunlight. The photogeneration of excess charge carrier pairs mainly occur in the bulk c-Si. In order to efficiently collect the photogenerated carriers which can be extracted and thus contribute to the externally retrievable current, reducing the carrier recombination in silicon, especially on the surface, is particularly important.

The carrier recombination originating from bulk silicon primarily comprises two types, i.e., the intrinsic and extrinsic ones. The intrinsic recombination, which mainly includes radiative [46] and Auger recombination [47], occurs in a pure and defect-free semiconductor, while the extrinsic recombination stems from contaminants and defects which act as recombination centers.

Total bulk recombination is the sum of the above three terms and is expressed as

$$U = U_{Rad} + U_{Aug} + U_{SRH} = \frac{\Delta n}{\tau_{Rad}} + \frac{\Delta n}{\tau_{Aug}} + \frac{\Delta n}{\tau_{SRH}} = \frac{\Delta n}{\tau_b} \tag{4}$$

where τ_{SRH} denotes the contribution from the extrinsic recombination of Shock-Read-Hall effect [48, 49].

Surface recombination is a special case of SRH recombination due to the localized states appearing at the surface. Different from the bulk SRH centers, these states usually form a set of states across the band gap instead of a single energy level. Surface recombination is usually evaluated in terms of surface recombination velocity (SRV) instead of lifetime. However, the principles are same to bulk SRH recombination. It mainly originates from the surface dangling bond, i.e., the missing of crystalline network. The chemically and mechanically created defect can also result in surface recombination. In a simple condition with a constant bulk lifetime τ_b and a small constant SRV (two surfaces are same to each other), the effective lifetime τ_{eff} is expressed as

$$\frac{1}{\tau_{eff}} = \frac{1}{\tau_b} + \frac{2S_{eff}}{W}, \tag{5}$$

where W is the thickness of the bulk silicon.

High carrier lifetime is a critical issue in achieving high-energy conversion efficiency in solar cells. There are many techniques that have been developed over the past two decades with the objective of measuring minority carrier lifetime in Si [50]. Two techniques, namely the transient photoconductance decay (TPCD) and the more recent quasi-steady-state photoconductance method (QSSPC), are the basis of the WCT-120 tester, which are widely used in this work to characterize the solar cell precursors.

Compared with the optical management by conventional methods like an antireflection layer, the minimization of electronic losses at the crystal silicon surface is a more delicate challenge for high-efficiency solar cell [51]. Recombination losses at silicon interface or surface can be minimized by using two passivation schemes, namely, via chemical passivation or filed passivation. The former aims to reduce the interface defects that originate from the dangling bonds by H atom, a thin dielectric, or semiconductor film. The latter is based on the fact that the electron/hole density at the interface will be significantly reduced by formation of a built-in potential from the introduction of a passivation layer. Based on these two strategies, various Si-based materials, such as hydrogenated amorphous silicon (a-Si:H) [52], silicon nitride (SiN) [53], thermal oxygenated silicon dioxide (SiO_2) [54], and amorphous silicon carbide (a-SiC_x) [55] have been investigated for passivation purposes. Among these materials, a-Si:H is still the best candidate for silicon heterojunction solar cells [56, 57] due to its excellent passivation properties obtained at low deposition temperatures, and also simple processing without the introduction of any atoms other than silicon and hydrogen.

The a-Si:H (including SiO_x, SiN_x and SiC_x) passivation layers were mainly deposited by using conventional CCP-PECVD or hot wire chemical vapor deposition (HWCVD) in literatures. A detailed review of c-Si surface passivation can be found in Refs. [2] and [51], where the millisecond level lifetime and below tens cm/s surface recombination velocity prevails. Schmidt et al. [59] used a-$Si_{1-x}N_x$ deposited at a temperature 300 °C ≤ T ≤ 400 °C and obtained recombination velocity $S \leq 10$ cm/s. Martin et al. [60] adopted a-SiC_x:H as a c-Si surface passivation layer yielding $S \leq 30$ cm/s. Dauwe et al. [61] achieved a low recombination velocities $S \leq 10$ cm/s by 200–250 °C deposited a-Si:H. Descoeudres et al. [52] adopted a VHF-PECVD (40.68 MHz) to deposit intrinsic silicon film for c-Si surface passivation and obtained a lifetime of 5.9 ms in N-type FZ Si wafer with a resistivity of 4.0 Ωcm. Furthermore, the authors proposed a new indicator of silane depletion fraction for good passivation effect. It was reported that the good interface passivation corresponds to the highly depleted silane plasma. Willem et al. [2] prepared the a-Si:H film using three different CVD techniques, namely, PECVD, VHF-PECVD, and HWCVD for passivation purpose and found that all three deposition methods yield excellent surface passivation with milliseconds level lifetime values.

However, the ICP-grown a-Si:H for c-Si surface passivation is rarely reported [30], not to mention that grown by LFICP.

3.2. c-Si surface passivation by semi-remote LFICP CVD a-/μc-Si:H

It was found that a-/μc-Si films deposited at direct LFICP region (with an antenna-substrate distance $d = 11$ cm) had an unsatisfied passivation effect for c-Si due to the damaged interface by severe ion bombardment. Therefore, the distance must be lengthened in order to yield the high-quality interface and good passivation. In this subsection, the adopted distance is $d = 33$ cm, where a semi-remote plasma is obtained.

3.2.1. Passivation effect of a-/μc-Si:H deposited from hydrogen-diluted silane

In this subsection, a-/μc-Si:H films deposited from only hydrogen-diluted silane at the antenna-substrate distance of d = 33 cm is used to passivate c-Si surface with different resistivity. First, the surface passivation effect of a-/μc-Si:H grown under different discharge powers has been investigated in the case of a higher substrate resistivity of 8 Ωcm. The effect of pre-deposition hydrogen plasma treatment, post thermal annealing, and the correlation between the lifetime and the thickness of the incubation layer at the interface have been analyzed. At this antenna-substrate distance, the resulting films possess fairly good passivation even without any post-deposition thermal annealing.

The a-/μc-Si:H thin films of thickness of 50 nm were deposited on the double-side polished, p-type CZ Si (comparatively higher resistivity of 8 Ωcm, thickness of 600 μm, (100) orientation, and area of about 3 × 3 cm^2). The pre-deposition hydrogen plasma treatment on the substrates was applied for various durations to investigate the effect of hydrogen plasma exposure on the surface recombination. The microstructures and properties of the thicker counterpart of the thin passivation layer were studied in Section 2.1. The effective minority carrier lifetime of the symmetrically passivated wafer was measured by using the quasi-steady state photoconductance decay (QSSPCD) technique on the facility of WCT-120.

3.2.1.1. Effect of pre-deposition H_2 plasma treatment on surface passivation

Prior to the passivation layer deposition, the substrates were treated by H_2 plasma for different durations (0–90 s) to improve the interface properties. The final structure for the minority carrier lifetime measurement is shown in inset of **Figure 22**. All the specific minority carrier lifetimes were recorded at an excess carrier of 10^{15}/cm^3.

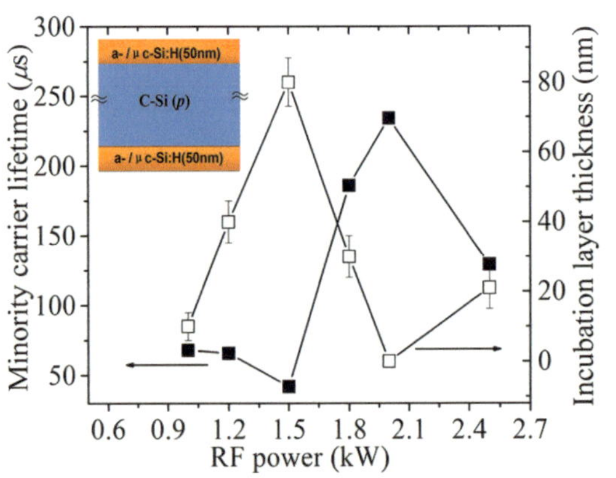

Figure 22. RF power-dependent minority carrier lifetime (filled squares) and incubation layer thickness (open squares). The inset shows the passivation scheme for the lifetime measurement.

τ_{eff} values are plotted in **Figure 23** as a function of H_2 plasma treatment time. τ_{eff} is 127 μs without H_2 plasma treatment. An intermediate treatment time (30–60 s) is beneficial to silicon surface passivation, which is evidenced by the improved lifetime value. However, overlong treatment (>60 s) is detrimental to the surface quality, resulting in a less lifetime value of 73 μs. The maximum value of carrier lifetime ~230 μs was obtained at a time of 30 s.

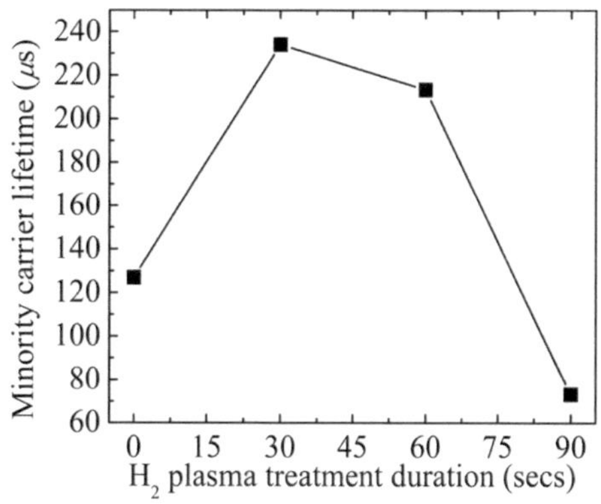

Figure 23. The minority carrier lifetime of the sample exposed to H_2 plasma for different time before the passivation layer deposition.

3.2.1.2. Influence of RF power on surface passivation

The RF power-dependent effective lifetime τ_{eff} is shown in **Figure 22** (filled squares). This curve can be separated into three stages: In the first stage (1.0–1.5 kW), τ_{eff} decreases from 67 to 42 μs; in the second stage (1.5–2.0 kW), τ_{eff} rapidly increases from 42 μs to its maximum of 234 μs; in the third stage (2.0–2.5 kW) τ_{eff} drops down to 129 μs. The observed pattern of τ_{eff} will be explained in terms of the properties of the thin films and the interface between the thin film and silicon substrate.

The passivation effect described by τ_{eff} is not only related to the deposited thin films itself but also the a-Si:H/c-Si interface quality. The low defect density in the thin films is presumably a consequence of hydrogen passivation. As mentioned above, the hydrogen content in these thin films is comparatively larger. However, in the current experiments, the maximum value of τ_{eff} is obtained at RF power of 2.0 kW, where the corresponding thin film contains the least hydrogen as shown in **Figure 5**. Therefore, the change of lifetime τ_{eff} should not simply be attributed to the change of hydrogen content with the RF power.

Another crucial mechanism to determine the passivation effect is the interface defect between the passivation layer and the substrate. It has been observed that during the deposition of a μc-Si:H layer on top of a substrate, a 30–50 nm thick interlayer referred as incubation zone [63] is produced. Its crystallinity and thickness strongly depend on the deposition condition and the substrate material. In our experiments, the incubation layer was directly observed in the high-magnification cross-sectional SEM images of the deposited thin films on double-side-polished Si substrate. **Figure 24** shows the high-magnification cross-sectional SEM images of the thin films deposited at the power of 1.2 (a), 1.5 (b), 1.8 (c) and 2.0 kW (d). The clear scratches and cracks come from the cross-section preparation process. Except the sample deposited at 2.0 kW, evident incubation layers were observed between the crystal Si and the films. Furthermore, the thickness of the incubation layers can be determined from the SEM measurement tool by making measurement multiple times at different locations and the results as a function of RF power are plotted in **Figure 22** (the open squares). The error bars shown in **Figure 22** represent the standard deviations from the mean thickness values from the multiple metering. At a power of 1.5 kW, the thickness has a maximum of 80 ′ 7 nm (>the standard passivation layer thickness of 50 nm), while the lifetime has a minimum of 42 μs. No evident incubation layer was observed at the power of 2.0 kW, at which the greatest minority lifetime value was obtained. Linking the curve of incubation layer thickness with that of the lifetime, one will find that the incubation layer thickness changes with the RF power in an opposite way to the minority carrier lifetime τ_{eff} dose. This observation suggests the fact that the incubation layer on the interface between the passivation layer and the crystal Si will worsen the passivation effect of the a-/μc-Si:H thin films.

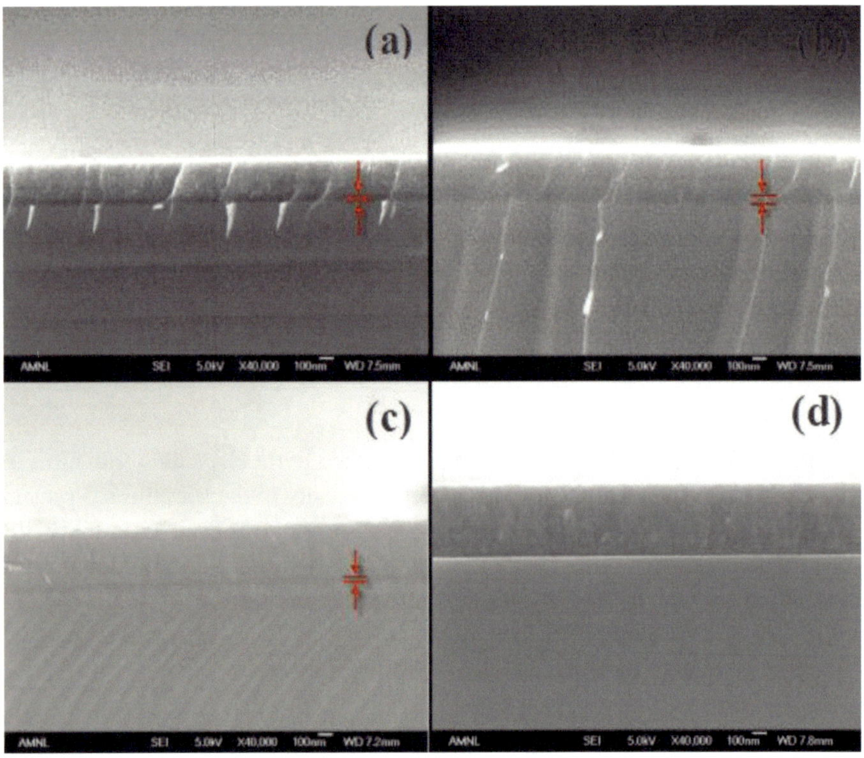

Figure 24. The cross-sectional SEM images of the thin films deposited at RF power of 1.2 (a), 1.5 (b), 1.8 (c), and 2.0 k W (d). Evident incubation layers were observed at the interfaces.

The CVD deposition of Si thin films involves a complex combination of several processes including the arrival and removal of gas molecules or precursors at the substrate surface, the decomposition into reactive species, and the migration of these species on the surface where they can lead to nucleation and continued deposition [64]. The adsorbed reactive species are likely to come to rest when a position of minimum energy position is found. The minimum energy position can be a defect at the substrate surface, which will result in a new nucleus, or an existing nucleus [65]. At certain process conditions, the formation of nuclei will be delayed for some time known as the incubation time [64]. From the point view of the growth mechanism, the incubation layer (initial growth) will be rich of defects, which will lead to the intense recombination of the photon-induced carriers and thus a decreasing lifetime value observed in **Figure 22**.

3.2.1.3. Influence of thermal annealing on surface passivation

In order to investigate the influence of the post-deposition annealing on the silicon surface, the samples were thermally annealed in vacuum and H_2 flow atmosphere. For simplicity, one sample with an as-deposited lifetime value of 196 μs was annealed just in vacuum in the temperature range of 230–500 °C. **Figure 25(a)** shows the evolution of lifetime τ_{eff} with annealing temperatures. The annealing duration at each temperature point is 30 min. One can see that the lifetime increases almost linearly in the range of 230–420 °C. Further increasing of temperature to 500 °C results in a significant decrease in lifetime value. The relatively low temperature (<500 °C) annealing in vacuum leads to the lifetime improvement by a factor of about 2.5. The corresponding FTIR transmission spectra of the sample before and after thermal annealing were measured in the wavenumber range of 550–950 cm^{-1} to clarify the H role in the surface passivation. The spectra of the as-deposited and the 420 and 500 °C annealed samples are shown in **Figure 25(b)**. The Si-H wagging/rocking mode at about 630 cm^{-1} slightly changes with temperature increasing from 230 to 420 °C. In comparison, this mode considerably recedes when temperature is increased from 420 to 500 °C, which is indicative of lower hydrogen content after annealing. The mode around 850–900 cm^{-1} also becomes weak after annealing at a temperature of 420 and 500 °C.

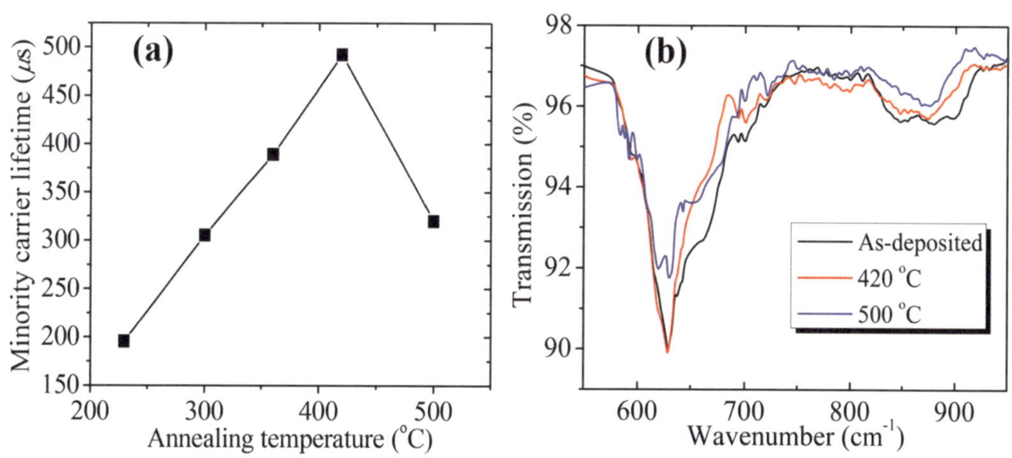

Figure 25. The minority carrier lifetime of the sample annealed at different temperatures in vacuum (a), and FTIR transmission spectra (550–950 cm^{-1}) of the as-deposited sample and that annealed at 420 and 500 °C (b). Increase in the annealing temperature from 420 to 500 °C leads to the decreasing hydrogen content in the film.

The vacuum annealing results discussed above represent the important role of the hydrogen atom in the Si surface passivation. Low-temperature (<420 °C) annealing enables the diffusion of hydrogen in the thin film towards the Si substrate, which effectively reduces the dangling bond density at the interface and enhances the minority carrier lifetime. At higher temperatures (>420 °C), hydrogen atom effuses leading to the re-generation of the dangling bonds and thus a decrease in the lifetime value. Systematic annealing experiments were performed on the samples deposited at different RF powers, and the ultimate lifetime values (not shown here) presented the same trend with that of the as-deposited samples shown in **Figure 22** (filled squares). These results show that the as-deposited interface features, especially the incubation layer properties discussed above, are the crucial factors in determining the ultimate lifetime values in our case. Therefore, we will focus on the sample with the highest as-deposited lifetime value. It is worth stressing that the ultimate lifetime value is dependent on the annealing circumstance. **Figure 26** exhibits the lifetime value evolution with annealing time at 420 °C in vacuum (filled squares), H_2 flow (filled circles), and N_2 flow. In the case of vacuum annealing, the lifetime value increases sharply in the 95 min and saturates at the value of 421 μs at about 185 min. On the contrary, the lifetime reaches the maximum of 524 μs in a short time of 11 min in the case of H_2 flow. It seems that the presence of hydrogen in the annealing circumstance is essential to improve surface passivation by present a-Si:H/μc-Si:H thin films, despite the presence of rich hydrogen in the film as evidenced by the FTIR result. With N_2 flow, lifetime reaches its maximum value of 425 μs in 5 min, followed by a drastic drop for time longer than 5 mins. The H_2 preference is confirmed by multiple annealing experiments. It is well known that N_2 is the commonly used protective gas for post-deposition annealing of passivation layer a-Si. But H_2 circumstance annealing is indeed better than N_2 in our experiments. The underlying mechanism needs a further study to clarify.

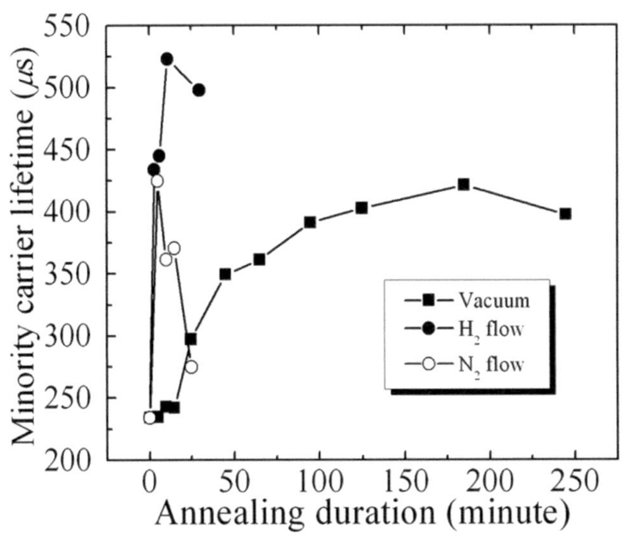

Figure 26. The minority carrier lifetime as a function of annealing duration at 420 °C at an atmosphere of vacuum (filled squares), H_2 flow (filled circles), and N_2 flow (open circles).

http://dx.doi.org/10.5772/63529

3.2.1.4. Injection level-dependent surface recombination

The effective minority carrier lifetime τ_{eff} is related to the effective surface recombination velocity S_{eff} through Equation 5. By neglecting the bulk lifetime ($\tau_b \to \infty$, we can get the maximum value of the surface velocity from this equation. The injection level-dependent S_{eff} with and without annealing is plotted in **Figure 27**. Comparing the results of the as-deposited and annealed, one can see that the low-temperature annealing leads to the decreasing of the surface recombination, especially in the range of the low injection levels. The as-deposited sample has a surface recombination of about 150 cm/s. The annealing at 230 °C for 11 min results in the decrease of S_{eff} above the injection level of 1 x 15/cm^3 and increase of S_{eff} below the same level. This abnormal increase in S_{eff} possibly stems from the excitation of the adsorbed atoms in the circumstance, which acts as the additional recombination centers to deteriorate passivation effect. Subsequent annealing at higher temperatures causes the continuous decreasing towards the injection level of 2.8 x 16/cm^3, where the Auger recombination plays a dominant role and the curves tend to converge. At the temperature of 420 °C, the lowest surface recombination velocity is 60 cm/s at the injection level of 1 x 15/cm^3. It should be emphasized that this S_{eff} value is the upper limit because of the adopting of an infinity assignment for the bulk lifetime.

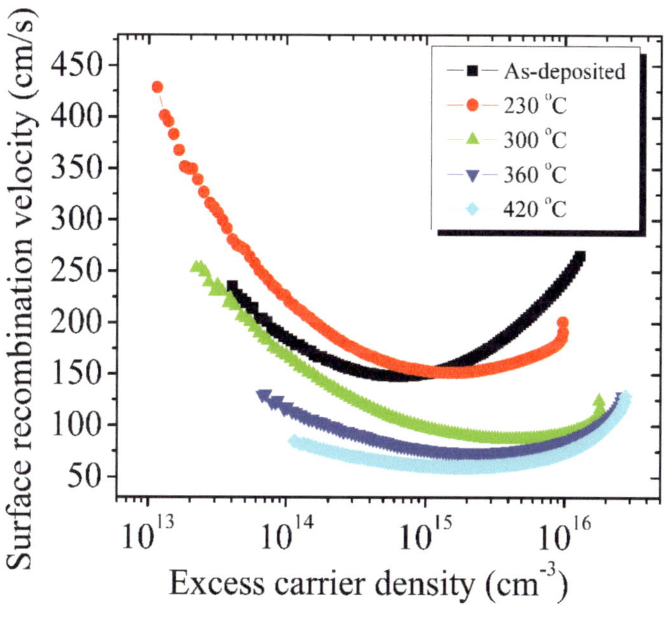

Figure 27. Injection level-dependent surface recombination velocity of the as-deposited sample and that annealed at different temperatures.

3.2.1.5. The high as-deposited lifetime values in the case of semi-remote LFICP

The present as-deposited lifetime value of 234 μs is much higher than that reported in literatures, where PECVD or HWCD methods were used to deposit the passivation a-/μc-Si-H layer. Jan-Willem et al. [62] summarized the passivation effects of a-Si:H deposited by various methods: 34 μs for PECVD, 43 μs for VHF PECVD, and 87 μs for HWCVD. In principle, passivation can be physically divided into two types, i.e., hydrogen-related chemical passivation and built-in potential related field passivation. In the former, defect density can be remarkably lowered due to the reduced number of dangling bonds which is terminated by hydrogen atoms. With regard to the filed passivation, a-/μc-Si:H has a wider band gap than crystal silicon substrate, which results in a barrier on the interface preventing the recombination of photogenerated carrier pairs through the surface/interface defects.

It has been shown that the pre-deposition hydrogen plasma exposure significantly influences the surface passivation (see **Figure 23**). In the post-deposition annealing, the interaction between hydrogen atom from the passivation layer and the c-Si substrate is crucial to the surface passivation. Mitchell et al. [66] studied the thermal activation of the hydrogen-related chemical passivation on silicon substrate and determined the thermal activation energy of E_A through the following formula:

$$1/\tau_{reac} = A\exp(-E_A/k_BT) \tag{6}$$

where τ_{reac} is the reaction time between hydrogen atoms and silicon atoms, E_A the activation energy of surface passivation, k_B the Boltzmann constant, and T the temperature. According to Equation (6), a low E_A value will point to a high reaction rate $1/\tau_{reac}$ or a short reaction time τ_{reac}. In earlier works [33, 39], we analyzed the effect of high-density hydrogen plasma on the crystallization mechanism of LFICP-grown μc-Si:H. Abundant atomic hydrogen resulting from the dissociation of SiH_4 and H_2 will result in high surface coverage by hydrogen, which reduces the activation energy E_A and the reaction time τ_{reac}. In addition, high-density excited hydrogen atoms will recombine with the precursor of Si-H in the way described by the reaction H + Si-H→Si- + H_2 (with the dangling bond Si-). This recombination process can produce effective local plasma heating on the growing surface, which will contribute to the hydrogen atom transport to the interface and bulk silicon and a considerable lifetime value even without any additional thermal annealing.

3.2.1.6. Incubation layer and its thickness control

Figure 28(a) [**(b)**] shows a low- (high-) magnification cross-sectional TEM micrograph of LFICP-grown μc-Si:H on silicon substrate. In **Figure 28(a)**, a clear incubation layer (dark area) between the film and substrate is discriminated like in SEM images. **Figure 28(b)** exhibits more details of the interface. The grey area in the film corresponds to the microcrystalline Si fractions, which are conical conglomerates of microcrystals. As observed above, the incubation layer is detrimental for the improvement of lifetime and its thickness is highly dependent on the process parameters, e.g., the applied RF power in present experiments. Therefore, a deep understanding on the mechanism to control the formation of incubation layer is extremely important. It was recognized that the ion bombardment played a crucial role in the formation of incubation layer [67]. The ion beam works in two possible ways, namely (i) a surface effect: very small grains do not survive the high dose of ion irradiation, thereby reducing the nucleation rate, (ii) a bulk effect: the ion beam induces defect-related grains growth. Accordingly, two strategies [67] to control the ion beam were proposed: the choice of the frequency of the power source generating the plasma, and the design of the electrode configuration to avoid its formation. In present antenna-substrate distance $d = 33$ cm, most of the ions/electrons energy is around several eVs, below the threshold value of 16 eV for defect formation for Si. Therefore, the lengthened distance d makes it possible to minimize the effect of the ion bombardment to an acceptable level.

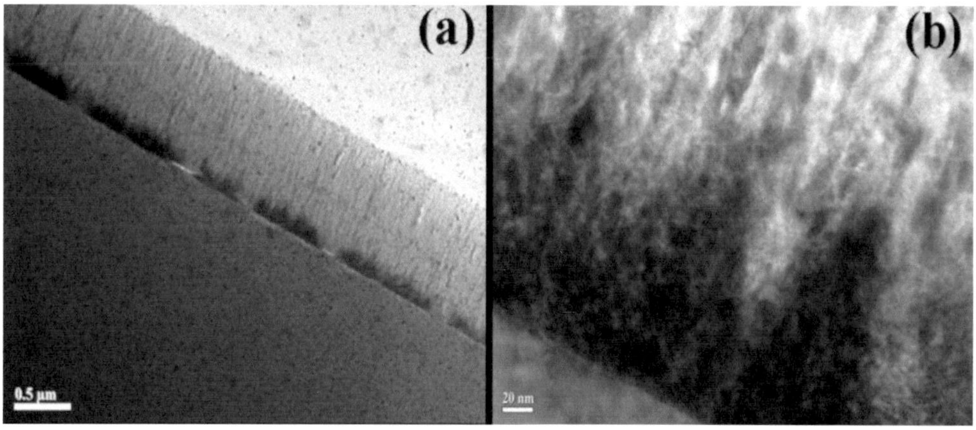

Figure 28. A typical cross-sectional TEM micrograph of LFICP-grown μc-Si:H films on Si substrate at low (a) and high (b) magnification.

From the point of view of a growth mechanism, the key to diminish the initial growth of incubation layer seems to be the rapid nucleation without retard on the interface. In present experiment, this occurs at the power of 2.0 kW, near the transition point (1.8 kW) from a-Si:H to μc-Si:H. This is consistent with some authors' findings that the good passivation effect is obtained with microcrystalline-oriented passivation layers, where the depletion fraction of SiH_4 is comparatively higher than that of amorphous regimes [52, 68]. In the current study, although the silane depletion was not experimentally measured at certain RF power, the high depletion fraction at the power of 2.0 kW is expected. The detailed relation between the depletion fraction and the incubation time in the growth of a-/μc-Si:H needs to be further clarified.

3.2.1.7. Passivation of low resistivity c-Si

It is more difficult to passivate a lower-resistivity Si surface due to the defect associated with more recombination. However, the lower-resistivity wafer is more meaningful in practical solar cell application. In this part, two types of wafers have been investigated: p-type (100), 2 Ωcm, and n-type (100), 3 Ωcm wafers. The process parameters such as discharge power, substrate temperature, dilution ratio, and post-deposition annealing have been widely explored to improve their surface passivation. **Figure 29** shows the effective minority carrier lifetime τ_{eff} in the n-type 3 Ωcm c-Si passivated with a-Si:H deposited at different substrate temperatures. τ_{eff} drastically drops from 164 μs to 25 μs with increasing substrate temperature from room temperature to 140 °C. This is possibly originated from the local epitaxial growth or reduced hydrogen content with increasing temperature.

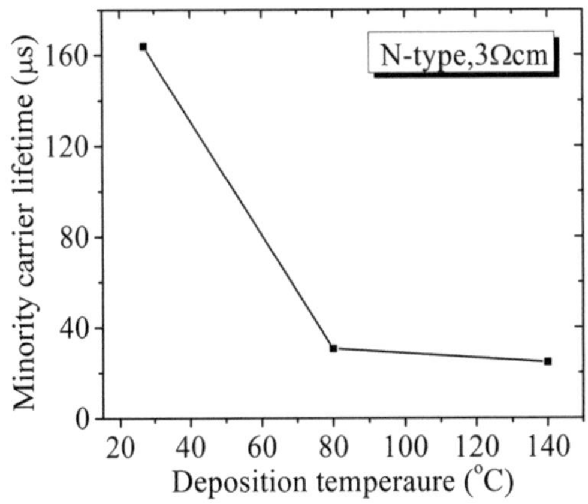

Figure 29. As-deposited lifetime values as a function of the substrate temperature. The other parameters are: SiH_4:H_2 = 12.5:12.5 (in sccm), discharge power 1.5 kW.

Figure 30 displays τ_{eff} in both types c-Si passivated by a-Si:H deposited from hydrogen-diluted silane with different dilution ratios (H_2/SiH_4). The discharge power and substrate temperature were kept at 2.0 kW and room temperature, respectively. The flow rate of SiH_4 and H_2 is 20:0, 12.5:12.5, 8.3:16.7, 5:20, corresponding to the hydrogen dilution ratio of 0, 1, 2, and 4, respectively. τ_{eff} reaches its maximum values at the dilution ratio of 1 in both p- and n-type c-Si. Based on the above analysis on passivation, the higher depletion of silane and reduced incubation layer are expected. **Figure 31** shows the discharge power-dependent lifetime values, which reached its maximum at RF power of 2.5 kW. Above 2.5 kW, τ_{eff} begins to drop again.

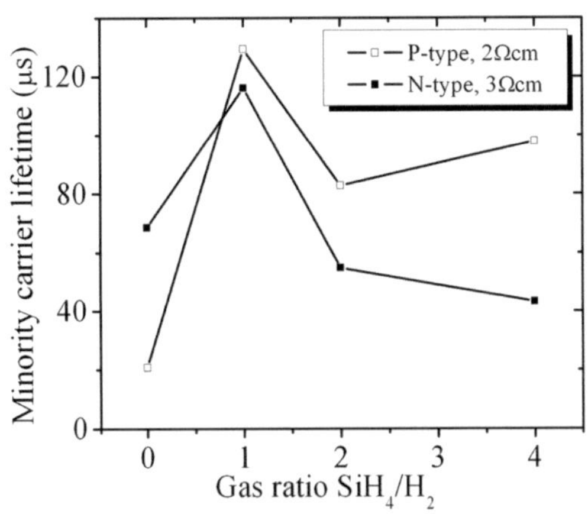

Figure 30. As-deposited lifetime values in both type c-Si passivated by a-Si:H grown from hydrogen-diluted silane with different dilution ratios.

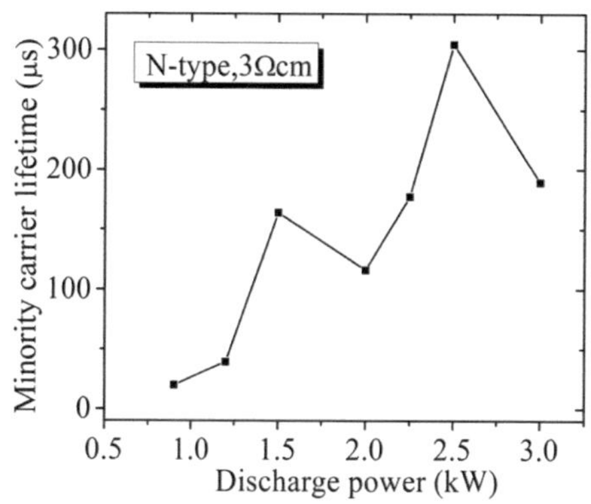

Figure 31. As-deposited lifetime values in N-type 3 Ωcm c-Si passivated by a-Si:H grown under different RF powers.

The optimized lifetime results of these two types of low-resistivity c-Si have been shown in **Figure 32**. At the specific excess carrier density of $10^{15}/cm^3$, the lifetime value is 305 and 150 μs for the n-type 3 Ωcm and p-type 2 Ωcm c-Si, respectively. It should be noted that these results are as-deposited values without post-deposition annealing. Unfortunately, post-deposition annealing cannot improve the lifetime values anymore in the present low-resistivity c-Si substrates. The hydrogen chemical annealing mentioned above has already raised the lifetime values to its maximum during the deposition process.

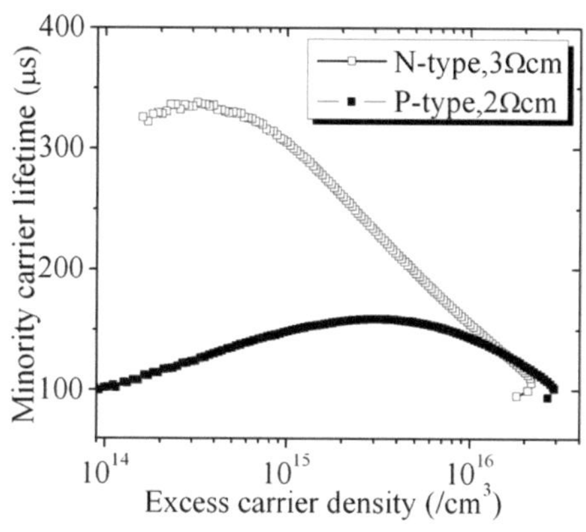

Figure 32. Injection level-dependent minority carrier lifetime in two types of low-resistivity c-Si wafers: p-type, 2 Ωcm and n-type, 3 Ωcm.

Although the as-deposited values without post-deposition long time annealing are much bigger than that reported in references, the present lifetime values are still lower than the ms level results in the literatures. The plasma density in the growth region has already been reduced a lot by increasing the antenna-substrate distance *d* to 33 cm. The semi-remote LFICP grown a-/μc-Si provides a very high as-deposited lifetime value in c-Si (low and high resistivity). In turn, these high as-deposited lifetime values reflect the still high-density plasma in the present growth area, which provides effective chemical annealing during the growth process. On the other hand, the high-density plasma makes the growth surface exposed to a pronounced ion bombardment, which will damage the interface between the substrate and the passivation layer. In terms of process parameters, the high-discharge power points to the high silane depletion. Meanwhile, the ion bombardment on the substrate is inevitably enhanced under the increasing discharge power. In this sense, the key to improve the passivation effect lies in how to balance the silane depletion and ion bombardment on the interface.

3.2.2. *Introduction of CO_2 into the reactive gas*

a-Si:H is easy for (partially) epitaxial growth on c-Si even at low deposition temperatures, which will severely reduce the carrier lifetime values and thus the photovoltaic performance of HJ solar cell [56]. Incorporation of O into the amorphous silicon network can effectively suppress the epitaxial growth [69]. In the aspect of optics, the inherent blue light loss in a-Si:H layer also restricts the solar cell optimization. As a consequence, wide-banded Si-based materials become a promising candidate. Thermally grown SiO_2 provides a state-of-art level of surface passivation with a reduced surface recombination velocity as low as 2 cm/s [54]. However, the thermally grown SiO_2 suffers from the inherent disadvantages such as extremely high process temperature ((1000 °C) and low deposition rate. As a consequence, non-stoichiometric silicon oxide (SiO_x) based on non-thermal deposition methods have been proposed as an alternative to SiO_2. Apart from conventional PECVD and ESR, VHF-PECVD (13.56–110 MHz) has also been used to grow SiO_x:H for c-Si passivation [70, 71]. Mueller et al. [71] investigated the RF frequency effect on the passivation behavior for n-type FZ silicon wafer (resistivity (0.5 Ωcm) and obtained the minority carrier lifetime of 480 μs at the optimum frequency of 70 MHz. Hydrogen-diluted SiH_4 acts as the feedstock gas, whereas the oxygen source varies from oxygen gas [72] to various composites such us N_2O [73] and CO_2 [74].

In this section, CO_2 is introduced into the reactive gases of SiH_4+H_2 and efficiently dissociated in the LFICP system. The discharge power is set at 1.8 kW, flow rate of both SiH_4 and H_2 are 12.5 sccm, and no external heating is applied to the substrate. Various flow rates of CO_2 are introduced into the deposition chamber. The microstructure evolution with the introduction of CO_2 in the obtained films is investigated. Two types of c-Si wafers: n-type, 3 Ωcm and p-type, 2 Ωcm, are passivated by these films. The underlying passivation mechanism is analyzed.

Figure 33 shows E_g and deposition rate as functions of CO_2 flow rate. The incorporation of CO_2 into the reactive gases results in the effective widening of band gap E_g and the pronounced decrease in the deposition rate. E_g increases by about 0.3 eV with CO_2 increased from 0 to 12.5 sccm. According to Watanabe [75], SiO:H is a two-phase material with an island of SiO in a matrix of a-Si:H. In their two-phase model, the oxygen-rich phase is effective in increasing E_g and the Si-rich phase contributes to high conductivity. The widened E_g is beneficial to the light transmittance through the window layer when SiO_x:H is used as the window layer. Actually, Sritharathikhun et al. has succeeded in fabricating SiO_x:H HIT solar cell on n-type [76] and p-type [77] silicon substrate and obtained the competitive photovoltaic performance efficiency of 17.9% for n-type substrate, and 15.3% for p-type substrate.

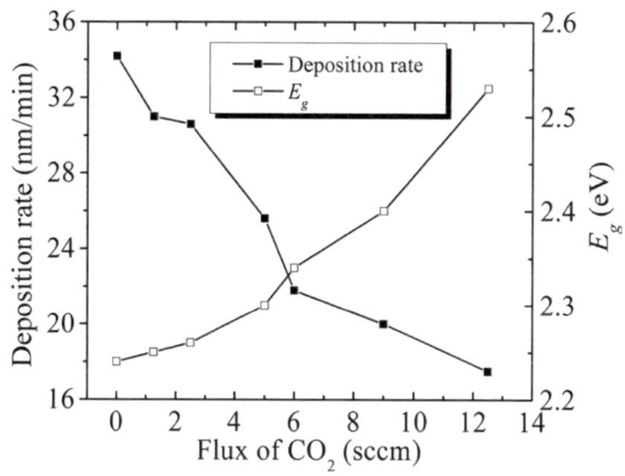

Figure 33. Deposition rate and optical band gap E_g evolution with CO_2 flow rate.

Figure 34 shows the FTIR absorbance spectra of the corresponding films. Besides the Si-H waging/rocking signal ((640 cm^{-1}) and Si-H stretching signal ((2100 cm^{-1}) usually observed in a-Si:H, some oxygen and carbon-associated modes appear. The Si-C stretching mode, which couples with the doublet signal of $(SiH_2)_x$ ((845–890 cm^{-1}) in the case of low CO_2 flow rate, grows stronger, and eventually dominates the latter. The Si-O-Si stretching mode located around 1100 cm^{-1} gradually shifts from 1107 to 1120 cm^{-1}. The former and latter are assigned to the nature of SiO_2 and SiO_x, respectively [78]. At minor CO_2 addition, the distinct peak at 1107 cm^{-1} is possibly due to the remaining oxygen in the deposition chamber. The consecutive increase of CO_2 makes the structure change from SiO_2 to SiO_x. Another effect of CO_2 addition is the shift of the Si-H stretching signal, indicating the incorporation of O into the Si-H bond. In literatures, carbon was regarded as being excluded because of the formation of volatile CO. However, the FTIR method tracks carbon in the form of Si-C bond in our experiment. Therefore, the obtained films can be strictly denoted as $SiC_{1-x}O_x$:H.

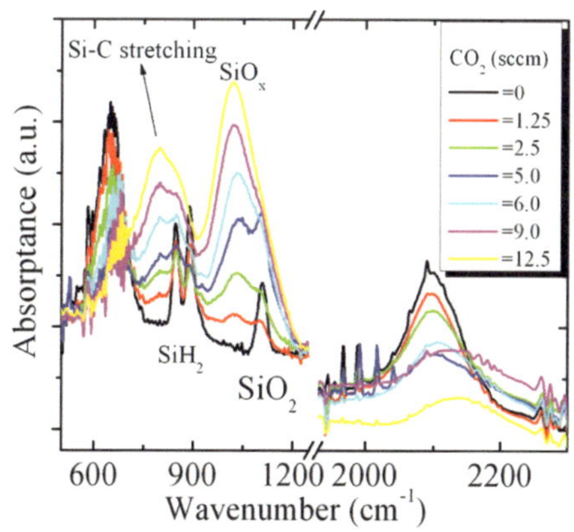

Figure 34. FTIR absorptance spectra of the deposited films with varying CO_2 flow rates.

About 60 nm thick $SiC_{1-x}O_x$:H films were deposited on both sides of the two types of c-Si substrate and the surface passivation was examined by the method of steady-state photoconductance decay method described above. **Figure 35** shows the effective lifetime values of these two substrates passivated by the $SiC_{1-x}O_x$:H films deposited with different flux of CO_2. At first glance, τ_{eff} degrades with the increasing addition of CO_2. This should be correlated to the increasing defect from the introduction of C and O into the network. However, incorporation of minor CO_2 is beneficial to the surface passivation. As an example, lifetime τ_{eff} has a highest value at flux of 2.5 sccm (1.25 sccm) in n-type (p-type) c-Si. The optimized injection level-dependent lifetime values before and after thermal annealing (275 °C, 30 min) have been presented in **Figure 36**. The obtained best results for n-type 3 Ωcm and p-type 2Ωcm are 313 and 166 μs at the specific carrier density of $10^{15}/cm^3$, respectively.

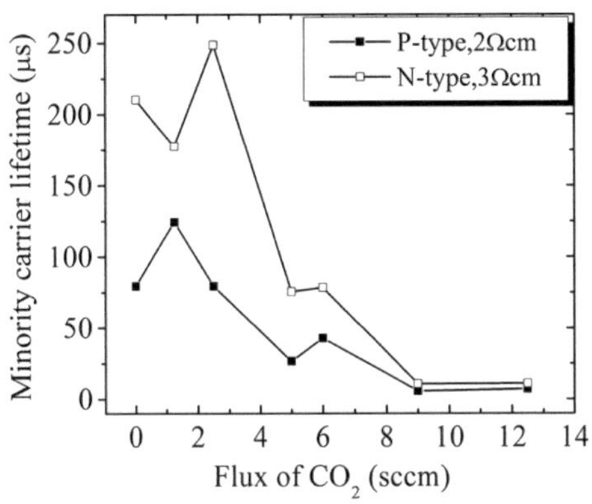

Figure 35. The as-deposited lifetime values in two types of c-Si substrates passivated by $SiC_{1-x}O_x$:H deposited with different flux of CO_2. The lifetime values are taken at the specific excess carrier density of $10^{15}/cm^3$.

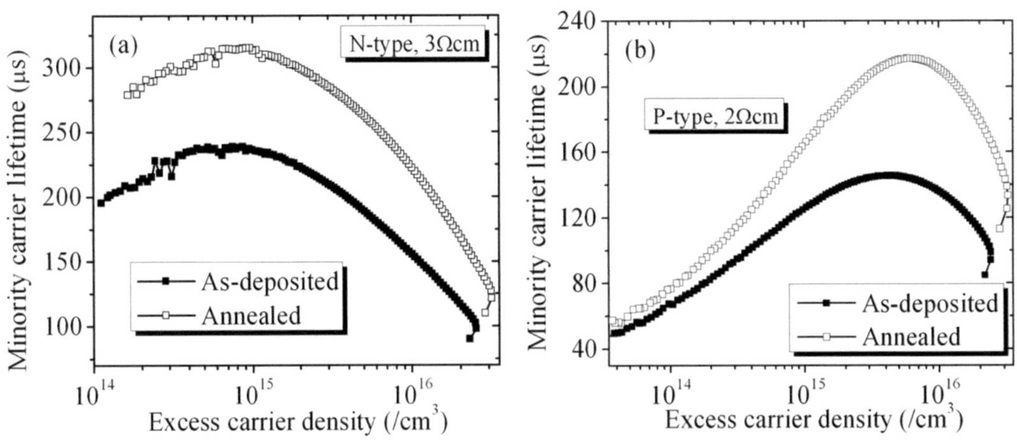

Figure 36. Injection level-dependent lifetime values in the substrate of n-type 3 Ωcm (a) and p-type 2 Ωcm (b) before and after the thermal annealing. The fluxes of CO_2 are 2.5 and 1.25 sccm in (a) and (b), respectively.

In order to clarify the passivation mechanism for the LFICP-grown $SiC_{1-x}O_x$:H, high-frequency (1MHz) C-V measurements were performed on a metal-semiconductor-insulator (MIS) structure shown in the inset of **Figure 37**. The insulator layer is $SiC_{1-x}O_x$:H material. The front contact is a small Al dot with a diameter about 0.5 mm and the rear contact is sheet Al. Al electrode was deposited by the method of magnetron sputtering. The amplitude of the AC signal was 10 mV. The work function difference of Al and silicon was kept at −0.85 V [79]. The recorded C-V curve is shown in **Figure 37**, from which a fixed charge density of $-2.7\times10^{11}/cm^2$ was deduced. However, this level is not high enough to contribute to the field passivation, which is usually accompanied by a fixed charge density of the order of $10^{12}/cm^2$ [74]. Therefore, the passivation effect of the present $SiC_{1-x}O_x$:H mainly stems from the chemical passivation like in a-Si:H case.

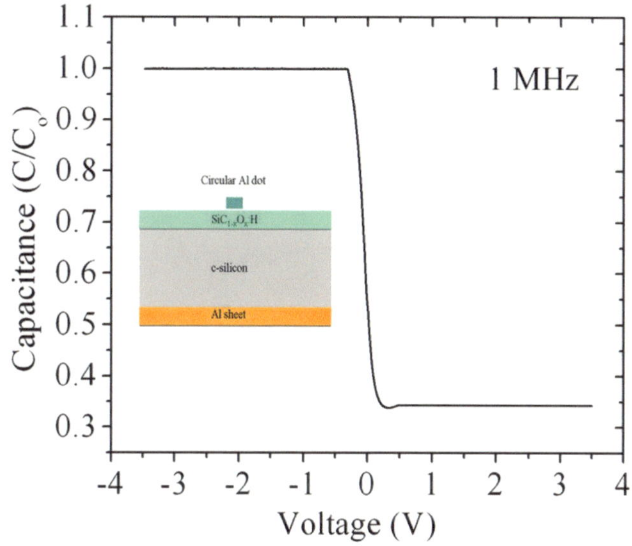

Figure 37. The C-V curve of the metal-semiconductor-insulator (MIS) structure, from which the fixed charge in the insulator can be deduced.

3.3. c-Si surface passivation by remote LFICP CVD a-Si:H

In subsection 3.2.2, reactive gas CO_2 was introduced into SiH_4+H_2 to improve the c-Si surface passivation. The introduction of CO_2 can widen the optical band gap of the passivation layer, and simultaneously suppress the partial epitaxial growth of the passivation layer on c-Si. The passivated sample exhibits high as-deposited lifetime values, whereas the lifetime values cannot be enhanced effectively after post-deposition annealing. It is believed that the frequent ion bombardment from the high-density plasma should be responsible for these two effects: it conveys energy to the growing surface acting as plasma annealing, and possibly results in damages to the interface. The damaged interface will define the hydrogen diffusion to the interface when annealing and results in a limited enhancement in lifetime. One possible solution is to keep the substrate away from the plasma generation space. In this section, the antenna-substrate distance *d* is further enlarged to *d* = 53 cm to form remote plasma, and thus further decrease the ion bombardment on the surface.

Figure 38 shows the lifetime τ_{eff} curves before and after the thermal annealing in 275 °C for different durations. In **Figure 38(a)**, the as-deposited τ_{eff} value is 3 μs, and τ_{eff} surges upon annealing. The similar behavior is observed in the case of the n-type 4 Ωcm c-Si. The saturation values of 295 and 716 μs are achieved in p-type 1 Ωcm and p-type 4 Ωcm c-Si, respectively. The lifetime values τ_{eff} reaches their maximums at annealing duration of 45 min. The present annealing behavior is different from that in the case of antenna-substrate distance *d* = 33 cm. Herein, τ_{eff} is improved by more than two orders of magnitude after thermal annealing. This fact supports the assumption of the ion bombardment effect accompanying with a lower distance *d* value.

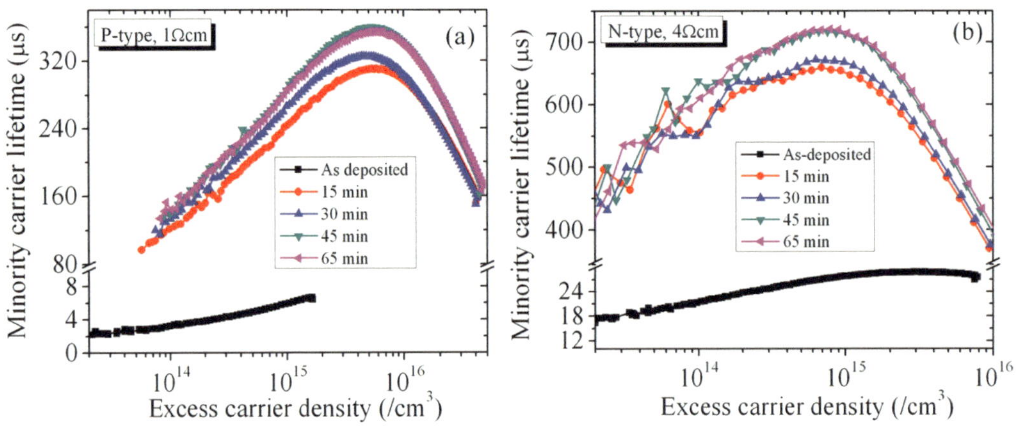

Figure 38. Injection level-dependent lifetime values in the p-type 1 Ωcm (a) and n-type 3 Ω cm (b) c-Si, before and after the thermal annealing. The thermal annealing was conducted at 275 °C in N_2 atmosphere for different durations.

In a narrow tunable working pressure range, we have compared the lifetime curve of the passivated samples under different pressures. **Figure 39** shows the comparison between the two discharge pressures of 0.2 and 1.0 Pa. The lifetime values are measured in the passivated c-Si substrates (n-type 4 Ωcm) before and after the thermal annealing treatment. The as-deposited lifetime values at 0.2 Pa are much greater than that at 1.0 Pa. Nevertheless, the vice verse is true after the thermal annealing. As mentioned above, a higher pressure means a smaller mean free path of the particles and a higher deposition rate. A faster growth can minimize the partial epitaxial growth of a-Si:H on c-Si surface. In addition, the lower-pressure mode possibly still causes some surface defects which limit the beneficial effect of thermal annealing.

http://dx.doi.org/10.5772/63529

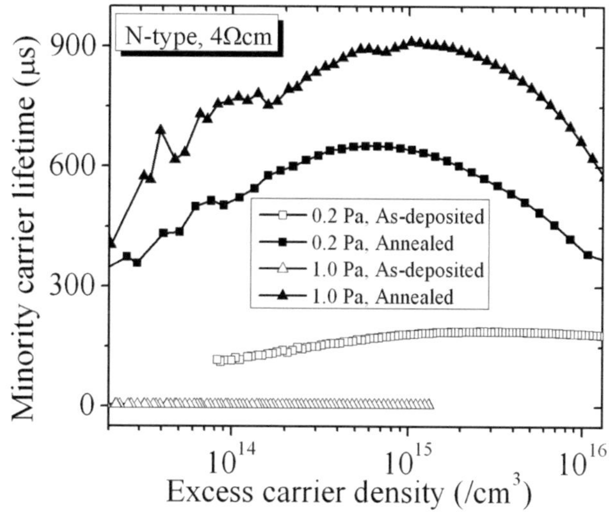

Figure 39. Injection level-dependent lifetime curves of the n-type 4 Ωcm c-Si passivated by a-Si:H deposited under different working pressures.

Figure 40 shows the lifetime curves in two types of c-Si substrates passivated by a-Si:H deposited under two temperature points of RT and 140 °C. For the p-type 2 Ωcm c-Si substrate, substrate temperature does not strongly influence the lifetime curve. In comparison, increasing temperature from RT to 140 °C deteriorates the lifetime value severely in n-type 4 Ωcm c-Si substrate. This difference is possibly from the different conductivity values which strongly influence temperature-dependent hydrogen diffusion on the surface or at the interface. The detailed mechanism needs further study to clarify.

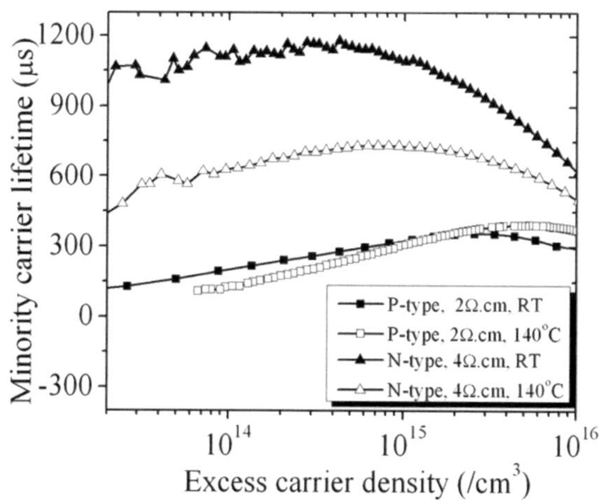

Figure 40. Injection level-dependent lifetime curve of two types of c-Si substrates passivated by a-Si:H deposited under different substrate temperatures.

Figure 41 shows the best lifetime results we obtained in three types of c-Si substrates. At the excess carrier density of $10^{15}/cm^3$, the best lifetime values are 1.3 ms, 620 μs, and 290 μs for the n-type 4 Ωcm, p-type 2 Ωcm, and p-type 1 Ωcm c-Si substrates, respectively.

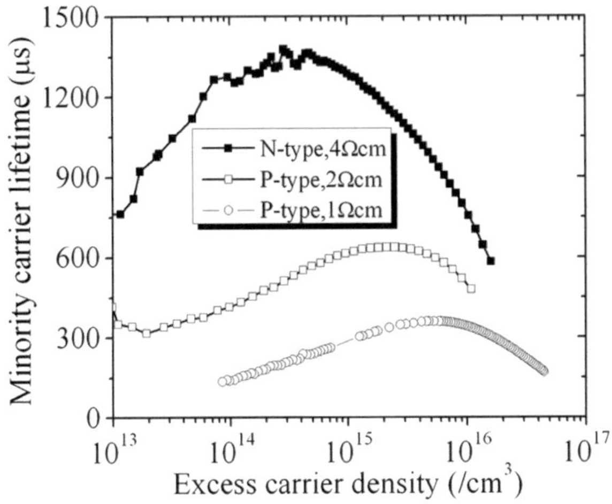

Figure 41. The best lifetime curves obtained in three types of c-Si substrates passivated by a-Si:H deposited in remote LFICP with an antenna-substrate distance of 53 cm.

3.4. c-Si passivation by stack configuration and implied V_{oc}

It is well known that the passivation effect is strongly dependent on the passivation layer thickness. In a practical HIT solar cell, the total thickness of deposited functional layer on one side is several tens of nanometers, less than the commonly used thickness of 50 nm in passivation studies. In addition, in a practical HIT solar cell, the passivation configuration is a stack of intrinsic and doped a-Si:H layers instead of above only intrinsic layer. Therefore, it is important to study the stack layer passivation configuration for the HIT solar cell applications. The passivation effect associated open-circuit voltage, namely, the implied V_{oc}, can be also determined from the QSSPC method. Using this method, we can estimate V_{oc} of the solar cell precursor (without the metallization procedure).

Intrinsic and doped a-Si:H films were symmetrically deposited on the p-type 1 Ωcm c-Si substrates to form the passivation configurations of n/i/P-c-Si/i/n or p/i/P-c-Si/i/p. This stack passivation scheme combines the chemical passivation (by intrinsic a-Si:H) and the field passivation (by doped a-Si:H). As an example, **Figure 42** shows the field passivation by p-a-Si:H in (a) and the combination of both passivation mechanisms in (b). Addition of a p-doped layer introduces an additional electrical filed to hinder the minority diffusion to the surface. The barrier Φ_{BSF} exponentially reduces the concentration of minority at the interface and thus the surface recombination. In **Figure 42(b)**, the interface defect density is reduced by the chemical passivation of addition of an intrinsic layer. Then, the solar cell precursor with the structure of n/i/P/i/p was fabricated to estimate the lifetime values. In these structures, the intrinsic and doped layer thicknesses are about 5 and 15 nm, respectively, estimated from the average deposition rate at corresponding deposition conditions. **Figure 43** shows the lifetime curves in different passivation configurations. Lifetime value τ_{eff} in p/i/P/i/p is much lower than that in n/i/P/i/n. It is reported that B dopant is prone to react with Si atom to form complex, increasing the interface defect density [80].

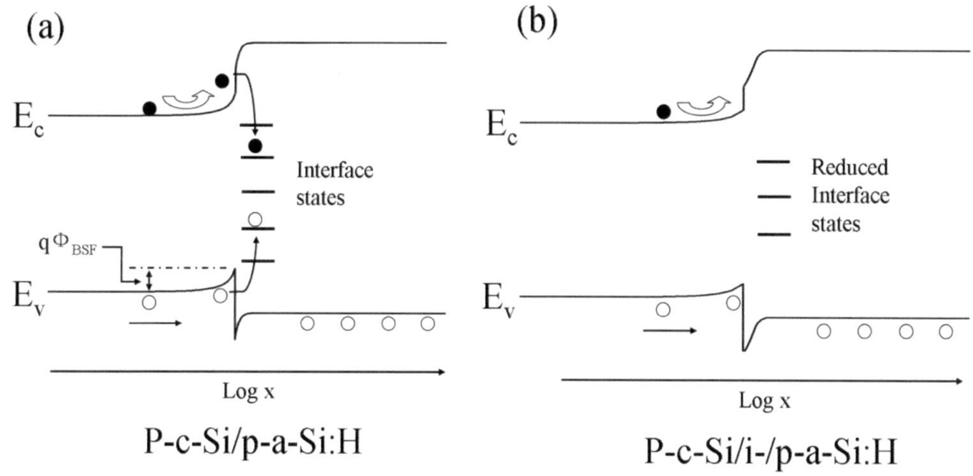

Figure 42. The filed passivation created by p-doped a-Si:H (a), and the combination of chemical surface passivation and field passivation in a stack of an intrinsic and p-doped layer (b).

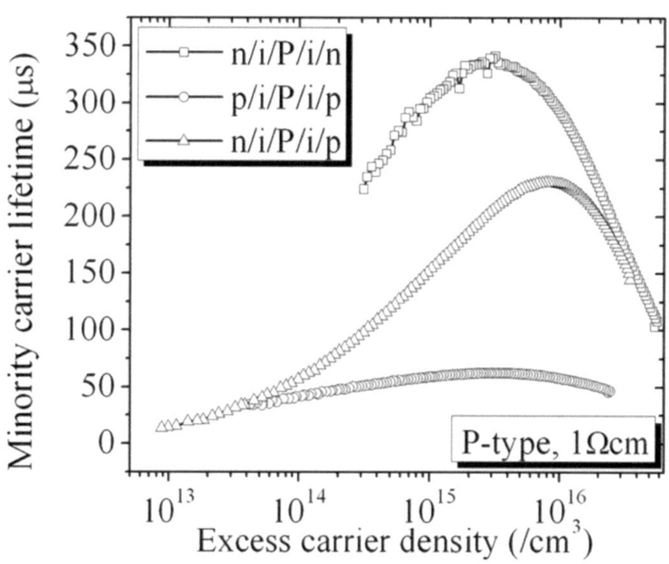

Figure 43. The injection level-dependent lifetime values in the p-type 1 Ωcm c-Si passivated by symmetric and asymmetric structures consisting of intrinsic and doped a-Si:H.

The relation between V_{oc} and effective lifetime τ_{eff} can be expressed as [81]:

$$V_{oc} = \frac{kT}{q} \ln \frac{2J_{ph}\tau_{eff}(N_A + \Delta n)}{qn_i^2 W} \tag{7}$$

where J_{ph} is the photocurrent density, N_A the background dopant density, and n_i the intrinsic carrier concentration in silicon. As expected, a high lifetime value corresponds to a high V_{oc}. In our p-type 1 Ωcm Si (thickness of 300 μm), the implied V_{oc} is about 670 and 620 mV corresponding to the solar cell precursor (n/i/P/i/p) lifetime of 150 and 64 μs, respectively.

4. Conclusions

As a low-temperature plasma source, LFICP has the inherent advantages of high-density plasma, low sheath potential, and low electron temperature, etc., which deserves an excellent CVD route. In LFICP CVD, the parameters of the RF power, hydrogen dilution, working pressure, and substrate temperature significantly influence the microstructures and properties of the films. The ion bombardment on the growing surface in LFICP CVD can be effectively controlled by the parameter of antenna-substrate distance *d*, providing an additional way to tune the microstructures and properties of Si-based materials.

The passivation effects of LFICP CVD Si-based materials are strongly dependent on the plasma configuration. In a region not far away the plasma generation area (semi-remote LFICP), the passivated c-Si shows a high as-deposited carrier lifetime due to the hydrogen-related chemical annealing during the deposition process. However, the additional thermal annealing does not lead to a considerable increase in the lifetime value. The addition of CO_2 into the reactive gases substantially influences the microstructure of the obtained thin films and improves the interface quality. In the region far away from plasma generation region (with a high antenna-substrate distance, remote-LFICP), a-Si:H passivated c-Si shows a low as-deposited and a high post-deposition annealing lifetime value. The highest lifetime value of 1.3 ms after thermal annealing is obtained in passivated 4 Ωcm c-Si substrate. After being passivated by the device-like scheme of undoped/doped silicon film stack, p-type substrate (1 Ωcm, 300 μm) shows an implied open voltage of 670 mV.

Acknowledgements

This work was jointly supported by AcRF Tier1 (No. RP 6/13 XS), A*STAR, Singapore, and the National Science Foundation of China (Grant No. 51302028 and 61404061).

Author details

H. P. Zhou[1,2*], S. Xu[2*] and S. Q. Xiao[3]

*Address all correspondence to: haipzhou@uestc.edu.cn; shuyan.xu@nie.edu.sg

1 School of Energy Science and Engineering, University of Electronic Science and Technology of China, Chengdu, Sichuan, China

2 Plasma Sources and Application Center, NIE, and Institute of Advanced Studies, Nanyang Technological University, Singapore

3 Key Laboratory of Advanced Process Control for Light Industry (Ministry of Education), Department of Electronic Engineering, Jiangnan University, Wuxi, Jiangsu, China

References

[1] Panasonic HIT Solar Cell Achieves World's Highest Energy Conversion Efficiency of 25.6% at Research Level Internet] 2014. Available from: http://news.panasonic.com/press/news/official.data/data.dir/2014/04/en140410-4/en140410-4.html

[2] Olibet S, Monachon C, Hessler A, Vallat-Sauvalin. E, Wolg De, Fesquet L, Damon-Lacoste J, Ballif C. Textured silicon heterojunction solar cells with over 700 mV open-circuit voltage studied by transmission electron microscopy. In: 23th European Photovoltaic Solar Energy Conference, Valencia, Spain; 2008. p. 1140

[3] Roschek T, Repmann T, Muller J, Rech B, Wagner H. Comprehensive study of micro-crystalline silicon solar cells deposited at high rate using 13.56 MHz plasma-enhanced chemical vapor deposition. Journal of Vacuum Science and Technology A. 2002; 20:492-498. DOI: 10.1116/1.1450585

[4] Ostrikov K. Reactive plasmas as a versatile nanofabrication tool. Review of Modern Physics. 2005; 77:489-511. DOI: http://dx.doi.org/10.1103/RevModPhys.77.489

[5] Amanatides E, Mataras D. Frequency variation under constant power conditions in hydrogen radio frequency discharges. Journal of Applied Physics, 2001; 89:1556-1566. DOI: 10.1063/1.1337597

[6] Rakhimova TV, Braginsky OV, Ivanov VV, Kim TK, Kong JT, Kovalev AS, Lopaev DV, Mankelevich YuA, Proshina OV, Vasilieva AN. Experimental and theoretical study of RF plasma at low and high frequency. IEEE Transactions on Plasma Science. 2006; 34:867- 877. DOI: 10.1109/TPS.2006.875849

[7] Rech B, Repmann T, van den Donker MN, Berginski M, Kilper T, Hüpkes J, Calnan S, Stiebig H, Wieder S. Challenges in microcrystalline silicon based solar cell technology. Thin Solid Films, 2006; 511:548-555. DOI: 10.1016/j.tsf.2005.12.161

[8] Xu S, Levchenko I., Huang SY, Ostrikov K. Self-organized vertically aligned single-crystal silicon nanostructures with controlled shape and aspect ratio by reactive plasma etching. Applied Physics Letters, 2009; 95:111505. DOI: 10.1063/1.3232210

[9] Denysenko I. B., Xu S., Long J. D., Rutkevych P. P., Azarenkov N. A. and Ostrikov K. Inductively coupled $Ar/CH_4/H_2$ plasma for low-temperature deposition of ordered carbon nanostructures. J. Appl. Phys. 2004; 95: 2713-2724. DOI: 10.1063/1.1642762

[10] Xiao SQ, Xu S, Ostrikov K. Low-temperature plasma processing for Si photovoltaics. Materials Science and Engineering R. 2014; 78:1-29. DOI: 10.1016/j.mser.2014.01.002

[11] Xu S, Ostrikov K, Li Y, Tsakadze EL, Jones IR. Low-frequency, high-density, inductively coupled plasma sources: operation and applications. Phys. Plasmas. 2001; 8: 2549-2557. DOI: 10.1063/1.1343887

[12] Ostrikov K, Xu S, Xu MY. Power transfer and mode transitions in low-frequency inductively coupled plasmas. Journal of Applied Physics. 2000; 88:2268-2271. DOI: 10.1063/1.1289055

[13] Cheng QJ, Xu S, Long JD, Ni ZH, Rider AE, Ostrikov K. High-rate, low-temperature synthesis of composition controlled hydrogenated amorphous silicon carbide films in low-frequency inductively coupled plasmas. Journal of Physics D: Applied Physics. 2008; 41:055406. DOI:10.1088/0022-3727/41/5/055406

[14] Cheng QJ, Xu S, Long JD, Huang SY, Guo J. Homogeneous nanocrystalline cubic silicon carbide films prepared by inductively coupled plasma chemical vapor deposition. Nanotechnology. 2007; 18:465601. DOI: 10.1088/0957-4484/18/46/465601

[15] Cheng QJ, Xu S, Ostrikov K. Rapid, low-temperature synthesis of nc-Si in high-density, non-equilibrium plasmas: enabling nanocrystallinity at low hydrogen dilution. Journal of Material Chemistry. 2009; 19:5134-5140. DOI: 10.1039/b904227

[16] Zhou HP, Xu LX, Xu S, Huang SY, Wei DY, Xiao SQ, Yan WS, Xu M. On the conductivity type conversion of the p-type silicon exposed to low frequency inductively coupled plasma of $Ar+H_2$. Journal of Physics D: Applied Physics. 2010; 43:505402. DOI: 10.1088/0022-3727/43/50/505402.

[17] Xu S, Huang SY, Levchenko I, Zhou HP, Wei DY, Xiao SQ, Xu LX, Yan WS, Ostrikov K. Highly efficient silicon nanoarray solar cells by a single-step plasma-based process. Advanced Energy Materials, 2011; 1:373-376. DOI: 10.1002/aenm.201100085

[18] Xiao SQ, Xu S, Zhou HP, Wei DY, Huang SY, Xu LX, Sern CC, Guo YN, Khan S. Amorphous/crystalline silicon heterojunction solar cells via remote inductively coupled plasma processing. Applied Physics Letters. 2012; 100:233902. DOI: http://dx.doi.org/10.1063/1.4721642

[19] Zhou HP, Wei DY, Xu S, Xiao SQ, Xu LX, Huang SY, Guo YN, Khan S, Xu M. Crystalline silicon surface passivation by intrinsic silicon thin films deposited by low-frequency inductively coupled plasma. Journal of Applied Physics 2012; 112:013708. DOI: http://dx.doi.org/10.1063/1.4733701.

[20] Cheng QJ, Xu S, Ostrikov K. Controlled-bandgap silicon nitride nanomaterials: deterministic nitrogenation in high-density plasmas. Journal of Material Chemistry. 2010; 20:5853-5859. DOI: 10.1039/c0jm01060j

[21] Zhou HP, Wei DY, Xu LX, Guo YN, Xiao SQ, Huang SY, Xu S. Low temperature SiN_x:H films deposited by inductively coupled plasma for solar cell applications. Applied Surface Science. 2013; 264:21-26. DOI: 10.1016/j.apsusc.2012.09.050

[22] Zhou HP, Wei DY, Xu S, Xiao SQ, Xu LX, Huang SY, Guo YN, Khan S, Xu M. Si surface passivation by SiO_x:H films deposited by a low-frequency ICP for solar cell applications. Journal of Physics D: Applied Physics. 2012; 45:395401-395408. DOI: 10.1088/0022-3727/45/39/395401

[23] Droz C, Sauvain EV, Bailat J, Feitknecht L. Relationship between Raman crystallinity and open-circuit voltage in microcrystalline silicon solar cells. J. Meier and A. Shah, Sol. Energy Materials and Solar Cells. 2004; 81:61-71. DOI: 10.1016/j.solmat.2003.07.004

[24] Vetterl O, Finger F, Carius R, Hapke P, Houben L, Kluth O, Lambertz A, Mück A, Rech B, Wagner H. Intrinsic microcrystalline silicon: A new material for photovoltaics. Solar Energy Materials and Solar Cells. 2000; 62:97-108. DOI: 10.1016/S0927-0248(99)00140-3

[25] Li J, Wang J, Yin M, Gao P, He D, Chen Q, Li Y, Shirai H. Deposition of controllable preferred orientation silicon films on glass by inductively coupled plasma chemical vapor deposition. Journal of Applied Physics. 2008; 103:043505-1- 043505-7. DOI: 10.1063/1.2885158

[26] Shen C, Shieh J, Huang J. Y, Kuo H, Hsu C. W, Dai B, Lee C, Pan C, Yang F. Inductively coupled plasma grown semiconductor films for low cost solar cells with improved light-soaking stability. Applied Physics Letters. 2011; 99:033510-1- 033510-3 DOI: 10.1063/1.3615650

[27] Huang J, Lin C, Shen C, Shieh J, Dai B. Low cost high-efficiency amorphous silicon solar cells with improved light-soaking stability. Solar Energy Materials and Solar Cells. 2012; 98:277-282. DOI: 10.1016/j.solmat.2011.11.023

[28] Sakikawa N, Shishida Y, Miyazaki S, Hirose M. High-rate deposition of hydrogenated amorphous silicon films using inductively coupled silane plasma. Solar Energy Materials and Solar Cells. 2001; 66:337-343. DOI: 10.1016/S0927-0248(00)00192-6

[29] Kosku N, Kurisu F, Takegoshi M, Takahashi H, Miyazake S. High-rate deposition of highly crystallized silicon films from inductively coupled plasma. Thin Solid Films. 2003; 435:39-43. DOI: 10.1016/S0040-6090(03)00374-2

[30] Dao VA, Duy NV, Heo J, Choi H, Kim Y, Lakshminarayan, Li J. Hydrogenated amorphous silicon layer formation by inductively coupled plasma chemical vapor

deposition and its application for surface passivation of p-type crystalline silicon. Japan Journal of Applied Physics. 2009; 48:066509-1-066509-4. DOI: 10.1143/JJAP.48.066509

[31] Mott NF, Davis EA. Electronic Processes in Non-Crystalline Solids. second ed., Clarendon, Oxford; 1979. 590 p.

[32] Matsuda A, Yamaoka T, Wolff S, KoyamaM, Imanishi Y, Kataoka H, Matsuura H, Tanaka K. Preparation of highly photosensitive hydrogenated amorphous Si-C alloys from a glow-discharge plasma. Journal of Applied Physics. 1986; 60:4205-4207. DOI: 10.1063/1.337528

[33] Zhou HP, Wei DY, Xu S, Xiao SQ, Xu LX, Huang SY, Guo YN, Yan WS, Xu M. Dilution effect of Ar/H_2 on the microstructures and photovoltaic properties of nc-Si:H deposited in low frequency inductively coupled plasma. Journal of Applied Physics. 2011; 110:023517. DOI: 10.1063/1.3605288

[34] Saha JK, Bahardoust B, Leong K, Gougam AB, Kherani N, Zukotynski S. Spectroscopic ellipsometry studies on hydrogenated amorphous silicon thin films deposited using DC saddle field plasma enhanced chemical vapor deposition system. Thin Solid Films. 2011; 519:2863-2866. DOI: 10.1016/j.tsf.2010.12.074

[35] Cavalcoli D, Rossi M, Cavallini A. Defect states in nc-Si:H films investigated by surface photovoltage spectroscopy. Journal of Applied Physics. 2011; 109:053719-1-053719-5. DOI: 10.1063/1.3553583

[36] Bagolini L, Mattoni A, Colombo L. Electronic localization and optical absorption in embedded silicon nanograins. Applied Physics Letters. 2009; 94:053115-1-053115-3. DOI: 10.1063/1.3078281

[37] Smit C, van Swaaij RACM, Donker H, Petit AMHN, Kessels WMM, van de Sanden MCM. Determining the material structure of microcrystalline silicon from Raman spectra. Journal of Applied Physics. 2003; 94:3582-3588. DOI: 10.1063/1.1596364

[38] Myong SY, Lim KS, Konagai M. Effect of hydrogen dilution on carrier transport in hydrogenated boron-doped nanocrystalline silicon-silicon carbide alloys. Applied Physics Letters. 2006; 88:103120. DOI: 10.1063/1.2177641

[39] Cheng QJ, Xu S, Ostrikov K. Temperature-dependent properties of nc-si thin films synthesized in low-pressure, thermally nonequilibrium, high-density inductively coupled plasmas. Journal of Physics and Chemistry C. 2009; 113:14759-14764. DOI: 10.1021/jp9047083

[40] Solomon I, Drévillon B, Shirai H, Layadi N. Plasma deposition of microcrystalline silicon: the selective etching model. 15th International Conference on Amorphous Semiconductors: Science and Technology; 6-10 September 1993; Cambridge: Journal of Non-Crystalline Solids. 1993; 164–166: 989-992.

[41] Kaxiras E, Joannopoulos JD. Hydrogenation of semiconductor surfaces: Si and Ge (111). Physical Review B.1988; 37:8842-8848. DOI: http://dx.doi.org/10.1103/PhysRevB.37.8842

[42] Smets AHM, Matsui T, Kondo M. Infrared analysis of the bulk silicon-hydrogen bonds as an optimization tool for high-rate deposition of microcrystalline silicon solar cells. Applied Physics Letters. 2008; 92:033506. DOI: 10.1063/1.2837536

[43] Spear WE, Le Comber PG. Substitutional doping of amorphous silicon. Solid State Communications. 1975; 17:1193-1196. DOI:10.1016/0038-1098(75)90284-7

[44] Street RA. Hydrogenated Amorphous Silicon. Cambridge University Press; 1991. 417 p. DOI: 10.1017/CBO9780511525247

[45] Winer K, Street RA, Johnson NM, Walker J. Impurity incorporation and doping efficiency in a -Si:H. Physics Review B. 1990; 42:3120-3128. DOI: http://dx.doi.org/10.1103/PhysRevB.42.3120

[46] Michaelis W, Pilkuhn MH. Radiative recombination in silicon p-n junctions. Physics State Solids. 1969; 36:311-319. DOI: 10.1002/pssb.19690360132

[47] Ashcroft NW, Mermin ND. Solid State Physics. Brooks/Cole Thomson Learning. 1976. 826 p.

[48] Shockley W, Read WT. Statistics of the recombination of holes and electrons. Physics Review. 1952; 87:835-842. DOI: http://dx.doi.org/10.1103/PhysRev.87.835

[49] Hall RN. Electron-hole recombination in germanium. Physics Review. 1952; 87:387-387. DOI: http://dx.doi.org/10.1103/PhysRev.87.387

[50] Schroder DK. Semiconductor Material and Device Characterization. John Wiley & Sons, Inc; 1990. 755 p. DOI: 10.1002/0471749095

[51] Aberle AG. Surface passivation of crystalline silicon solar cells: a review. Progress in Photovoltaics. 2000; 8:473-487. DOI: 10.1002/1099-159X(200009/10)8:5<473:AID-PIP337>3.0.CO;2-D

[52] Descoeudres A, Barraud L, Bartlome R, Choong G, Wolf SD, Zicarelli F, Ballif C. The silane depletion fraction as an indicator for the amorphous/crystalline silicon interface passivation quality. Applied Physics Letters. 2010; 97:183505. DOI: http://dx.doi.org/10.1063/1.3511737

[53] Kerr MJ, Schmidt J, Cuevas A, Bultman JH. Surface recombination velocity of phosphorus-diffused silicon solar cell emitters passivated with plasma enhanced chemical vapour deposited silicon. Journal of Applied Physics. 2001; 89:3821-3826. DOI: http://dx.doi.org/10.1063/1.1350633

[54] Kerr MJ, Cuevas A. Very low bulk and surface recombination in oxidized silicon wafers. Semiconductor Science Technology. 2002; 17:35-38. DOI: http://dx.doi.org/10.1088/0268-1242/17/1/306

[55] Martín I, Vetter M, Orpella A, Voz C, Puigdollers J, Alcubilla R. Characterization of a-SiCx:H films for c-Si surface passivation. Applied Physics Letters. 2002; 81:476-479. DOI: http://dx.doi.org/10.1557/PROC-715-A24.5

[56] Fujiwara H, Kondo M. Effects of a-Si:H layer thicknesses on the performance of a-Si:H/c-Si heterojunction solar cells. Journal of Applied Physics. 2007; 101:054516. DOI: 10.1063/1.2559975

[57] Wang Q, Page MR, Iwaniczko E, Xu Y, Roybal L, Baure R, To B, Yuan HC, Duda A, Fasoon F, Yan YF., Levi D, Meier D, Branz HM, Wang TH. Efficient heterojunction solar cells on p-type crystal silicon wafers. Applied Physics Letters. 2010; 96:013507. DOI: 10.1063/1.3284650

[58] Lee JY, Glunz SW. Investigation of various surface passivation schemes for silicon solar cells. Solar Energy Matters & Solar Cells. 2006; 90:82-92. DOI: 10.1016/j.solmat.2005.02.007

[59] Schmidt J, Kerr M, Cuevas A. Surface passivation of Si solar cells using plasma-enhanced chemical vapor deposited SiN films and thermal SiO2/thermal SiN Stacks. Semiconductor Science Technology. 2001; 19:164-170. DOI: 10.1088/0268-1242/16/3/308

[60] Martin I, Vetter M, Orpella A, Puidollers J, Cuevas A, Alcublla R. Surface passivation of p-type crystalline Si by plasma enhanced chemical vapor deposited amorphous SiCx:H films. Applied Physics Letters. 2001; 79:2199-2201. DOI: 10.1063/1.1404406

[61] Dauwe S, Schmid J, Hezel R. Very low surface recombination velocities on p- and n-type silicon wafers passivated with hydrogenated amorphous silicon films. Proceedings of the Photovoltaic Specialists Conference; 19–24 May 2002; New York: IEEE; 2002. p. 1246-1249.

[62] Willem J, Schuttauf A, Van der Werf KHM, Kielen IM, Van Sark WGJHM, Rath JK, Schropp REI. High quality crystalline silicon surface passivation by combined intrinsic and n-type hydrogenated amorphous silicon. Applied Physics Letters. 2011; 98:153514. DOI: 10.1063/1.3662404

[63] Vetterl O, Finger F, Carius R, Hapke P, Houben L, Kluth O, Lambertz A, Mück A, Rech B, Wagner H. Intrinsic microcrystalline silicon: a new material for photovoltaics. Solar Energy Materials Solar Cells. 2000; 62:97-108. DOI: 10.1016/S0927-0248(99)00140-3

[64] Rem JB, Holleman J, Verweij JF. Incubation time measurements in thin-film deposition. Journal of Electrochemical Society. 1997; 144:2101-2106. DOI: 10.1149/1.1837748

[65] Tickle AC. Thin-Film Transistors: A New Approach to Microelectronic. John Wiley & Sons, New York; 1969. 144 p.

[66] Mitchell J, Macdonald D, Cuevas A. Thermal activation energy for the passivation of the n-type crystalline silicon surface by hydrogenated amorphous silicon. Applied Physics Letters. 2009; 94:162102. DOI: 10.1063/1.3120765

[67] Rath J K. Low temperature polycrystalline silicon: a review on deposition, physical properties and solar cell applications. Solar Energy Materials Solar Cells. 2003; 76:437-487. DOI: 10.1016/S0927-0248(02)00258-1

[68] Strahm B, Howling A, Sansonnens L, Hollenstein C. Fast equilibration of silane/hydrogen plasmas in large area RF capacitive reactors monitored by optical emission spectroscopy. Plasma Sources Science Technology. 2007; 16:679-696. DOI: 10.1088/0963-0252/16/4/001

[69] Fujiwara H, Kaneko T, Kondo M. Application of hydrogenated amorphous silicon oxide layers to c-Si heterojunction solar cells. Applications of Physics Letters. 2007; 91:133508. DOI: 10.1063/1.2790815

[70] Sritharathikhun J, Banerjee C, Otsubo M, Sugiura T, Yamamoto H, Sato T, Limmanee A, Yamada A, Konagai M. Surface passivation of crystalline and polycrystalline silicon using hydrogenated amorphous silicon oxide film. Japan Journal of Applied Physics. 2007; 46:3296-3300. DOI: 10.1143/JJAP.46.3296

[71] Mueller T, Schwertheim S, Fahrner WR. Crystalline silicon surface passivation by high-frequency plasma-enhanced chemical-vapor-deposited nanocomposite silicon suboxides for solar cell applications. Journal of Applied Physics. 2010; 107:014504 DOI: 10.1063/1.3264626

[72] Bulkin PV, Swart PL, Lacquet BM. Electron cyclotron resonance plasma enhanced chemical vapour deposition and optical properties of SiOx thin films. Journal of Non-Crystalline Solids. 1998; 226:58-66. DOI: 10.1016/S0022-3093(98)00362-7

[73] Colberg SK, Barkmann N, Streich C, Schütt A, Suwito D, Schäfer P, Müller S, Borchert D. 26th European PV Solar Energy Conference and Exhibition; 5-9 September 2011; Hamburg: EU PVSEC Proceedings; 2011. p. 1770-1773.

[74] Jana T, Mukhopadhyay S, Ray S. Low temperature silicon oxide and nitride for surface passivation of silicon solar cells. Solar Energy Materials and Solar Cells. 2002; 71:197-211. DOI: 10.1016/S0927-0248(01)00058-7

[75] Watanabe H, Haga K, Lohner T. Structure of high-photosensitivity silicon oxygen alloy-films. 15th International Conference on Amorphous Semiconductors: Science and Technology; 6-10 September 1993; Cambridge: Journal of Non-Crystalline Solids; 1993. p. 1085-1088.

[76] Sritharathikhun J, Yamamoto H, Miyajima S, Yamada A, Konagai M. Optimization of amorphous silicon oxide buffer layer for high-efficiency p-type hydrogenated microcrystalline silicon oxide/n-type crystalline silicon heterojunction solar cells. Japan Journal of Applied Physics. 2008; 47:8452-8455. DOI: 10.1143/JJAP.47.8452

[77] Banerjee C, Sritharathikhun J, Yamada A, Konagai M. Fabrication of heterojunction solar cells by using microcrystalline hydrogenated silicon oxide film as an emitter.

Journal of Applied Physics D: Applied Physics. 2008; 41:185107. DOI:10.1088/0022-3727/41/18/185107

[78] Chen MS, Santra AK, Goodman DW. Structure of thin SiO_2 films grown on Mo(112). Physics Review B. 2004; 69:1124-1133. DOI: http://dx.doi.org/10.1103/PhysRevB.69.155404

[79] Sze SM. Semiconductor Devices: Physics and Technology, John Wiley & Sons; 2002. 576 p.

[80] Wolf SD, Kondo M. Nature of doped a-Si:H/c-Si interface recombination. Journal of Applied Physics. 2009; 105:103707-103707-6. DOI: 10.1063/1.3129578

[81] Sinton RA, Cuevas A. Contactless determination of current–voltage characteristics and minority-carrier lifetimes in semiconductors from quasi-steady-state photoconductance data. Journal of Applied Physics. 1996; 69:2510-2512. DOI: S0003-6951(96)02743-X

Chapter 9

Applications of CVD to Produce Thin Films for Solid-State Devices

A.M. Torres-Huerta, M.A. Domínguez-Crespo and
A.B. López-Oyama

Additional information is available at the end of the chapter

http://dx.doi.org/10.5772/63964

Abstract

Thin films of Pt-YSZ and Pd-ZrO_2 cermets by chemical vapor deposition (CVD) from metallorganic precursors (MOCVD) were evaluated as electrode in solid-state devices. Morphology and structural characteristics were studied by X-ray diffraction (XRD), scanning electronic microscopy, atomic force microscopy (AFM), and transmission electronic microscopy (TEM). Electrochemical performance was determined using Tafel and electrochemical impedance spectroscopy methods. Metallorganic precursors were used (metal-acetylacetonates), and argon and oxygen were used as the carrier and reactive gases, respectively. The particle average size was less than 20 nm, with high and uniform particle dispersion according to TEM measurements.

Keywords: thin films, CVD, platinum metals, zirconia, cermets

1. Introduction

Chemical vapor deposition (CVD) is a fast and economic method to obtain controlled deposits of high quality; the main drawbacks are the possible incorporation of impurities due to non-exhausted remains and poor adhesion of the film in some cases [1]. Chemical vapor deposition (CVD) from metallorganic precursors (MOCVD) is a rapidly developing method for producing films and coatings of ceramic materials for a variety of applications [2]. Among solid electrolyte cells, commonly used for numerous applications including gas sensors, gas pumps, solid oxide fuel cells, and electrochemically promoted catalysts, platinum electrode deposited on yttria-stabilized zirconia (YSZ) is one of the most widely studied examples [3] because of

YSZ conducts at temperatures above 300°C via O^{2-} diffusion and platinum has always been of particular interest due to its electrocatalytic properties [4]. The electrochemical performance of electrode layers depends largely on the microstructure of the three-phase boundaries (TPBs) [5] in which the ionic conductor, the electronic conductor, and the oxygen gas are in contact, being the best pathway to reduce oxygen to oxygen ion at the surface [6].

Nanoparticles exhibit very interesting features at the nanometer scale due to the quantum confinement, where the changes induced to the microstructure allow them to increase the specific surface area and, in the case of electrodes, more TPBs are arisen. These characteristics are desirable for high electrochemical performance. The higher sintering temperatures are related to the formation of undesirable reaction products highly resistive at the electro/ electrolyte interface. By reaching the nanometer size, the sintering temperature can be low facilitating the ceramic processing, due to the more TPBs which as a geometrical parameter required for high electrochemical performance [5].

The appearance of the material at nanometric scale can be achieved by modification of the preparation methods despite its identical chemical composition. The material undergoes a phase transformation by changing temperature and pressure or by combining different materials. The induced changes onto the microstructure modify the electrode behavior [4]. One approach to increase the TPBs areas, and in consequence the yield of the electrodes, is to prepare a cermet material that spreads the solid electrolyte into the electrode [5, 7–10]. The development in thin-film technology and the application of ionic-electronic composites as electrodes have significantly reduced the ohmic loss caused by the electrolyte [5]. The rate and selectivity of catalytic reactions occurring on metal and metal-oxide porous catalyst-electrode films deposited on solid electrolytes can be reversibly affected in a very pronounced and controlled manner by polarizing of the catalyst-electrode, which can exceed the corresponding rate of ion transport through the solid electrolyte by several orders of magnitude [11]. Noble metals are historically known to be extremely non-reactive, but the possibility for obtaining coatings of platinum metals by means of CVD on the substrates made of different materials was demonstrated widely [12]. Metal, oxide, and metal-oxide composite systems can be obtained by means of CVD using the process of thermal decomposition of a mixture of the volatile initial metal complexes with organic ligands (precursors) on a heated surface [12]. Additionally, chemical vapor deposition (CVD) is one of the most important synthesis processes for making nanoparticles of metals, semiconductors, and magnetic materials. In some cases, Pt is deposited as continuous films whereas in other cases, it is preferred in the form of a dispersion of particles with a high surface area. On the other hand, ceramics based on zirconium dioxide (zirconia, ZrO_2) are promising materials for multifunctional applications. In previous works [13–16], ZrO_2, IrO_2, Pt, and IrO_2-YSZ thin films have been synthesized by MOCVD process and, in the case of ZrO_2, a thermodynamic study to explain the phase and the conditions that promote the formation of pure ZrO_2 from a β-diketonate precursor (Zr-acetylacetonate, $Zr(acac)_4$) and oxygen (O_2) as carrier/reactive gas was reported [17], and now, the structural, nanocrystalline nature and morphological characterizations of Pt-YSZ composites are completed and reported; furthermore, in order to determine the performance of these composites as electrodes in YSZ solid-state devices, electrochemical tests were also carried out.

Additionally, the synthesis of Pd-ZrO_2 cermets by MOCVD and their characterization are reported. X-ray diffraction (XRD), atomic force microscopy (AFM), scanning electron microscopy (SEM), Raman spectroscopy, and electrochemical characterizations were conducted in order to determine the structure, composition, morphology, oxidation state, and electrocatalytic properties of these materials. Metallorganic precursors were used (metal-acetylacetonates), and argon and oxygen were used as the carrier and reactive gases, respectively.

2. Experimental procedure

The experimental setup has been described elsewhere [15]. Amorphous quartz and YSZ were used as substrates; the composite electrodes were produced by a horizontal hot wall MOCVD. The employed precursors were $Zr(acac)_4$, $Y(acac)_3$, $Pt(acac)_2$, and $Pd(acac)_2$ [Sigma-Aldrich, ≥97%]. Argon and oxygen gases were used as the carrier and reactive gases, respectively. Before the MOCVD process, the quartz substrates were cleaned with a soapy solution and thereafter rinsed with tap water, deionized water, and acetone. The optimal experimental conditions for producing nanocomposite electrodes were identified as follows: 450–600°C deposition temperature (T_{dep}), 1.0 Torr total pressure (P_{tot}), 190–220°C precursor temperature (T_{prec}), 130 sccm O_2, and 180 sccm Ar flow rates (FR). The structural analysis and the particle size determination were carried out by X-ray diffraction (XRD, Siemens D5000 and Bruker D8 Advanced) using a monochromatic Cu-$K_{\alpha 1}$ radiation at 35 kV, 25 mA and a scan rate of 0.02° min^{-1}; the average crystallite size was calculated from the diffraction peak broadening using the Scherrer equation; and the morphology, texture, and surface roughness of the films were examined by scanning electron microscopy (SEM, JEOL JSM-6300), transmission electron microscopy (TEM, JEOL-2000 FX-II), and atomic force microscopy by AFM, Nanosurf easyscan 2 AFM/STM in contact mode with a Si cantilever and nominal force of 20 nN, respectively.

Finally, the electrochemical tests were conducted to determine the performance of these films as electrodes. The method for this purpose was a three terminal one: the working electrode (WE, 1.7 cm^2 in area) on one side of the YSZ pellet and the reference and the counter electrodes on the opposite side (RE, 0.5 cm^2 in area, CE, 1.2 cm^2 in area, respectively). Impedance measurements were carried out over the frequency range of 10^5–10^{-2} Hz using a Solartron 1255 frequency response analyzer from 300 to 800°C in air. The signal applied to the cell was generally 10 mV rms. Potentiodynamic polarization was obtained from -400 mV to +400 mV at a scan rate of 1.67 mV s^{-1}.

3. Results and discussion

Figure 1a and **b** shows XRD spectra of platinum thin films obtained at 500°C and 550°C, and quartz is shown as reference (**Figure 1c**). Pt crystallites in face-centered cubic structure (ICDD 04-0802) grow in (1 1 1) orientation at 500°C while at 550°C, is polycrystalline, and at both temperature, the films are purely metallic and shiny. The broadening of the reflections was associated with a small crystallite size, determined by Scherrer equation:

$$t = \frac{0.9\lambda}{B\cos\theta_B} \tag{1}$$

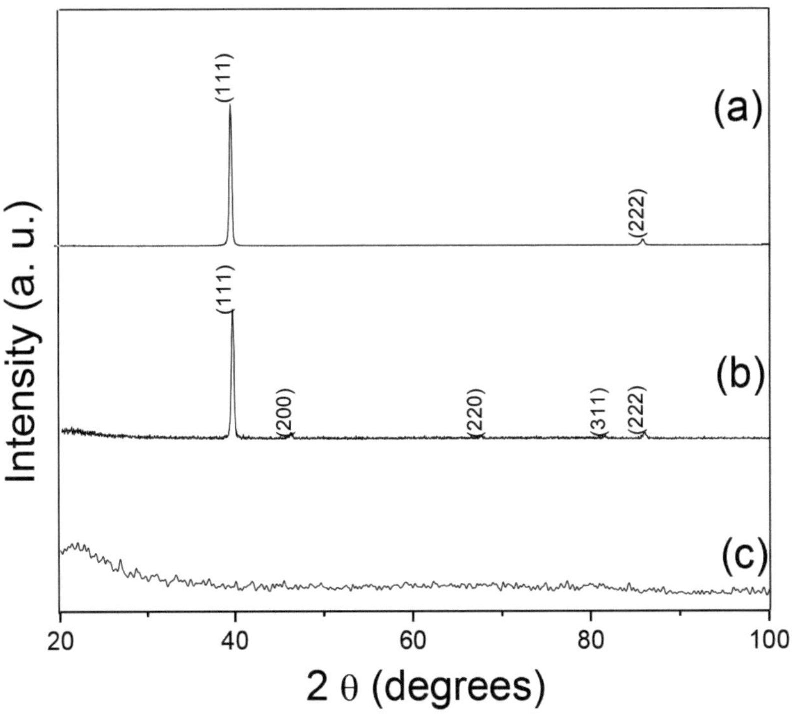

Figure 1. Pt thin-film diffractograms (a) 500°C, (b) 550°C and (c) amorphous quartz.

Here, t is the mean crystalline size, λ is the wavelength of the X-ray, B is the full width of a peak at half of the maximum of the peak (FWHM) in the diffraction spectra (measured in radians), and θ is the Bragg angle.

Then, the mean Pt crystallite size was 48 nm. Pt-ZrO_2 composite was obtained at 600°C (**Figure 2a**); ZrO_2 signals are weak, while Pt reflections are better defined than that for ZrO_2. It was reported before that ZrO_2 at these particular experimental conditions was not obtained below 600°C [16]; the Pt-ZrO_2 composites had to be synthesized at this temperature, taking in account the experimental conditions for ZrO_2, that is, only oxygen was used as the carrier and reactive gas. The next step was the deposition of yttria-stabilized zirconia (YSZ), shown in **Figure 3b**. These films were transparent. The XRD spectrum displays the signals corresponding to (1 1 1), (2 0 0), and (3 1 1) planes for cubic YSZ (ICDD 30-1468); the Zr/Y weight ratio to obtain YSZ was 0.87:0.13, according to Wang et al. [18]; the mean crystallite size was 15 nm. The appearance of YSZ cubic microstructure could be the result of these nanocrystalline sizes [19, 20]. The formation of cubic structure of YSZ film at low temperatures is somewhat alike with the formation of metastable tetragonal structure as a result of nanocrystalline size effect in the transformation of zirconium phase [18, 19, 21]. Finally, once the individual thin films

http://dx.doi.org/10.5772/63964

were synthesized, the composite Pt-YSZ was deposited by combining the different experimental conditions (**Figure 3c**). The signals of both Pt and YSZ are clearly observed. In this case, the mean Pt crystallite size was 21 nm, while the YSZ size remains the same, 15 nm. The decrement in Pt crystallite size could be attributable to the film microstructure, which is avoiding the agglomeration, sintering, and, consequently, the growth of Pt particles. The XRD spectrum for amorphous quartz is shown in **Figures 1c**, **2b**, and **3a** in order to identify the signals arising from thin films. Experimental deposition conditions are summarized in **Table 1**.

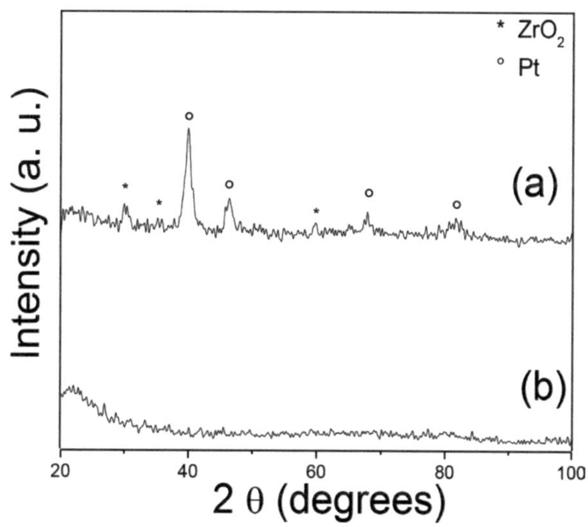

Figure 2. Pt-ZrO_2 diffractograms (a) 600°C and (b) amorphous quartz.

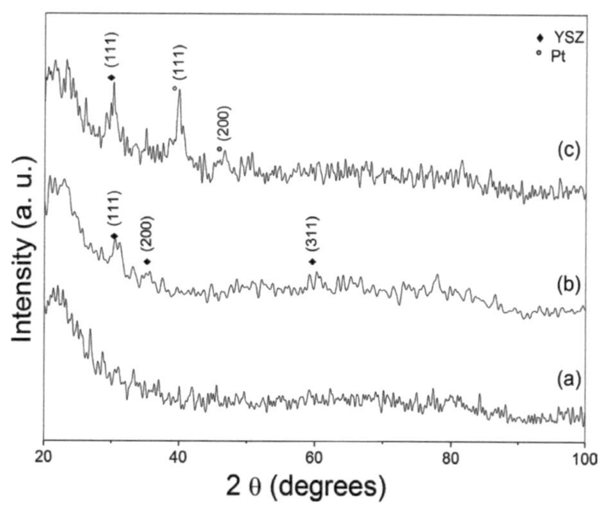

Figure 3. Diffractograms of (a) amorphous quartz, (b) YSZ, and (c) Pt-YSZ at 600°C.

Thin film	T_{prec} (°C)	T_{dep} (°C)	Gas flow (cm^3/min)
Pt	180	500	180
ZrO_2	220	600	130
Pt-ZrO_2	220	550	130
Y_2O_3-ZrO_2	220	600	130
Pt-YSZ	220	600	130

Table 1. Experimental deposition conditions for Pt-YSZ film.

A typical view of the composite Pt-YSZ film is given in **Figure 4**. The film is porous as a result of its columnar growth (**Figure 4d**, cross-sectional view). Thickness of the film was estimated to be about 1 μm with a growth rate of 50 nm/min.

Figure 4. SEM images of Pt-YSZ thin films (a–c) surface and (d) cross-section.

Figure 5 depicts the impedance spectrum at 300–800°C for Pt-YSZ nanocomposite electrode films prepared on YSZ pellets. According to the associated capacitance value (10^{-5} F), reactions of charge transfer at the electrode/electrolyte interface were assigned to the small semicircle at high frequency. It seems that a diffusion mechanism rise at the low-frequency region because of the large semicircle with capacitance values between 10^{-4} and 10^{-3} F, attributed to the oxygen diffusion at the finite thickness in the film. The bulk and grain boundary responses are missing due to the frequency scale. According to West et al. [22], $C = \sim 10^{-7}$ F corresponds to electrode-electrolyte interphase and $C = \sim 10^{-4}$ F is associated with the oxygen electrochemical reaction into the TPB; furthermore, a depressed semicircle could indicate an associated impedance to the oxygen path through the electrode [22]. The Pt-YSZ nanocomposite electrodes displayed lower charge-transfer resistance than diffusion resistance even at higher temperatures. This suggests easy interfacial charge-transfer reactions probably controlled by the diffusion of oxygen through the structure of the Pt-YSZ electrode.

http://dx.doi.org/10.5772/63964

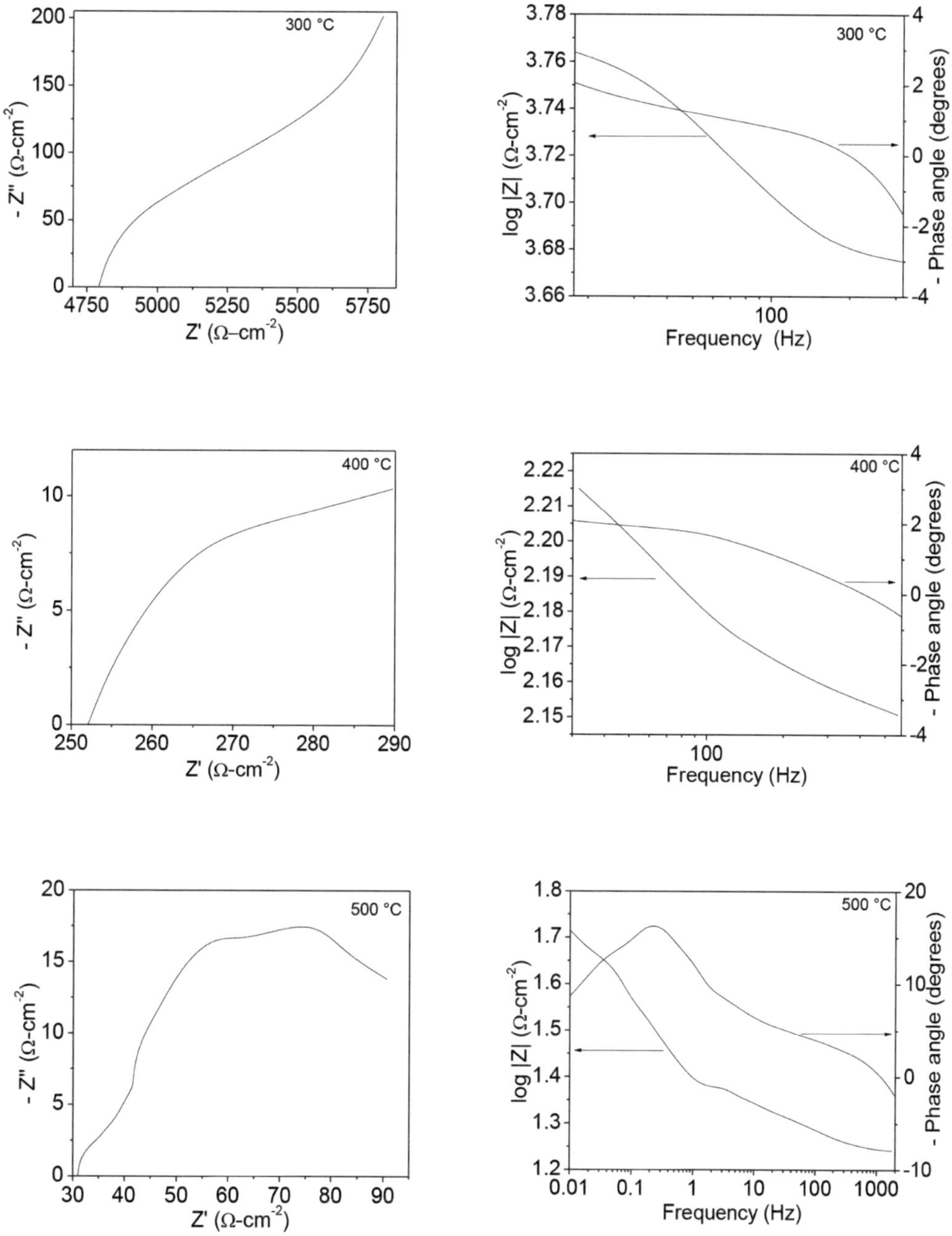

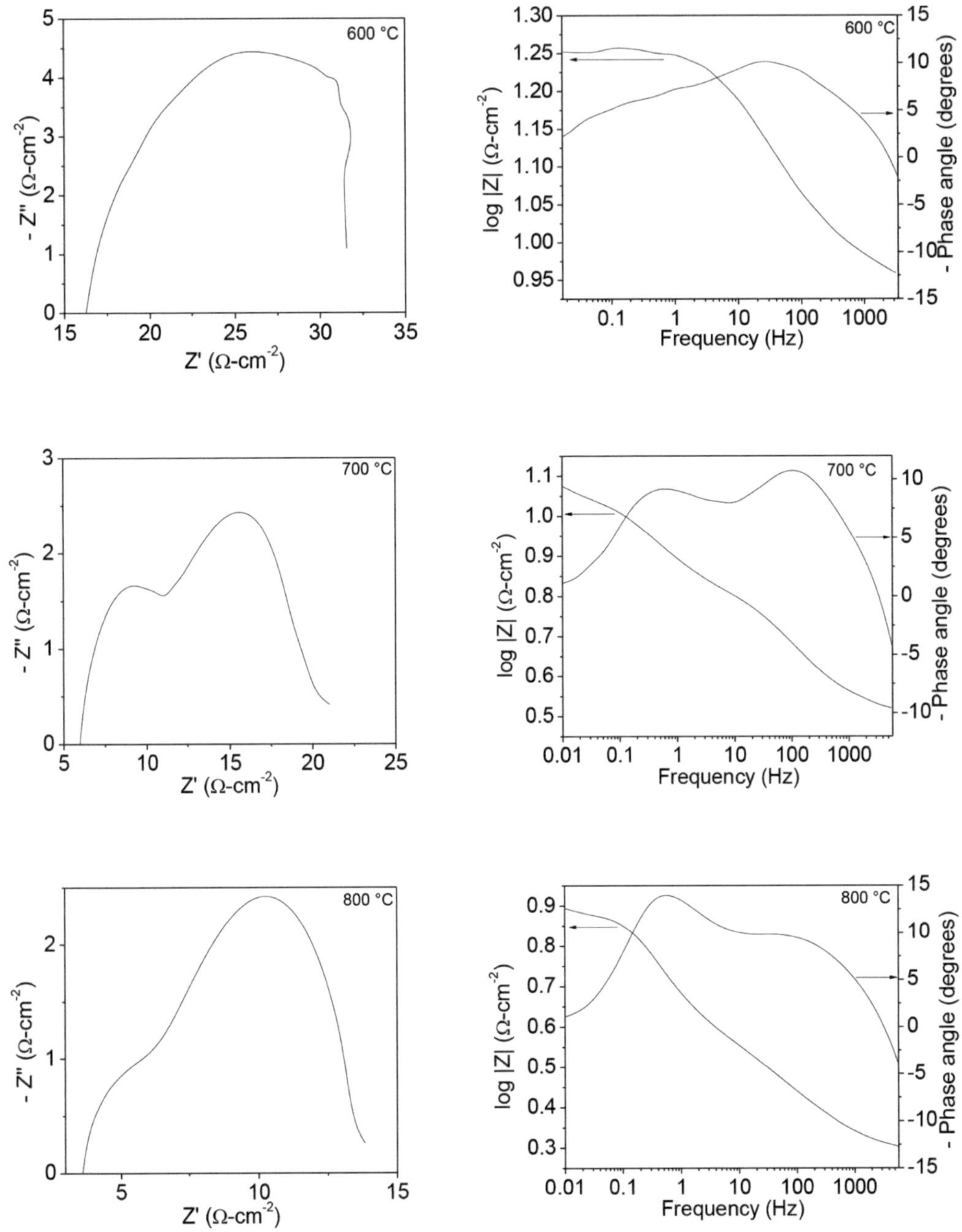

Figure 5. Impedance spectrum at 300–800°C for Pt-YSZ nanocomposite electrode films prepared on YSZ pellets.

From polarization plots (Tafel plots) can be observed that as temperature increases, the current increases through the electrode-electrolyte interphase (**Figure 6**). In the anodic branch where higher current was obtained than that obtained in cathodic branch, except at 600 and 700°C, at 600°C, the current in the cathodic branch is slightly higher than in anodic branch; and at

http://dx.doi.org/10.5772/63964

700°C, both values of currents (anodic and cathodic branches) seem to be the same; and the cathodic response looks similar at these temperatures (600 and 700°C). At 800°C, the cathodic response is lower than that observed at 600 and 700°C, but the current increases as temperature increases in the anodic branch. This behavior could be attributed to a decrement of TPB provoked by particle agglomeration, maybe Pt or YSZ, or by a reaction between Pt and YSZ as Badwal and de Bruin suggested [23]. **Table 2** shows the interchange current and Tafel slopes from **Figure 6**. According to the Tafel slopes, one order of reaction was obtained. Besides YSZ, no oxidized phases were found in Pt-YSZ electrodes after heating in air, which indicates a high thermal stability (**Figure 7**).

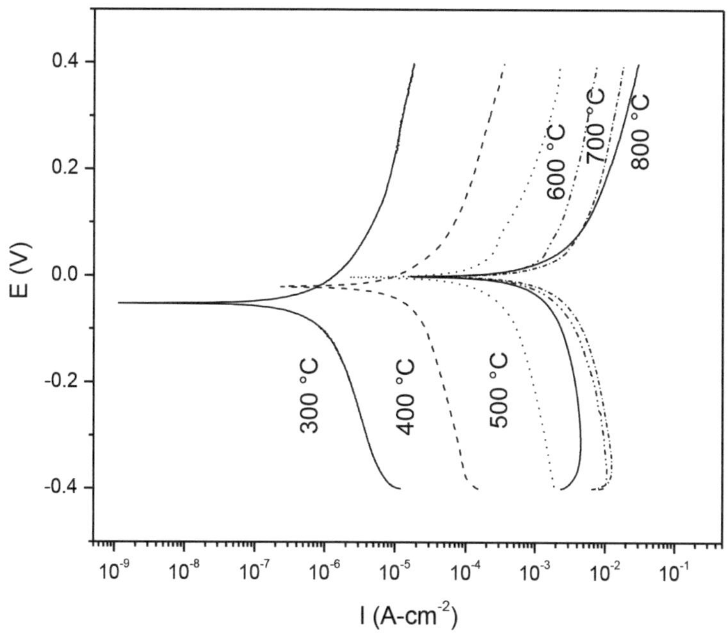

Figure 6. Tafel plots of Pt-YSZ at 300–800°C.

Temperature (°C)	Log I_o (A/cm^2)	Cathodic part m_c (mV/decade)	Anodic part m_a (mV/decade)
300	-6.22	-242	190
400	-4.76	-204	214
500	-3.62	-173	254
600	-2.77	-177	165
700	-2.43	-167	177
800	-2.67	-122	211

Table 2. Interchange current (I_o) and Tafel slopes (m), from Pt-YSZ.

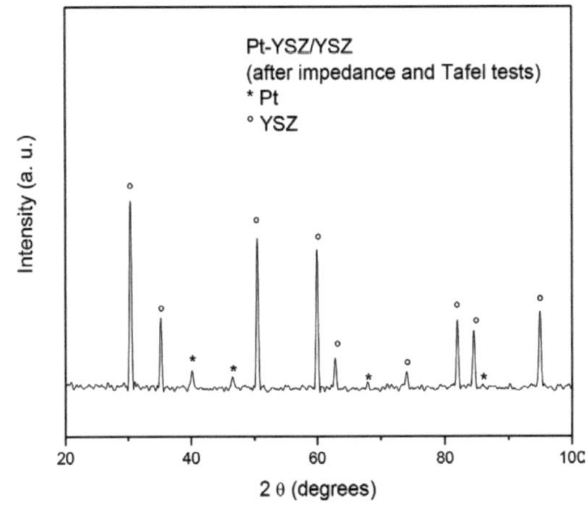

Figure 7. XRD pattern of Pt-YSZ/YSZ after impedance and Tafel tests.

The synthesis of cermets Pd-ZrO_2 by MOCVD and their characterization were conducted in order to determine the structure, composition, morphology, oxidation state, and electrocatalytic properties of these materials. Metallorganic precursors were used (metal-acetylacetonates), and argon and oxygen were used as the carrier and reactive gases, respectively. It is expected that these materials could be applied in water purification. The composite thin films were prepared using a ratio of Pd and Zr metals of 10–90 wt%, respectively. The deposition temperature (T_{dep}) was varied from 450, 500, and 550°C, and pressure (P) was controlled at 1.0 Torr. According to XRD patterns (**Figure 8**), palladium metallic reflections are difficult to observe and only in the films at 500°C, these signals are detected. Pd-ZrO_2 coatings on stainless

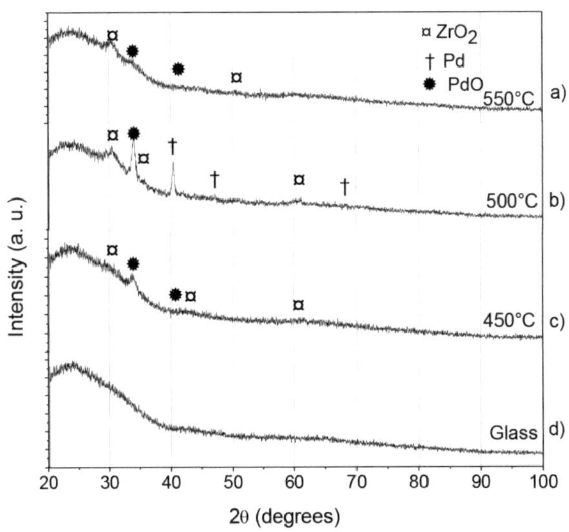

Figure 8. XRD patterns of Pd-ZrO_2 on glass substrate deposited at (a) 450°C, (b) 500°C, and (c) 550°C for 30 min and 80–20% volume Ar-O_2 and (d) of glass.

steel were deposited to characterizing their electrochemical behavior. It was reported that an increasing in Pd concentration favors the zirconia tetragonal phase as well as a higher crystal size [24]. The Pd-ZrO_2 coatings deposited on glass substrates were porous, uniform and thin (**Figure 9**). Apparently, the films exhibit good adherence, but an adherence test is needed to verify this appreciation; composites obtained at 450°C show higher porosity. **Figure 10** shows the AFM images for Pd-ZrO_2 at different temperatures; it can be suggested a columnar growth, characteristic for this type of technique and thin film. The average roughness is present in **Table 3**. Pd-ZrO_2 coatings on stainless steel, at 450 and 500°C, were uniform, and it is observed at the domes, which could indicate a columnar growth (**Figure 11**). The coatings display a shift in the corrosion potential (**Figure 12**) from 64 mV to 273 mV as consequence of deposition temperature.

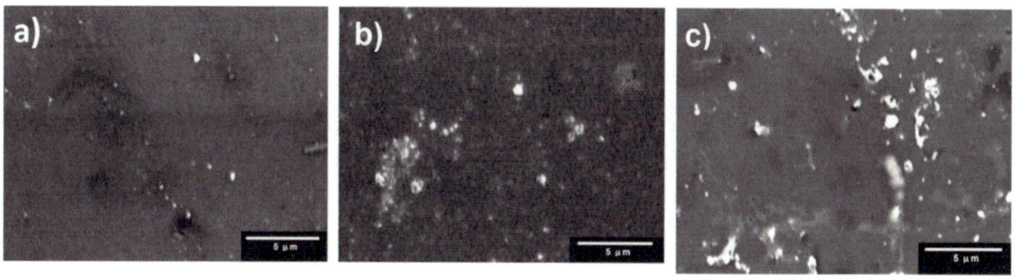

Figure 9. Scanning electron micrographs of Pd-ZrO_2 on glass substrate deposited at (a) 450°C, (b) 500°C, and (c) 550°C for 30 min and 80–20% volume Ar-O_2.

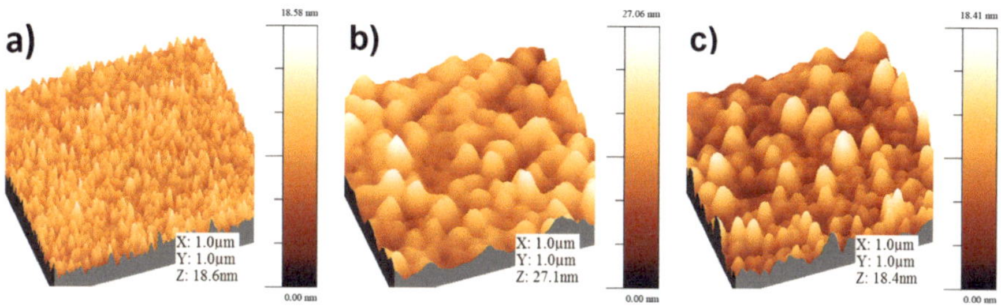

Figure 10. AFM micrographs of Pd-ZrO_2 on glass substrate deposited at (a) 450°C, (b) 500°C, and (c) 550°C for 30 min and 80–20% volume Ar-O_2.

Temperature (°C)	Rms (nm)
AISI 304L	6.1
450	6.3
500	15.9
550	3.7

Table 3. Average roughness for Pd-ZrO_2 thin films.

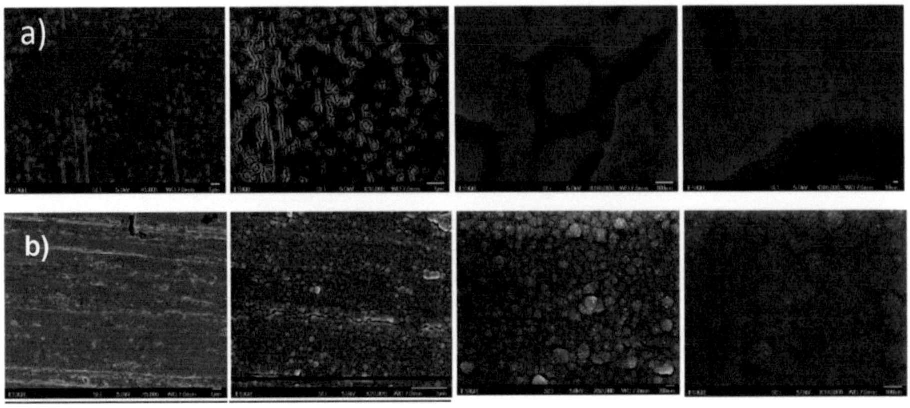

Figure 11. SEM micrographs of Pd-ZrO_2 on stainless steel at (a) 450°C and (b) 500°C.

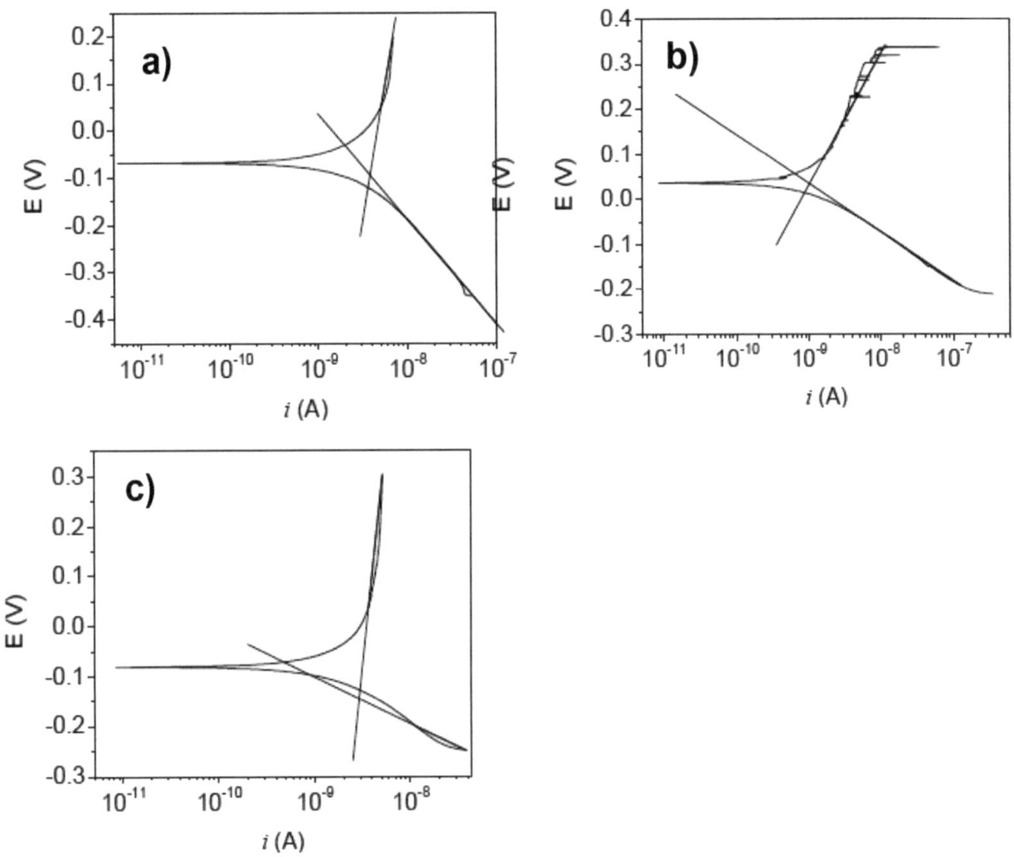

Figure 12. Tafel plot for Pd-ZrO_2 coatings on stainless steel at (a) 450°C, (b) 500°C, and (c) 550°C.

4. Conclusions

Metal, oxide, and metal-oxide composite systems can be obtained by means of CVD using the process of thermal decomposition of a mixture of the volatile initial metal complexes with organic ligands (precursors) on a heated surface. The possibility for obtaining coatings of platinum metals by means of CVD on the substrates made of different materials was demonstrated widely. The platinum-YSZ particle average size was less than 20 nm, with high and uniform particle dispersion according to TEM measurements.

Acknowledgements

Authors are grateful for the financial support by CONACYT, SNI, and IPN in the projects SIP2016-0541, and they thank J. Ferretiz-Anguiano, J.R. Vargas-García, and H. Dorantes for their technical support.

Author details

A.M. Torres-Huerta[1*], M.A. Domínguez-Crespo[1] and A.B. López-Oyama[1,2]

*Address all correspondence to: atohuer@hotmail.com and atorresh@ipn.mx

1 National Polytechnic Institute, CICATA-Altamira, Altamira, Tamaulipas, Mexico

2 CONACYT, Cd. México, Mexico

References

[1] C. Thurier, P. Doppelt. Coordination Chemistry Reviews 252 (2008) 155–169.

[2] S.P. Krumdieck, O. Sbaizero, A. Bullert, R. Raj. Surface and Coatings Technology 167 (2003) 226–233.

[3] G. Fóti, A. Jaccoud, C. Falgairette, C. Comninellis. Journal of Electroceramics 23 (2009) 175–179.

[4] A. Jaccoud, G. Fóti, R. Wüthrich, H. Jotterand, Ch. Comninellis. Topics in Catalysis 44 (2007) 409–417.

[5] A.J. Darbandi, T. Enz, H. Hahn. Solid State Ionics 180 (2009) 424–430.

[6] C. Lee, J. Bae. Solid State Ionics 179 (2008) 2031–2036.

[7] K. Sasaki, J.P. Wurth, R. Gschwen, K. Godickemeier, L.J. Gauckler. Journal of the Electrochemical Society 143(2) (1996) 530.

[8] T. Tsai, S.A. Barnett. Solid State Ionics 93 (1997) 207.

[9] T. Kenjo, M. Nishiya. Solid State Ionics 57 (1992) 295.

[10] M.J.L. Ostergard, C. Clausen, C. Bagger, M. Mogensen. Electrochimica Acta 40(12) (1995) 1971.

[11] N. Kotsionopoulos, S. Bebelis. Journal of Applied Electrochemistry 35 (2005) 1253–1264.

[12] I.K. Igumenov, N.V. Gelfond, P.S. Galkin, N.B. Morozova, N.E. Fedotova, G.I. Zharkova, V.I. Shipachev, E.F. Reznikova, A.D. Ryabtsev, N.P. Kotsupalo, V.I. Titarenko, Yu.P. Dikov, V.V. Distler, M.I. Buleev. Desalination 136 (2001) 273–280.

[13] A.M. Torres-Huerta, J.R. Vargas-Garcia. Journal of Metastable Nanocrystalline Materials 20–21 (2004) 393–398.

[14] A.M. Torres-Huerta, J.R. Vargas-Garcia. Journal Metastable Nanocrystalline Materials 24–25 (2005) 399–402.

[15] A.M. Torres-Huerta, J.R. Vargas-Garcia, M.A. Dominguez-Crespo. Solid State Ionics 178 (2007) 1608–1616.

[16] A.M. Torres-Huerta, M.A. Domínguez-Crespo, E. Ramírez-Meneses, J.R. Vargas-García. Applied Surface Science 255 (2009) 4792–4795.

[17] A.M. Torres-Huerta, J.R. Vargas-Garcia, M.A. Dominguez-Crespo, J.A. Romero-Serrano. Journal of Alloys and Compounds 483 (2009) 394–398.

[18] H.B. Wang, C.R. Xia, G.Y. Meng, D.K. Peng. Materials Letters 44 (2000) 23–28.

[19] Y. Jiang, J. Gao, M. Liu, Y. Wang, G. Meng. Solid State Ionics 177 (2007) 3405–3410.

[20] H.Z. Song, H.B. Wang, S.W. Zha, D.K. Peng, G.Y. Meng. Solid State Ionics 156 (2003) 249.

[21] J.S. Lee, T. Matsubara, T. Sei, T. Tsuchiya. Journal of Materials Science 32 (1997) 5249.

[22] A.R. West, D.C. Sinclair, N. Hirose. Journal of Electroceramics 1 (1997) 65.

[23] S.P.S. Badwal, H.J. de Bruin. Physica Status Solidi A 54 (1979) 261.

[24] E.M. Holmgreen, M.M. Yung, U.S. Ozkan. Journal of Molecular Catalyst A: Chemical 270 (2007) 101–111.

Chapter 10

Plasma-Enhanced Chemical Vapor Deposition: Where we are and the Outlook for the Future

Yasaman Hamedani, Prathyushakrishna Macha,
Timothy J. Bunning, Rajesh R. Naik and
Milana C. Vasudev

Additional information is available at the end of the chapter

http://dx.doi.org/10.5772/64654

Abstract

Chemical vapor deposition (CVD) is a technique for the fabrication of thin films of polymeric materials, which has successfully overcome some of the issues faced by wet chemical fabrication and other deposition methods. There are many hybrid techniques, which arise from CVD and are constantly evolving in order to modify the properties of the fabricated thin films. Amongst them, plasma enhanced chemical vapor deposition (PECVD) is a technique that can extend the applicability of the method for various precursors, reactive organic and inorganic materials as well as inert materials. Organic/inorganic monomers, which are used as precursors in the PECVD technique, undergo disintegration and radical polymerization while exposed to a high-energy plasma stream, followed by thin film deposition. In this chapter, we have provided a summary of the history, various characteristics as well as the main applications of PECVD. By demonstrating the advantages and disadvantages of PECVD, we have provided a comparison of this technique with other techniques. PECVD, like any other techniques, still suffers from some restrictions, such as selection of appropriate monomers, or suitable inlet instrument. However, the remarkable properties of this technique and variety of possible applications make it an area of interest for researchers, and offers potential for many future developments.

Keywords: plasma, polymers, vapor deposition

1. Introduction

Chemical vapor deposition (CVD) is a multifaceted procedure which is currently used for several applications such as the fabrication of coatings, powders, fibers, and uniform components. Metals, composites of nonmetallic materials such as carbon, silicon, carbides, nitrides, oxides, and intermetallics can be deposited through CVD. Considering one of the main applications of CVD, which is the synthesis of thin films and coatings, this procedure has in general overcome some of the problems Facing the chemical synthesis of thin films, and simplified the process by having a single-step uniform fabrication. In the CVD technique, a precursor gas flows into a chamber, over the heated substrates to be coated, and deposition of thin films on the surface occurs due to the chemical reaction in vapor phase as shown in **Figure 1A**. This procedure could be defined as an atomistic process where the primary Species deposited are atoms or molecules or a combination of both. Other common deposition procedures include physical vapor deposition technique (PVD), which uses evaporation, sputtering, and other physical processes to produce Vapors of materials instead of chemical processes. The PVD technique, however, has some drawbacks, such as low deposition rate and low pressure requirements. Moreover, this process might require subsequent annealing, which could be a drawback. Although these two techniques can be used separately, however, there are procedures that have utilized The benefits of the combination of CVD and PVD techniques, such as a modified PEVCD method.

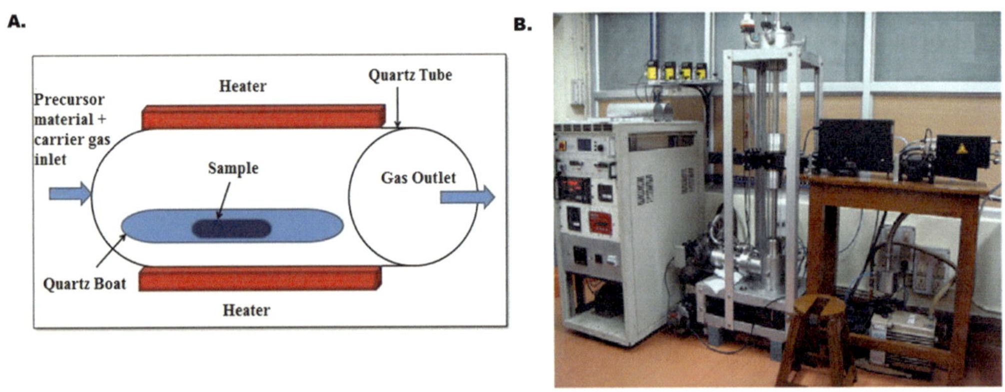

Figure 1. (A) Schematic of chemical vapor deposition technique demonstrating the various components of the instrument. (B) Microwave PECVD unit (reprinted with permission from [1], © Elsevier).

There are many hybrid techniques, which arise from CVD and are constantly evolving to modify the properties of the fabricated thin films [2], namely initiated CVD (iCVD), which utilizes an initiator to start the polymerization process because the initiator thermally decomposes to form free radicals and they are absorbed on the surface and therefore lead to start the radical polymerization, Oxidative CVD (oCVD), which can be considered as an iCVD process and utilizes an oxidant molecule as the initiator for starting the polymerization [3], metallo-organic CVD (MOCVD), which operates at temperatures lower than CVD (300–800°C) and is

http://dx.doi.org/10.5772/64654

used for achieving epitaxial growth and deposition of semiconductor materials [4, 5], Where deposition occurs at low pressure (10 –1000 Pa) to achieve and homogeneity of the deposited films [6], atmospheric pressure CVD (APCVD) in which deposition occurs at atmospheric pressure, which leads to lower equipment cost, avoids use of vacuum systems and simplicity of utilization and process control parameters [7] and plasma-enhanced CVD (PECVD), which uses electrical energy for producing a plasma, and the produced plasma activates the reaction by transferring the energy of its species to the precursors and induces free radical formation followed by radical polymerization [2]. Among them, PECVD can extend applicability of the vapor deposition process to various precursors, including reactive organic, inorganic, and inert materials. In this chapter, we will briefly demonstrate the characteristics of the PECVD technique, its advantages and disadvantages over other techniques, an overview of studies utilizing PECVD for both organic and inorganic materials deposition, as well as the future direction and potential applications.

CVD	PVD	PECVD
ADVANTAGES • Avoids the line-of-sight • High deposition rate • Production of thick coating layers • Co-deposition of material at the same time **DISADVANTAGES** • Requirement of high temperatures • Possibility of toxicity of precursors • Mostly inorganic materials have been used	**ADVANTAGES** • Atomic level control of chemical composition • Not requiring the usage of special precursors • Safer than CVD due to the absence of toxic precursors or by-products **DISADVANTAGES** • Line-of-sight deposition • Low deposition rate • Production of thin coating layers • Requirement of annealing time	**ADVANTAGES** • Avoids the line-of-sight issue to certain extent • High deposition rate • Low temperatures • Both organic and inorganic materials as precursors • Unique chemical properties of the deposited films • Thermal and chemical stability • High solvent and corrosion resistance • No limitation on substrates: complicated geometries and composition **DISADVANTAGES** • Instability against humidity and aging • Existence of compressive and residual stresses in the films • Time consuming specially for super-lattice structures • Existence of toxic, explosive gases in the plasma stream • High cost of equipment

Table 1. Comparison of the characteristics, advantages and disadvantages of the CVD, PVD, and PECVD techniques.

The CVD process, its modifications, and PVD each have their own unique advantages and applications. CVD may successfully overcome issues such as line-of sight deposition, which PVD commonly faces. This can improve the ease of coating both nonuniform and straight surfaces. From an economical perspective, CVD also offer advantages due to the ease of synthesis of thick coatings. The process also allows flexibility during deposition such as codeposition of different materials, inclusion of plasma, or initiators to improve reactivity and operation at atmospheric pressure. On the other hand, CVD in its unmodified form has disadvantages, such as the requirement of higher temperatures (over 600°C). The high temperature is a constraint both due to the energy required to heat the gas phase, which increases the cost of the process and the limitation of materials and substrates that can be used due to their instability at higher temperatures. The other disadvantage of the process is the utilization of chemical precursors with high vapor pressure, such as halides, metal-carbonyl precursors, and hybrid ones which may be an issue due to some of their associated toxicities, as well as the limitation of types of materials that may be used as precursors. Moreover, the by-products of the CVD process maybe toxic and their neutralization can make the process expensive. However, PECVD, which has seen a rapid development in the last few years, has eliminated these problems to a great extent as discussed in **Table 1**.

1.1. History of PECVD

The initial experiments using plasma in the CVD process dates back to 1950s and 1960s when the decomposition of organic compounds in the presence of an electron beam was first observed [8–10]. In the electro-optical systems, when a surface was exposed to an electron beam, it was covered by a thin film. Upon observing this effect, researchers reasoned that the formation of the thin film was due to the interactions between the electron beam and organic vapor present in the vacuum system. They concluded that the mechanism of formation of the thin layer was the free radical polymerization of the organic molecules, upon exposure to the electron beam followed by adherence to the target surface [8–10]. After roughly a decade, Buck et al. suggested that the observed mechanism of film formation can be beneficial for applications where thin insulating films were required [11]. Subsequently, relative electron bombardment of appropriate chemical compounds was used for the fabrication of polymeric or metallic thin films. Christy et al. in 1960 used a defocused, low energy electron beam to form solid films composed of silicon oil. The factors affecting the film growth rate, such as oil pressure, strength of electron beam current, and temperature were also investigated. He showed that the produced films had excellent electrical insulating properties [11]. In 1961, Baker et al. have Demonstrated the production of metallic films from decomposition of organometallic vapors by an electron beam [12]. Among the initial reports on the use of PECVD, there was one in 1962, in the Electronics Laboratory of General Electric Company for the utilization of the direct current glow discharge system for the production of thin films [13]. Laendle et al. reported the use of an organosilicon compound and its decomposition in a low-energy plasma to fabricate a silicon oxide film [10]. The decomposition reported was related to the collisions that occurred between the excited or ionized gas atoms with the organosilicon molecules, which in turn resulted in the formation of free radicals, followed by their attachment on the substrate and finally due to continuous bombardment, fabrication of stable Si-O-Si network and the resultant thin film. In 1964, Davern et al. reported the use of the same process for the fabrication of silicon oxide films at room temperature, but with the use of radio frequency (RF) plasma production source to reduce the undesired sputtering observed in the previous report. A continuous bombardment of the surface was used to prevent the deposition of decomposition products. In the following year, they also investigated the use of the silicon dioxide films as a diffusion masking against other diffusants such as N-type semiconductors including GaAs [14]. In 1965, the deposition of silicon oxide thin film capacitor was demonstrated in an AC glow discharge plasma. The produced film was amorphous, coherent and had good adherence. These capacitors also demonstrated thermal stability, resistance to humidity, and breakdown strength characteristics [15]. In 1971, Reinberg developed the "Reinberg reactor," which involved the use of a parallel plate, capacitively coupled RF reactor for depositions at lower temperatures for semiconductor encapsulation and optical coatings [16, 17]. This reactor was capable of inducing radial, laminar flow of the reactant gases over the substrate, resulted in enhancing the uniformity of the desired coating [16–18]. During this time, the process was also referred to as the chemical ion-plating and was commonly used for the deposition of silicon nitride films for encapsulation, and metal carbides [16]. Subsequently, the "plasma-assisted" or plasma-enhanced CVD as it is commonly known as evolved over the past few decades, and some of the process parameters and applications are discussed further on.

2. Methods

The CVD process can operate in two different reactor schemes: a closed reactor and an open one. In the closed reactor CVD, which is the most common one, the species are placed in a container and the container is closed; however, in the open reactor CVD, or flowing-gas CVD, the chemicals are being entered the system continuously. However, in both systems, there is a reactant supply section, in which the reactant, which may be in a gas, liquid, or solid phase, must be transported into the reactor. In the case of gaseous reactants, they can be transferred into the reactor through pressure controllers, however, in the case of liquid or solid reactants, the reactants need to be heated up to their evaporation temperature, which can sometimes be problematic to the process. Then the reactants can be transported by an inert carrier gas to the reactor [5].

Depending on the activation energy used in the CVD process, such as temperature, photon, or plasma it can be categorized into several groups, namely, thermal CVD, laser/photo CVD, and plasma CVD. Thermal CVD typically requires high temperature in the ranges of 800–2000°C, which can be produced by methods such as hot plate heating, and radiant heating. In the case of laser or photo CVD procedures, photons are produced using either a high power laser or ultraviolet (UV) radiation. Laser CVD involves the use of a laser for the generation of a strong beam of photons for the activation of CVD and photo CVD where the reaction is activated by the use of UV radiations for the dissociation of chemical bonds. However, in plasma CVD, or plasma-enhanced CVD (PECVD), the reaction is activated by the use of inert gas plasma. The utilization of plasma decreases the deposition temperature to a great degree in comparison with other methods such as thermal CVD. Deposition can occur at room temperature and the thermal effects are avoided and this opens up the opportunity of coating polymers and other materials with low melting temperatures [19]. As previously mentioned, PECVD is a variant of the CVD process in which an inert gas plasma is used for deposition of thin films. Thus far, CVD has been mostly applied to inorganic materials. The plasma, which is used in the PECVD technique, allows the usage of a wide range of precursors, both organic and inorganic due to the improved reactivity of the precursors [20, 21]. Plasma is a partially or fully ionized gas and generally is a mixture of electrons, charged particles, and neutral atoms, and therefore, the plasma state has extremely high energy; however, the plasma has no net charge, i.e., neutral. The energy that is available in a plasma discharge is used for various applications, one of these applications is the deposition of thin films and coatings.

In the PECVD process, an external energy source is required for the ionization of atoms and molecules (creation of the plasma), a pressure reduction system (maintain the plasma state), and finally, the existence of a reaction chamber. One method for plasma production is the heating of the gas; however, ionization temperatures are extremely high, and this can be limitation of this method. The other way of plasma production is the utilization of electrical energy, at various frequency discharges which can be subcategorized into audio frequency (10 or 20 kHz), radio frequency (13.56 MHz), and microwave frequency (2.45 GHz), and this can result in various types of plasma being produced. However, when a plasma produced by any of these techniques is utilized, the deposition species undergo fragmentation which forces them to become submonomers or free radicals and ions. The plasma induces radical or plasma polymerization in the monomers. Neutral molecules will be ionized or excited when the electrons and ions in the plasma interact with them, so they will become chemically reactive. The collision of a charged ion with a neutral atom will result in improved chemical reactivity and production of implantation reactions of atoms, radical generation, and polymer-forming reactions or an etching reaction [19, 22].

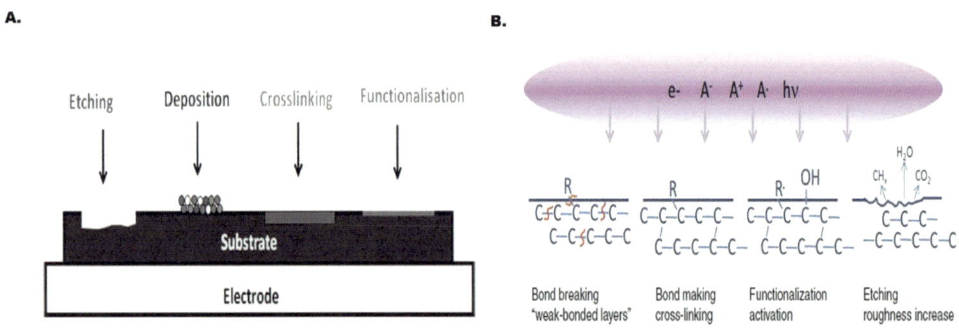

Figure 2. (A) Various processes that polymer surfaces can undergo in the time of exposure to the plasma stream (reprinted with permission from [16], © Springer). (B) Schematic of various modifications on polymer surfaces during plasma treatments (reprinted with permission from [27], © 2014 Walter de Gruyter).

The monomers that are used in this procedure are mainly in the gas or liquid states, which can be easily evaporated; however, there are studies which use solid monomers as well. Utilization of solid monomers requires inclusion of sublimation apparatus by which the solid monomer can sublime for deposition and this capability allows the use a vast range of materials as monomers [20, 23–26]. When a gaseous or liquid precursor with high vapor pressure is introduced into the PECVD reaction chamber, dissociation and activation of the precursor occur and in the presence of the plasma, which allows the deposition to happen at much lower temperatures compared to CVD. When the plasma comes in contact with the surface of a polymer substrate, modification of the surface can occur in different ways, namely *etching*, where the plasma treatment leads to the removal of materials from the surface; *deposition*, where precursors in the plasma stream are deposited as a plasma polymerized thin layer on the surface; *cross-linking and functionalization*, which involves modifications of the plasma polymers on the surface as shown in **Figure 2** [27–29].

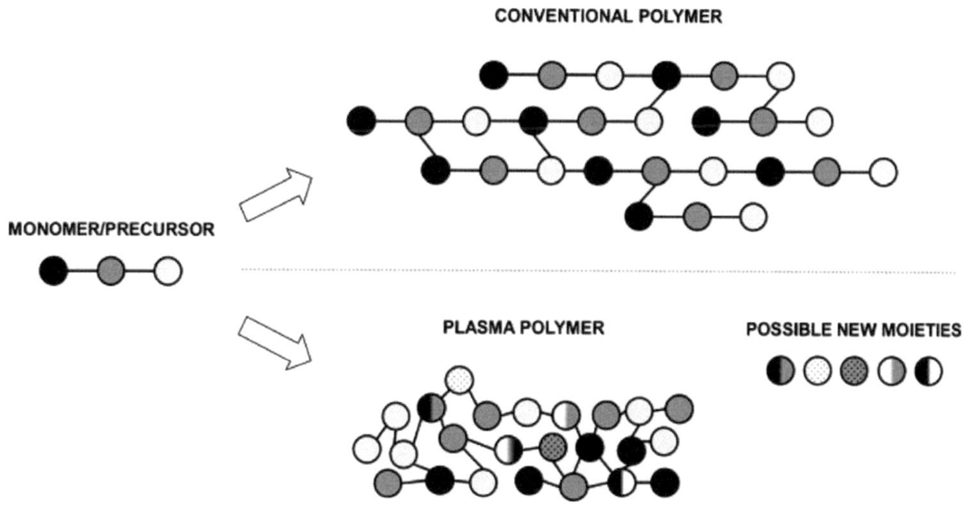

Figure 3. Schematic of structural differences among conventional polymers and plasma polymers (reprinted with permission from [30], © 2015 Elsevier).

During the PECVD process, the plasma polymerization of the precursors, which are induced in the plasma stream, is of the random radical recombination type. There is always a radical that initiates the process of polymerization, due to the radicalization of the precursor, which at the end of the process, results in the formation of polymer. And due to the presence of the radicals in each chain of the produced polymer, cross linking of the polymer also occurs; therefore, the final resulting plasma polymer coating has a cross-linked structure [29, 30, 32]. Due to the existence of a greater degree of cross-linked and branched structures, the overall structure of the plasma polymeric thin film is not similar to the conventional polymers, as shown in **Figure 3** [30, 33]. Another evidence for this difference is that the structure of subunits of the plasma polymers, which are referred to as "monomers" for conventional polymers, is not conserved during the process; therefore, they cannot be characterized by their repeating units [33–35]. However, the structure of the subunit can be reserved only under specific conditions, in which all the external and internal plasma parameters are optimized.

2.1. Variation of pressure for plasma production

The internal plasma parameters that affect the final polymerized film, include homogeneity of discharge, distribution of various species in the plasma, and energy of the species, and the external plasma parameters include reactor geometry, applied voltage, frequency, total pressure, and flow rate [32, 36]. Some of the parameters can be modified to create different variants of this technique, namely temperature, length of the deposition, pressure, inert gas flow rate, method of plasma production, and the power of the source. The most important ones among them are pressure and method of plasma production. Based on the pressure during deposition, the PECVD process can be classified as low pressure (LP- PECVD) and atmospheric pressure (AP-PECVD), with different specific applications [34]. Both of these techniques are similar to PECVD in the main steps of the procedure such as passage, transportation and absorption of the gaseous species over the surface, and production of the reaction products at the end [6]. Initial PECVD studies were performed mostly at low pressure to avoid the plasma instability due the mean-free path of the plasma species. But the utilization of AP-PECVD, which means having the plasma at an ambient pressure, has become an area of interest in the recent years. AP-PECVD offers two main advantages, namely, avoiding the cost of expensive vacuum pumping systems and transfer chambers from air to vacuum. The ease of controlling the plasma parameters such as current, gas flow, power, and voltage during the AP-PECVD process also makes it attractive and the schematic for the process is shown in **Figure 4**. Moreover, it has also been observed that APCVD demonstrates superior adherence of the coating on the substrate due to plasma activation [29, 34, 37]. Also, in LP-PECVD, there can be limitation on the size of the substrate to be treated, which is dependent on the size of the vacuum chamber, but it is not the case in AP-PECVD [38]. LP-PECVD has some other disadvantages too, such as lower deposition rates compared with AP-PECVD. For some of the LP-PECVD techniques, applying higher temperatures is required to bring about more uniformity with less defects for the produced layer, but this can also be considered as a disadvantage for the procedure [6]. Besides, it has been demonstrated that the chemical structure of the materials is more effectively retained in the AP-PECVD due to the lower energy of plasma in comparison with LP-PECVD [37, 39]. Treating materials such as textiles, scaffolds, membranes, and hydrogels is an area where the use of AP-PECVD over LP-PECVD has had a considerable effect [40, 41].

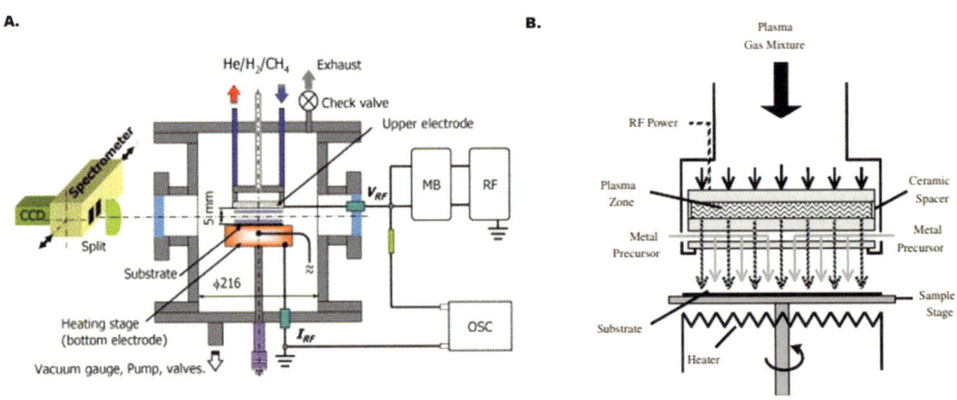

Figure 4. (A) Atmospheric RF discharge used for the deposition of single walled carbon nanotubes (reprinted with permission from [42], © Elsevier). (B) Schematic of APCVD apparatus used for the deposition of zinc oxide (reprinted with permission from [43], © 2007 Elsevier).

The primary advantage of the low pressure plasmas is the high purity of the coatings produced compared to the atmospheric pressure plasmas. When utilizing LP-PECVD, the ratio of mass transport velocity to the velocity of reaction on the surface will be different. When the pressure decreases during LP-PECVD, the diffusion of the gas decreases and therefore the velocity of mass transport will also decrease correspondingly and the produced layer is expected to be more uniform [6]. However, in the AP-PECVD, the ratio of mass transport velocity and the velocity of reaction on the surface is equal to one and the two velocities are of the same order of magnitude. Moreover, another issue facing with AP-PECVD is the problem of maintaining a glow discharge under the atmospheric pressure, which however, has been developed to some extent [38]. One of the major differences between the LP-PECVD and AP-PECVD is in the plasma excitation step. In LP-PECVD, the collision between the monomer with an electron is responsible for activating the monomers and the excitation mechanism can be denoted as follows (1):

$$M_i + e^- \longrightarrow M_j^* + M_k^* + e^- \tag{1}$$

Although this process exists in the AP-PECVD as well, but at atmospheric pressure, there are other reactions in which unstable high-energy atoms or molecules from the plasma gas exist too, so the aforementioned collision may not be the main collision responsible for activating the monomers. And therefore, the energy transfer in APCVD is more efficacious (2).

$$M^* + A_2 \longrightarrow M + A^+ + A + e^- \tag{2}$$

2.2. Methods of plasma production

The different methods of plasma production, include microwave frequency (MW) plasma, also known as glow-discharge, radio frequency (RF) plasma, and arc plasma. By utilizing two electrodes in the reaction chamber, the high frequency power is applied to the gas to be ionized. When the microwave frequency is used, plasma is produced by the application of a high frequency electric field of magnitude of 2.45 GHz at a low pressure to a carrier gas such as oxygen, argon, or nitrogen, and the schematic of this process is shown in **Figure 5A**. At such high frequencies, electrons due to their small masses can accelerate toward the higher energy levels; however, this is not the case for the heavier ions, therefore, the temperature of electrons increases, but for heavier ions, it will remain low, and so this is a nonisothermal plasma. In the microwave plasmas, the electrons with their high energy levels are responsible for collision with the gas molecules to dissociate them and make them reactive for further plasma polymerization and one such reactor is shown in **Figure 1B** [1]. This technique has been used for deposition of conductive films from a metal target. There are also many reports on the utilization of lower frequency discharges such as RF [23, 24]. Usually, in the case of an RF source, a matching box is used alongside the RF generator to reduce the reflected power, as shown in **Figure 5B**. The power setting differs based on the material deposited. The RF source is mainly used for nonconductive materials. Unlike for the LP-PECVD case in which microwave frequency or a microbeam plasma generator is used, production of a stable plasma in an AP-PECVD utilizing an RF source does not require a dielectric material between the electrodes [44]. Arc plasmas typically operate at moderate frequencies of 1 MHz lower than the microwave plasma sources.

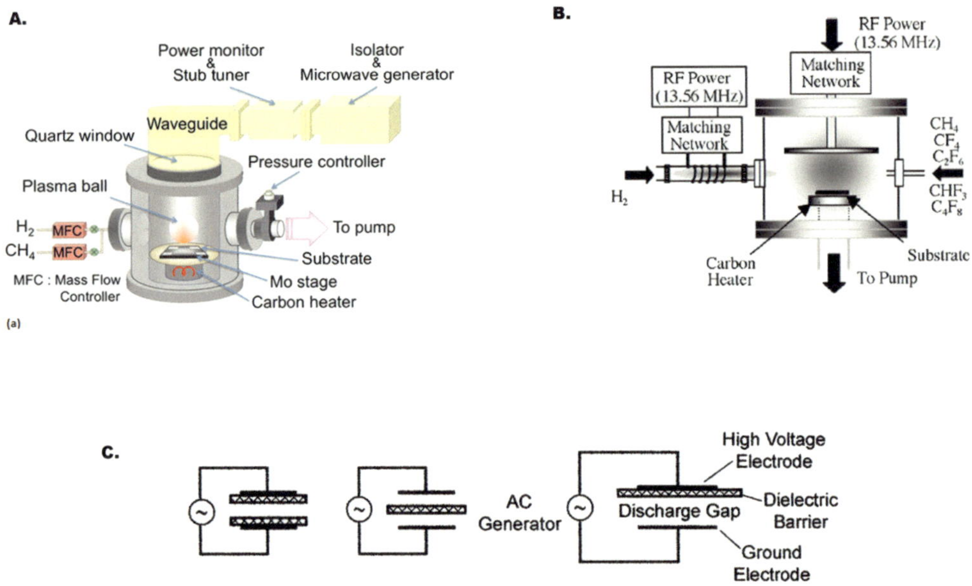

Figure 5. (A) Schematic of carbon nanotube growth procedure with microwave PECVD apparatus (reprinted with permission from [45], © InTech). (B) Schematic of the radio frequency PECVD apparatus for the production of carbon nanowalls (reprinted with permission from [46], © 2005 Elsevier). (C) Schematic of a dielectric barrier discharge system (reprinted with permission from [47], © 2003 Springer).

However, arc plasmas are extremely hot and require higher power, so they are not usually utilized and some of the efforts to avoiding this include limiting the discharge time, which is the time applied for the electrons to accelerate in the electric field, by applying pulsed voltages or using a dielectric material between the electrode to reduce the accumulation of the electrons, which is called dielectric barrier discharges (DBD), as shown in **Figure 5C** [19, 34]. Arc plasma CVD techniques are mainly used for deposition of diamonds where the process occurs at a higher pressures compared to MW PECVD, and therefore, the number of collisions increases, and, consequently, the temperature is higher. Due to high temperatures, total dissociation of hydrogen molecules occurs and proves to be advantageous over the MW PECVD and thermal CVD due to the decreased production of atomic hydrogen species [19]. Due of the high temperature of the arc plasma deposition, substrate cooling is necessary, but it is difficult process; therefore, the high temperature of this procedure counts as a disadvantage of it; however, high rate of deposition and thickness of the produced layer are some of the advantages. The plasma at different frequencies (microwave, radio frequency, and arc plasma) can be produced by various methods such as inductively coupled or capacitive-coupled parallel plates and/or electron cyclotron resonance (ECR). For example, an MW plasma can be produced by the latter method, which is the case when the frequency of the applied electric field becomes matched with the natural frequency of the electrons when they orbit the magnetic field. The RF frequency plasma can be generated in the capacitive-coupled parallel plate configuration plasma reactor in which one of the electrodes is electrically grounded while the RF power is applied to the other electrode as shown in **Figure 5A** [46]. By considering the temperature of the species present in the plasma stream, the process can be divided into two categories. One is the case in which all the plasma species: ions, electrons, and neutral species all exist at the same temperatures, which means that plasma is at the local thermal equilibrium (LTE), and the other is when this equilibrium does not exist and the species are at different temperatures (non-LTE), and electrons may be at a higher temperature in comparison with the other particles. In a LTE plasma state, the density of electrons is usually higher than the non-LTE state, and therefore, the mean free path of the electrons is lower, and, consequently, the number of collisions increases. In the case of AP-PECVD, the plasma is expected to be in the LTE state since the higher pressure also leads to increased number of collisions; therefore, the temperature of the molecules and ions increases [43]. This case is called an arc plasma, which operates at lower frequencies compared to microwave and higher frequencies compared to RF sources (1 MHz) [19, 34].

PECVD can be considered as a nonline of sight process, because the plasma stream can surround the substrate, and therefore, leads to uniform deposition and alleviates some of the issues associated with ionic or e-beam sputtering processes. It has been well-documented in the literature that the resulting chemistry of the thin films produced by PECVD is unique and cannot be obtained by common wet deposition techniques. These fabricated thin films also demonstrate high solvent and corrosion resistance along with thermal and chemical stability [24]. The process results in efficient and controlled usage of precursors and the generation of very little by-products. The process allows the utilization of unconventional precursors and can be used for deposition on surfaces with complicated geometries [48]. Moreover, because the interactions are occurring just at the surface of the substrates, mechanical properties of the substrates will not change and the film that is formed is completely uniform [49]. The plasma polymers fabricated by this technique have high modulus, due to the dense cross-linking in their polymer chains. There are some reports that show these produced polymer layers have a high permeability coefficient of oxygen and nitrogen which is mainly related to the deposition mechanism as well as the polymerization mechanism of plasma polymers [22]. Besides, if the plasma polymers contain ionic groups, they will have a wide variety of applications in various fields which require electrical conductivity, antistatic esterification, and biocompatibility.

A major difference between the films produced by PECVD and CVD is that, the ones produced through PECVD have higher content of hydrogen, which is due to the utilization of plasma in the deposition process [50]. The synthesized plasma polymers have some disadvantages as well, such as instability against aging and humidity which is due to free radicals that exist in the plasma polymers as well as the existence of polar side groups that can absorb water, and therefore may render them unsuitable for dielectric thin films [49, 51]. Some ways of overcoming this problem such as heat treatment of plasma polymers in a vacuum which will decrease the radical concentration in the deposited thin films has been reported [35, 49, 51]. Another disadvantage is that undesirable compressive and residual stresses in the deposited film can occur at lower frequencies, which could lead to subsequent cracking and cause problems in many applications, specifically metallurgy [35]. Residual stresses can also affect thick oxide films where fracture has been observed [52]. Techniques for controlling the residual stress, which ultimately resulted in limiting the thickness of the produced layer, such as rapid thermal analysis (RTA) have been utilized [53]. However, this treatment may have other impacts on the characteristics of the produced film too; therefore, the behavior of the PECVD oxide layer following the heat treatment should be analyzed [54]. Also, the ion bombardment during the process could damage some sensitive substrates. Moreover, the problem of line-of-sight does exist to some extent and therefore, the regions that are directly exposed to the plasma stream will be more affected than the other regions. One other major issue is that due to the presence of several species in the plasma stream (with varying mass and charge), the film growth can be influenced by different species. The mechanism of this growth depends on the chemistry of the depositing species and the conditions used for deposition. For example, in DLC film deposition, it has been shown that both ions and neutral species are a part of the mass of the produced film [55]. However, in the production of silane or polymeric films, the role of radicals, as well as the influence of plasma pressure, has been considered [56, 57]. Therefore, development in the understanding the roles of different plasma species on the film growth, will enhance the film production, as well as control of the process [58].

3. Applications

The initial application of PECVD was in microelectronics, namely electrical insulation films, conductive films, semiconductors, diffusion masks, and capacitors. Over a decade after the establishment of PECVD process, researchers started to utilize this technique for the fabrication of integrated circuits (ICs) [15, 52, 59], solar cells [43, 60, 61], transistors [62], photovoltaic and photonic applications [48, 59, 63, 64] such as notch filters and antireflective coatings [27]. Deposition of inorganic materials by PECVD also has applications in food packaging, scratch resistance, and biomedical applications [2, 65, 66]. PECVD has been extensively used in the development of organic polymers and organometallic compounds. Plasma polymers are used to avoid corrosion in many dielectric and optical devices. The film properties can be easily controlled to produce homogenous organic thin films on large substrates (silicon and glass) with varying mechanical, optical, chemical, thermal, and electric properties. Due to its relatively moderate nature, this technique has also recently been used for various biological applications. Biological applications include immobilization of the biomolecules (e.g. enzymes and proteins); biometallization of gold, silver and platinum nanoparticles with the aid of DNA, amino acids or proteins; Degradation and molecular weight reduction of biopolymers such as chitosan in order to enhance their properties; Changing surface functionalities by addition of functional groups such as amine, carboxylic and silane groups; Fabrication of organic/ inorganic hybrid film for enhancing the dielectric properties. Such surface treatment of

http://dx.doi.org/10.5772/64654

implants and biomaterials for applications in biosensors, device fabrication and anti-biofouling coatings [20, 25, 40, 67].

3.1. Silicon-based applications

PECVD is recognized as a unique method of organic thin film deposition and is commonly used for the deposition of inorganic and organic, doped, and undoped films. Silicon composite films is one area where thin films are deposited from various precursors such as silane (SiH_4), siloxane, and silazane mixed with other gases in varying ratios [68, 69]. The two predominant forms of silicon which get deposited are hydrogenated amorphous (a-Si:H) and microcrystalline (μc-Si:H) silicon [70–72]. The films deposited through PECVD, are used in semiconductor devices, solar cells, and optically active device applications due to their optical, mechanical, and electrical properties. Recently, PECVD has also been utilized for the processing of flexible and printable electronic devices, due to its high process efficiency, large-scale patternability, lower cost, and environmentally friendly nature [73, 74]. AP-PECVD at low gas temperatures due to absence of vacuum equipment, seems beneficial in the surface processing of flexible materials and has been demonstrated in the lower temperature deposition of SiO_x films on polymer substrates that are used as diffusion barrier and stable hydrophilic surfaces. The applications of flexible semiconductors can be seen in wearable electronics and organic electronics [75, 76].

3.1.1. Semiconductor devices

The physical properties of PECVD-deposited films are known to be superior since they are highly cross-linked, uniform, and resistant to thermal and chemical changes and have low rates of corrosion. Plasma polymers find extensive use in dielectric films and optical coating due to lower cost and higher efficiency [77, 78]. Plasma polymerized organic-inorganic hybrid thin films with ratio controlled precursors has been used to optimize dielectric properties. Also, the characteristics of the thin film were modified by the gases' mixing ratio, plasma power, and chamber pressure [35, 79]. Silicon oxynitride (SiON) and silicon nitride (SiN) are some of the commonly used precursors for the deposition of insulating thin films for MIM (metal-insulator-metal) capacitors and allow easy manipulation of film properties [78]. It was observed that the varying the ratios of silane (SiH_4), nitrous oxide (N_2O), and ammonia (NH_3), lead to the deposition of thin films of varying compositions. Properties including deposition rate, refractive index, and extinction index of the film were observed to improve with the increase in SiH_4/N_2O (NH_3) ratio. Although both SiN and SiON have their advantages, SiN films are used for semiconductor applications due to their higher capacitance density, breakdown voltage and particle performance (**Figure 6A–C**) [63, 73, 80]. Silicon carbide, which has a high band gap and chemical inertness, is another material commonly deposited using PECVD. SiC is used widely as a material for MEMS devices due to its stability in harsh and high temperature environments [36, 81, 82]. PECVD, being a metal-compatible deposition method has been used for the deposition of α-SiC at 450°C using methane and silane precursor gases. Low frequency RF (40 Hz) power and pressure of range 700–1000 Torr, along with changes in the precursor ratios were utilized for the production of films of varying electrical and mechanical properties. The results indicate the potential of the α-SiC films as structural components for devices that need stable residual film stress at high temperatures of operation [36]. Although the thin films of SiC have demonstrated promise in the development of high temperature withstanding MEMS devices, the increase in surface roughness in deposited thicker films remains an obstacle. Increasing the carbon content of the film in mixed frequency condition and decreasing the RF power was observed to decrease the surface roughness, making it a potential solution. Such methods pave the way for future studies on the influence of PECVD process parameters on the growth rate, morphology, and microstructure in the field of semiconductors.

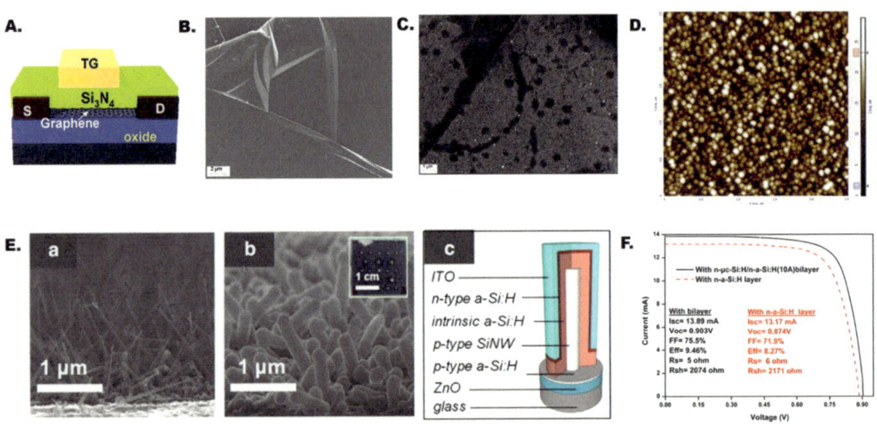

Figure 6. (A) Schematic representation of graphene FET device with silicon nitride coating. (B and C) SEM images of silicon nitride and silicon oxide deposited on graphene (A, B and C: reprinted with permission from [80] © 2010 ACS). (D) 2D surface morphology of n-µc-Si: H layer used in single junction a-Si solar cells [83]. (D, a) SEM images of p-type Silicon nanowires using PECVD on a glass substrate (Sn catalyzed, 600°C). (b) Deposition of a conformal intrinsic and n-type a-Si:H after a temperature drop of 175°C using PECVD. (c) Schematic representation of a PIN radial junction. (Reprinted with permission [84], © 2011 Elsevier). (E) I–V characteristics of solar cells with n-a-Si: H layer and n-µc-Si:H/n-a-Si: H bilayer (F and G: reprinted with permission [83], © 2015 Solar Energy).

3.1.2. Solar cells

Various deposition techniques such as PVD, CVD, oxidation, plating, and spin-coating have been used extensively for the deposition of silicon thin films with good photoelectric properties [85]. Currently, the photovoltaic market is dominated by crystalline silicon cells, which accounted for nearly 95% of world's photovoltaic cell and module production in 2004 [85, 86]. PECVD is widely used in the microelectronics and solar cell production, specifically for the deposition of thin films from a mixture of gas-phase species onto a solid substrate [87]. There has been a continuous effort to establish the process conditions to improve the film quality and the rate of deposition. PECVD at 60 MHz excitation frequency is now used extensively in large area thin film technology and industrial applications [83, 88, 89].

An increase in growth rate (going up from 1.5 to 4 nm s^{-1}) was obtained by switching from a low pressure (2–4 Torr) conventional RF plasma (13.56 MHz) to a 60 MHz excited plasma-assisted CVD was observed by Matsuda et al. and by several other research groups [88, 89]. The crystallinity of the deposited silicon plays a vital role in the film properties. It has been observed that even high quality amorphous silicon solar cells have very low efficiency in trapping light when compared to crystalline counterparts, which has led to various studies to improve solar power conversion (**Figure 6F**) [83]. Several groups have worked on the transformation of the silicon thin film from amorphous to nanocrystalline in the presence of nonthermal SiH_4/H_2 plasmas. The interactions between the hydrogen atoms of the plasma and the solid silicon matrix determine the crystallinity and other properties of the thin films. The presence of H atoms in between the strained Si-Si bonds led to the disordered-to-ordered transitions [70, 71]. The deposition parameters such as chamber pressure (P_r), gas mixture composition, and flow rates, RF power density (P_w), and substrate temperature can be varied to obtain the desired film properties, including the crystallinity [69, 70]. Microdoping is another process used for the betterment of optical, electrical and structural properties of a microcrystalline silicon film for solar cell applications. The effect of boron microdoping through PECVD at high temperature and very high frequency (VHF, 60 MHz) PECVD at 200°C with a varying diborane flow rate of 0.00–0.30 sccm has been studied [90, 91]. Improved dark current flow and conductivity changes up to 10 orders of magnitude were observed upon boron doping was observed in these studies and could subsequently be used to improve solar conversion efficiency.

Another area where PECVD has been used extensively is the deposition of a passivation layer, which involves a shielding outer-layer deposited as a microcoating. In microelectronics and photovoltaics, a passivating layer can reduce surface recombination, which is a significant cause of losses in solar cells. Hydrogen atoms play an important role in the termination and passivation of silicon dangling bonds [70, 92]. Hydrogenated amorphous and nanocrystalline silicon films produced by PECVD are now widely being used in electronic and optoelectronic devices [70, 89]. Also, nanostructures have been used extensively to improve efficiency of solar cells in the recent times. One example is the use of silicon nanowires to improve the light-trapping in low-cost silicon photovoltaic cells (**Figure 6E**) [84, 93]. The growth of radial junctions of hydrogenated amorphous silicon over p-doped crystalline silicon nanowires on a glass substrate using PECVD (single pump) and Sn catalysts has demonstrated a considerable increase in the dark current and conversion efficiencies [84]. Silicon solar cells produced on flexible plastic substrates at an optimum deposition temperature of ~200°C, typically are cheap and robust [61, 94, 95].

3.1.3. Optically active films

Thin films with dye molecules are used in laser cavities, optical filters, and optical gas sensors [96–98]. Such thin films are synthesized using wet and sublimation techniques. PECVD avoids the harsh (chemical and temperature) and multistep process of these fabrication methods and can be used to deposit optically active thin films with tailorable properties. The deposition involves a partial polymerization of dye molecules that are evaporated over a substrate while exposed to an arc plasma. Ultrathin highly planar films (~100 nm) with controlled gradation of color were synthesized using the PECVD process. Such films can be used in development of photonic materials and devices such as sensors, and wavelength couplers [99, 100]. Plasma polymerization of fluorine-containing materials yield many properties such as low dielectric constant and refractive index, low surface energy, low coefficient of friction, low permeability constant, and good biocompatibility [23, 101]. Fabrication of thin films using plasma homo- and copolymerization of octafluorocyclobutane (OFCB, C_4F_8) and hexamethyldisiloxane (HMDSO, $C_6H_{18}Si_2O$) using PECVD (RF at 13.56 MHz, power 20–45 W and pressure 0.001–1 Torr) were studied for their optical properties [23, 102]. Rare earth-doped optical materials is an area that has also garnered interest in the recent times, Pitt et al. report the synthesis of silicon-rich SiO_2 along with erbium doping. PECVD was used for the synthesis of such films due to low processing temperatures, ability of controlling stoichiometry, and growth conditions for the film. Microclusters of Si were found to be embedded in the erbium doped SiO_2 films demonstrating strong absorption peaks in the visible region along with the strong near IR fluorescence due to the erbium ions [97, 103].

3.1.4. Quantum dots

Silicon nanostructures have been found to exhibit luminescent properties, leading to many studies to understand them [63, 104]. Si-rich silicon nitride films have been grown by low frequency (440 kHz) PECVD using SiH_4 and NH_3 as reactant gases at a plasma power and a chamber pressure of 1000 W and 1500 mTorr, respectively. The total gas flow being 800 sccm, the ratio of NH_3/SiH_4 flows was changed between 2 and 10, and the flow rates were changed at a temperature of 370°C to get a different Si concentration in films. These kinds of films can be used in photovoltaic devices [105, 106].

3.2. Carbon-based nanostructures

Carbon-based nanostructures can be classified into 2D structures such as nanowalls, graphene, and 3D structures such as nanotubes and nanowires. Based on their morphological features and orientation, such nanostructures are used for different applications. For example, conical structures give higher mechanical and thermal stability when compared to narrow ones and the large surface area of nanowalls prove helpful in electrochemical devices and gas storage [107, 108]. The main drawback in the synthesis of these nanostructures is the absence of a controlled and deterministic method for depositing large-scale assemblies, which can be used in nanoelectronics, biological applications (probes), field emission displays, and radiation sources. PECVD can be used to address these difficulties and modifications of the deposition conditions produces different morphologies, location, and orientation-specific carbon nanostructures such as carbon nanotubes, diamond-like carbon (DLC) films deposited using microwave PECVD [108, 109]. However, the use of PECVD for the synthesis of DLC films is limited because of the issue of being unable to reach harder tetrahedral amorphous carbon state, due to the presence of hydrogen in the gaseous precursor [109].

Carbon nanotubes (CNTs) formed from graphene sheets have excellent electronic and mechanical properties. CNTs in the range of 1 nm in diameter and a few micrometers in length can be classified broadly as single-walled or multiwalled CNTs. Single-walled nanotubes (SWNTs) are known for their semiconductor properties and find applications in high performance electronics. SWNTs have been grown by using RF PECVD (13.56 MHz, power ~75 W, low pressure) with methane as a precursor and metallic seed catalyst as shown in **Figure 5**. MW PECVD process with frequency of 2.45 GHz, pressure of 20 Torr, and power of 5 kW could also be used to produce SWNTs [45, 110]. Various parameters and factors involved in the process influence the preferential growth of the CNTs (**Figure 7**) [108, 110, 112]. Vertically oriented SWNTs have been synthesized, and this alignment is attributed to the strong electric field in a plasma sheath [44, 113]. Such CNTs can be used in various applications such as field emission devices and nanoscale electrochemical actuators [111, 113–115]. The formation of iron silicide (FeSi) roots on the Si substrate was observed to be helping in the formation of a base growth mode, which led to well-aligned CNTs deposition (**Figure 7B** and **C**) [114]. Recently, Bo et al. have reported the catalyst-free PECVD growth of graphene. The precursor was a volatile natural extract from the tea tree plant (*Melaleuca altornifolia*) [116]. The deposition rate was high and produced high quality graphene films without the presence of a catalyst. The graphene films at the end of this sustainable bottom up process, when tested yielded a high water contact angle of 135°, proving its potential use as a highly hydrophobic coating. Also, due to its mem-resistive behavior, such graphene could be used in the construction of memory devices [116].

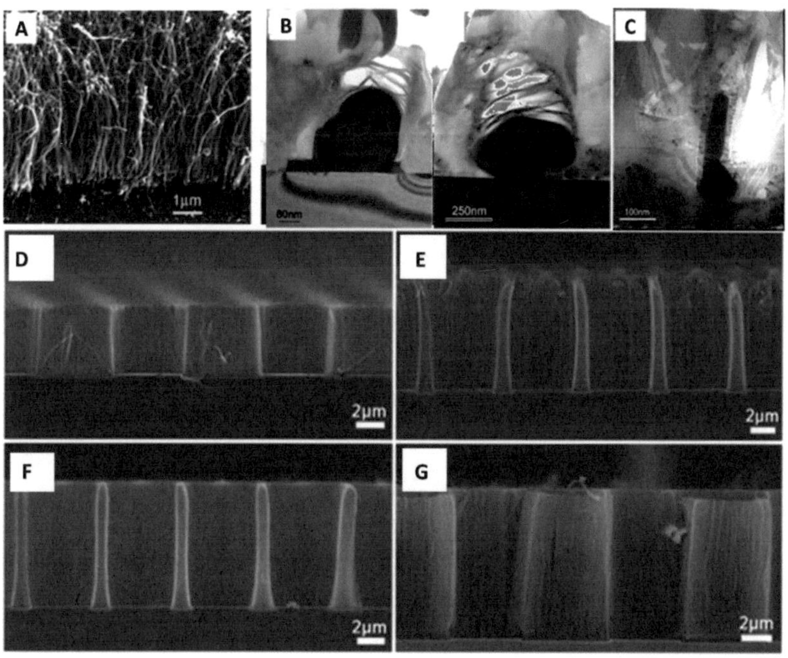

Figure 7. (A) SEM image of CNTs grown using microwave PECVD (4 nm Fe film as a catalyst). (B) TEM images of various possible shapes of catalyst at the CNT growth, i.e., not well attached to substrate and completely enclosed). (C) Elongation of catalyst particle in vertical direction moving from (A) and (B) (reprinted with permission from [113], © 2005 ACS). (D–G) CNTs deposited from different mixture ratios of H_2/C_2H_2. The flow rates of H_2/C_2H_2 were 140/2, 160/5, 140/5, and 180/10 sccm, respectively (reprinted with permission from [110], © 2015 Creative Commons Attribution (CC BY)).

Carbon nanowalls (CNWs) have been deposited using acetylene as precursors with iron as the catalyst. The growth parameters such as temperature, pressure, and gas flow rate were varied to get various samples of CNWs. The samples which displayed best field emission properties were the ones deposited at higher temperature and pressure (650°C with 120 sccm H_2 flow rate and a pressure of 300 Pa) [46, 112]. Graphene an allotrope of carbon is an example of 2D carbon nanostructure with extraordinary properties such as being 100 times stronger than steel and conducting heat and electricity with a nearly transparent sheet like morphology [117]. Graphene deposition for dielectrics is a challenge because of discontinuity in film formation. This can be compensated by the deposition of SiN by PECVD, which provides excellent coverage of graphene for preservation of their carrier mobility for application in FETs [80]. Their unique structural features, i.e., vertical orientation of the substrate without any agglomeration, sharp and exposed edges, and the controllable intersheet connectivity provide numerous advantages. These nanosheets can be deposited using PECVD on a variety of substrates using gas, solid, or liquid precursors [117]. Perpendicular orientation of the graphene sheets demonstrates better properties in comparison with conventional graphene films and is utilized for energy storage, sensors, and environmental applications. There are several unanswered questions in the development of carbon-based nanostructures by PECVD. Some of the unknowns involve the chemical species, or ions that are responsible for nanostructure growth, and their influence on the adhesion to the substrate, the rate determining step in their growth, and the effects of pressure and process parameters on the growth. Understanding some of these processes will help in the manipulation of the parameters to acquire required shape and orientation of carbon nanostructures and can lead to efficient large area deposition for various applications [108].

3.3. Titanium-based applications

Titanium has many applications due to its strength and low density and is used as an alloy, for thin film coatings, as substrates and photocatalysts. Its applications are wide-range and can be predominantly seen in medical prostheses (dental and orthoimplants), antimicrobial coatings, and electronic devices. RF PECVD has been most commonly used in the production of titanium thin films [118–123]. As a photocatalyst, titanium and its oxide (TiO_2) films have numerous applications such as antimicrobial activity, waste water treatment, and for biomedical implants. The antimicrobial activity of titanium-oxide-coated cotton textile substrates (**Figure 8A–C**) and glass substrates were tested against *E. coli*. The coatings were uniform and bacterial inhibition increased correspondingly with the increase in the refractive index of the film [118]. Titania powder has also been used for photocatalytic activities and this property was harnessed in the degradation of certain organic compounds, which could further be adapted for waste water treatment, but powders have lower efficiency due to crystalline defects. Although crystalline TiO_2 is more efficient, its surface area is not as high as powder. Hence, studies have been conducted to improve the surface area while retaining the crystallinity aspects of TiO_2. An excitation frequency of 40 kHz was employed in a cold-wall-type PECVD (precursor: titanium tetra-isopropoxide (Ti (O-*i*-$C_3H_7)_4$), and oxygen gas carrier flow rate: 50 sccm and reduced pressure conditions ~30 Pa) was used to produce crystalline TiO_2 particles, which had a high surface area (**Figure 8E**) and good photocatalytic properties [120].

The combination of PECVD deposited TiO_2 films with silver nanoparticles have been studied (**Figure 8D**) to characterize the films for their photolytic activity and antimicrobial property (silver is known for its antimicrobial activity). These films demonstrated good photochemical activity at nearly room temperature and could be used for heat-sensitive substrates [119]. PECVD can also be used for other material-based precursors and here are some examples and applications of such depositions: the mixtures of Ti $(O\text{-}i\text{-}C_3H_7)_4$ and oxygen can be used to produce amorphous TiO_2 thin films by PECVD. The dielectric properties of TiO_2 films were put to use by assuming that a TiO_2/SiO_2 bilayer would behave as two capacitors in series. The rate of deposition was observed to decrease with oxygen concentration and RF power and whereas an increase with equivalence ratio. The film thickness was seen to influence the dielectric constant of the film and it was pretty consistent, even in the presence of an interfacial SiO_2 layer [42].

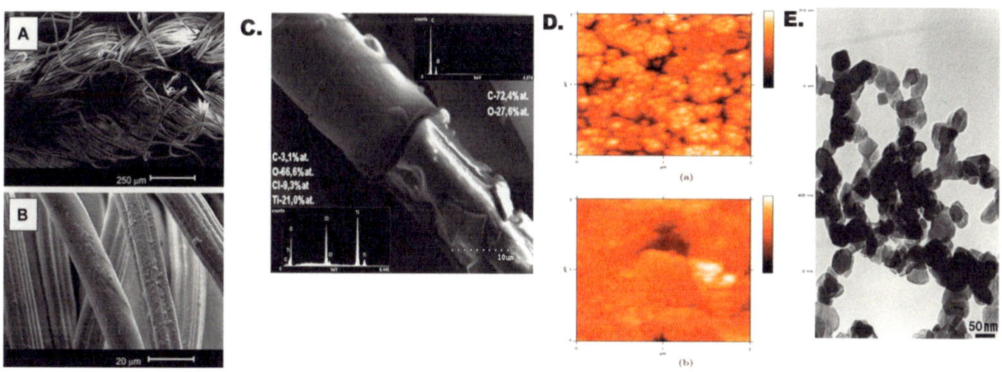

Figure 8. (A and B) Low and high magnifications SEM images of cotton knitwear deposited made of titanium oxide with RF PECVD (100 W). (C) SEM image of a single cotton fiber and it has been stripped partially. The EDX results in the two windows are for coated and uncoated fiber (reprinted with permission from [118], © 2005 Elsevier). (D) AFM pictures of the TiO_2/PECVD films made at (a) 300°C, (b) 40°C using PECVD (2 × 2 μm) (reprinted with permission from [119], © 2009 Springer). (E) TEM image of the TiO_2 powder produced using PECVD following treatment (800°C for 1.5 h) (reprinted with permission from [120], © 1999 Springer).

3.4. Biological applications

PECVD is a technique which operates at lower temperatures, suitable for the deposition of organic precursors, unlike other processes. Synthesized thin films are biocompatible, adherent, and flexible for applications such as implants [2, 65]. PECVD offers several advantages in case of deposition of ultrathin layers of macromolecular materials such as alcohols, carboxylic acids, acid chlorides, simple amines, anhydrides, and ethylene glycols [124–130]. The chemical reactivity of the deposited layer is generally based on the functional group density and various degrees of cross-link density. The chemical reactivity of these functional layers also determines shelf life of the resultant device. These plasma-polymerized thin films are utilized in various biological applications such as antimicrobial coatings, biomaterials, tissue engineering, implant coatings, and bioelectronics [2, 66, 67, 131].

3.4.1. Antimicrobial and antifouling

Antimicrobial surfaces are in demand owing to increase in bacterial colonization issues in medical devices, implants, and healthcare products. The lifetime of implanted prosthesis is brought down drastically due to bacterial infections, especially in immunocompromised patients. The use of antimicrobial surfaces has gained importance lately, because they do not lead to resistance (common example *Staphylococcus aureus* which is multidrug resistant). The PECVD process was an effective way of designing these surfaces because it can be used efficiently to control the coating properties and content of nanoparticles embedded. Biocidal activity of silver nanoclusters embedded in a PECVD deposited film is believed to be a result of effective progressive release of Ag ions from the nanoparticles in to the media surrounding it [119, 131–133]. In a recent study, polydimethylsiloxane (PDMS) surfaces were coated with PFM (pentafluorophenyl methacrylate) using PECVD with a continuous RF power (15 W), and pulsed plasma polymerization (DC of 0.5) for 3–5 min to produce a flexible and highly reactive surface with ester groups [134]. The incubation of these surfaces in an amine sugar (glucosamine) produced a controlled reductive surface capable of reducing silver salts to form micro- and nanostructured silver coatings. Gilabert et al. have demonstrated the antibacterial activity of such coatings due to their hydrophobic behavior which eliminated bacterial adhesion. The initial rapid release followed by a sustained release proved effective against *P. aeruginosa* and *S. aureus* bacterial adhesion but had no cytotoxic effects on the mammalian cell line (COS7 cells) when tested (**Figure 9**) [132]. Silver coatings are also commonly used on implants to avoid bacterial infections and biofilm formation [134].

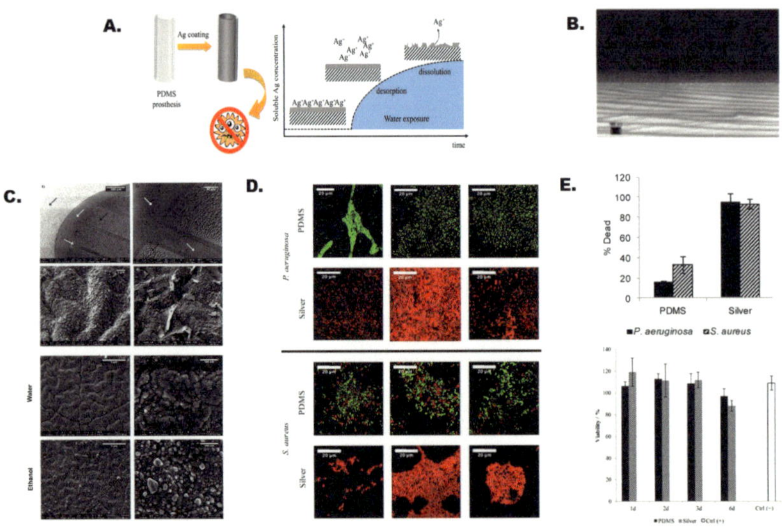

Figure 9. (A) Schematic representation of synthesis of highly hydrophobic silver nanoparticle surface and silver ion release (reprinted with permission from [132], © 2015 ACS). (B) Silver nanoparticles on silicon oxide layer, a product of APCVD plasma jet (SEM micrograph) (reprinted with permission from [131], © 2012 Wiley). (C) SEM images depicting the surface morphology of silver modified PDMS samples. (D) Confocal images showing the bacterial viability on noncoated (PDMS) and coated (silver) surfaces. (E) Antibacterial activity of coated (silver) and noncoated (PDMS) surfaces on two different strains (*P. aeruginosa* and *S. aureus*) and cell viability of COS-7 cell-line (C–E: reprinted with permission from [132], © 2015 ACS).

3.4.2. Wettability and cell adhesion

In biological systems, the hydrophobic and hydration forces play an important role in the mediation of solute, for instance, in the adsorption and adhesion of proteins and cells (**Figure 9B**) [135, 136]. Production of bioabsorbable materials is one main area where PECVD has been utilized [137–139]. The design and synthesis of various functional hybrid film systems for SiO_x film coatings with good mechanical properties at high deposition rates were reported. These SiO_x surfaces are known for their hydrophilic and smooth properties that can be modified using plasma treatment [139, 140]. The deposition of SiO_x films on poly (lactic acid) (PLA) substrate was synthesized by Chaiwong et al. using octamethylcyclotetrasiloxane (OMCTS: $Si_4O_4C_8H_{20}$) as the precursor and oxygen as carrier gas. Oxygen plasma treatment after deposition enhanced the wettability of the surface leading to an increased BSA protein adsorption and cell adhesion (preosteoblasts, fibroblasts) [139].

3.4.3. Biosensors

Disease diagnosis using biosensors requires functionalization of sensors' surfaces with specific antigens to detect required biomolecules. Typically, biosensors use functionally modified polymer surfaces to immobilize proteins. Use of chemical synthesis methods is time consuming, complicated, and quality of film is difficult to control during the process. PECVD, being a single-step process, provides a venue to overcome some of these obstacles. Multiple studies have used PECVD for the deposition of various end-groups such as amines, carboxylic, mercapto, epoxy, and polyethylene glycol to address the problem of nonspecific binding of single-stranded DNA or other proteins of interest. Parameters such as higher RF power lead to increase in thickness of the film due to increased rate of deposition [140, 141]. Monomers such as amino silanes, allylamine, and allyl organophosphates have been commonly used for the deposition of thin films with reactive functional groups [58, 127].

3.4.4. Bioimplants

Hydroxyapatite (HA) is a major mineral component of natural bone that can readily integrate with human tissues within just a short period after implantation. Therefore, it is widely used in bone and dental implants. Although it has multiple advantages, it has poor mechanical strength and load bearing capacity. Hence, it is used predominantly as a coating on metallic implants [142–144]. Some studies have shown that the inclusion of HA along with titanium in a plasma sputtering process promotes the formation of calcium titanate, which largely contributes to the adhesive strength between the alloy and the film. The structure and morphology of the coatings were controlled using plasma parameters such as DC substrate bias. It was observed that at higher DC (negative) bias voltages, the calcium oxide (CaO), and calcium titanate ($CaTiO_3$) phase concentrations in the bioceramic coating increased. This means that the change (increase) in DC bias in turn increased the ion flux (CaO^+) leading to the increased presence of CaO and $CaTiO_3$ phases, and hence the biocompatibility (positive biomimetic response). Plasma-assisted RF magnetron sputtering deposition is also an efficient way of coating TiAlV orthopedic alloys with HA or any calcium phosphate-based mineral (**Figure 10A**) [144].

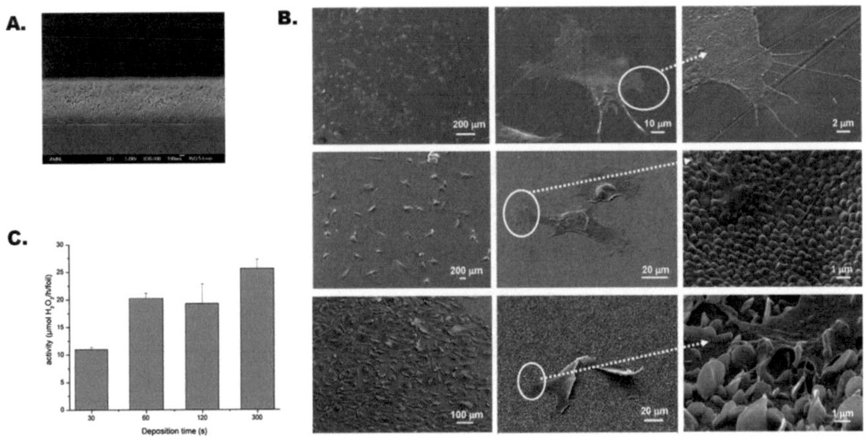

Figure 10. (A) SEM image of RF plasma deposited (RF power = 500 W). Calcium phosphate bioceramic film (1.2 μm) on an alloy (Ti_6Al_4V) as a bioimplant coating (reprinted with permission from [144], © 2007 Wiley). (C) Activity of immobilized GO_x in polyacetylene coating on plastic foil (AP-PECVD) after different times of deposition (reprinted with permission from [40], © 2011 Wiley). (B) SEM images of cell and PECVD deposited material interactions. Cell spreading based on surface roughness confirms the dependence of the cell behavior on surface topography (reprinted with permission from [145], © ACS).

3.4.5. Food preservation and bioactive coatings

Spoilage is a major problem in the food industry and various studies are trying to address this issue. Quintieri et al. used PECVD to coat bovine lactoferrin (BLF) and lactoferricin B (LfcinB) against spoilage by *Pseudomonas* strains. In one study, BLF and LfcinB were evaluated for their antimicrobial activity, when immobilized on plasma deposited films of ethylene/acrylic acid (pdEthAA) [146]. RF plasma with different input power, two fragmentation levels in plasma feed, and different –COOH group densities were used during the coatings, to vary the amount of LfcinB and BLF immobilization on the films [146]. Heyse et al. have reported the use of APCVD to form bioactive coatings (using enzymes, proteins, and other organic precursors) [40]. Though the temperature range during deposition did not exceed 32°C, the radical chemistry due to the plasma had harsh effects on the enzymes. There was considerable effort to minimize the effect, by atomizing the enzyme solution and encompassing it in water shell ("nano shuttle" formation) which subsequently protected the enzyme from the harsh plasma conditions. These enzymes were further trapped in the growing polymer network of organic polymers acetylene or pyrrole giving rise to bioactive films. The films formed using enzymes such as glucose oxide (GO_x) (**Figure 10C**) and lipase were successfully used in sensor and catalysis applications. They tested it on other sensitive and hyperthermophilic proteins such as allophycocyanin and Apase [40].

The PECVD technique, as discussed has been used for a broad range of applications from photovoltaics, biological, to semiconductors. Most recently, this technique has been used in several instances where barrier properties were required for the encapsulation of organic light-emitting devices (OLED) by silicon oxide (SiO_x) and nickel oxide (NiO_x) PECVD-deposited films, organic electronic fabrication, and anticorrosive coatings. Room temperature deposition

http://dx.doi.org/10.5772/64654

of polymer-like carbon films (PLC) for gate dielectric applications in organic thin film transistors has also reported [148]. Improvements in the encapsulation process of various materials utilizing PECVD can lead to potential advancements in the research area of PECVD-based encapsulation. [147]. Other areas such as anticorrosion coating of metallic substrates such as copper via plasma polymeric films of dicyclopentadiene (DCPD) films is an area of future research [149]. Tailoring the surface properties of organic materials to become suitable for biomedical applications is another area in which PECVD holds promise for future applications [150].

Acknowledgements

The authors would like to acknowledge the Provost's office at the University of Dartmouth and AFOSR.

Author details

Yasaman Hamedani[1], Prathyushakrishna Macha[1], Timothy J. Bunning[2], Rajesh R. Naik[2] and Milana C. Vasudev[1*]

*Address all correspondence to: mvasudev@umassd.edu

1 Department of Bioengineering, University of Massachusetts at Dartmouth, Dartmouth, MA, USA

2 Air Force Research Laboratory, Wright Patterson AFB, Dayton, OH, USA

References

[1] Das D, Banerjee A. Further improvements of nano-diamond structures on unheated substrates by optimization of parameters with secondary plasma in MW-PECVD. Surface and Coatings Technology. 2015;272:357–65.

[2] Vasudev MC, Anderson KD, Bunning TJ, Tsukruk VV, Naik RR. Exploration of plasma-enhanced chemical vapor deposition as a method for thin-film fabrication with biological applications. ACS Applied Materials & Interfaces. 2013;5(10):3983–94.

[3] Yagüe JL, Coclite AM, Petruczok C, Gleason KK. Chemical vapor deposition for solvent-free polymerization at surfaces. Macromolecular Chemistry and Physics. 2013;214(3):302–12.

[4] Eichfeld SM, Hossain L, Lin Y-C, Piasecki AF, Kupp B, Birdwell AG, et al. Highly scalable, atomically thin WSe2 grown via metal–organic chemical vapor deposition. ACS Nano. 2015;9(2):2080–7.

[5] Pierson HO. Handbook of Chemical Vapor Deposition (CVD): Principles, Technology and Applications. Park Ridge, NJ: Noyes Publication; 1992.

[6] Stoffel A, Kovács A, Kronast W, Müller B. LPCVD against PECVD for micromechanical applications. Journal of Micromechanics and Microengineering. 1996;6(1):1.

[7] Li X-L, Ge J-P, Li Y-D. Atmospheric pressure chemical vapor deposition: an alternative route to large-scale MoS2 and WS2 inorganic fullerene-like nanostructures and nanoflowers. Chemistry – A European Journal. 2004;10(23):6163–71.

[8] Poole KM. Electrode contamination in electron optical systems. Proceedings of the Physical Society Section B. 1953;66(7):542.

[9] Ennos AE. The source of electron-induced contamination in kinetic vacuum systems. British Journal of Applied Physics. 1954;5(1):27–31.

[10] Alt LL, Ing SW, Laendle KW. Low-temperature deposition of silicon oxide films. Journal of the Electrochemical Society. 1963;110(5):465.

[11] Christy RW. Formation of thin polymer films by electron bombardment. Journal of Applied Physics. 1960;31(9):1680–3.

[12] Baker AG, Morris WC. Deposition of metallic films by electron impact decomposition of organometallic vapors. Review of Scientific Instruments. 1961;32(4):458.

[13] Christy RW. Conducting thin films formed by electron bombardment of substrate. Journal of Applied Physics. 1962;33(5):1884–8.

[14] Ing SW, Davern W. Use of low-temperature deposited silicon dioxide films as diffusion masks in GaAs. Journal of the Electrochemical Society. 1964;111(1):120–2.

[15] Ing SW, Davern W. Glow discharge formation of silicon oxide and the deposition of silicon oxide thin film capacitors by glow discharge techniques. Journal of the Electrochemical Society. 1965;112(3):284–8.

[16] Wertheimer MR. Plasma processing and polymers: a personal perspective. Plasma Chemistry and Plasma Processing. 2013;34(3):363–76.

[17] Reinberg AR. Extended Abtracts. The Electrochemical Society Meeting, San Francicso, CA, May 12–17, 1974.

[18] Reinberg AR. Inventor, Patent No. US3757733 A, 1973.

[19] Pierson HO. Handbook of Chemical Vapor Deposition (CVD). 2nd Ed. Norwich, NY: William Andrew Publishing; 1999. p. iii.

[20] Vasudev MC, Koerner H, Singh KM, Partlow BP, Kaplan DL, Gazit E, et al. Vertically aligned peptide nanostructures using plasma-enhanced chemical vapor deposition. Biomacromolecules. 2014;15(2):533–40.

[21] Anderson KD, Slocik JM, McConney ME, Enlow JO, Jakubiak R, Bunning TJ, et al. Facile plasma-enhanced deposition of ultrathin crosslinked amino acid films for conformal biometallization. Small. 2009;5(6):741–9.

[22] Inagaki N. Plasma Surface Modification and Plasma Polymerization. Boca Raton, FL: CRC Press (Taylor and Francis group); 1996.

[23] Jiang H, Eyink K, Grant JT, Enlow J, Tullis S, Bunning TJ. PECVD siloxane and fluorine-based copolymer thin films. Chemical Vapor Deposition. 2008;14(9–10):286–91.

[24] Singamaneni S, LeMieux MC, Lang HP, Gerber C, Lam Y, Zauscher S, et al. Bimaterial microcantilevers as a hybrid sensing platform. Advanced Materials. 2008;20(4):653–80.

[25] Anderson KD, Young SL, Jiang H, Jakubiak R, Bunning TJ, Naik RR, et al. Plasma-enhanced copolymerization of amino acid and synthetic monomers. Langmuir. 2012;28(3):1833–45.

[26] Enlow JO, Jiang H, Grant JT, Eyink K, Su W, Bunning TJ. Plasma polymerized ferrocene films. Polymer. 2008;49(19):4042–5.

[27] Schulz U. Coating processes for plastic optics. Advanced Optical Technologies. 2014;3(1):29–39.

[28] Rossmann K. Improvement of bonding properties of polyethylene. Journal of Polymer Science. 1956;19(91):141–4.

[29] Wertheimer MR, Martinu L, Klemberg-Sapieha JE, Czeremuszkin G. In Mittal KL, Pizzi A, editors. Adhesion Promotion Techniques in Advanced Technologies. New York: Marcel Dekker; 1998.

[30] Nisol B, Reniers F. Challenges in the characterization of plasma polymers using XPS. Journal of Electron Spectroscopy and Related Phenomena. 2015;200:311–31.

[31] Osada Y, Biederman H, editors. Plasma Polymerization Processes. Amsterdam: Elsevier Science and Technology Books; 1992. ISBN: 9780444887245.

[32] Peri SR, Habersberger B, Akgun B, Jiang H, Enlow J, Bunning TJ, et al. Variations in cross-link density with deposition pressure in ultrathin plasma polymerized benzene and octafluorocyclobutane films. Polymer. 2010;51(19):4390–7.

[33] Kramer PW, Yeh YS, Yasuda H. Low temperature plasma for the preparation of separation membranes. Journal of Membrane Science. 1989;46(1):1–28.

[34] Merche D, Vandencasteele N, Reniers F. Atmospheric plasmas for thin film deposition: a critical review. Thin Solid Films. 2012;520(13):4219–36.

[35] Yasuda H. Plasma Polymerization. New York: Academic Press; 1985.

[36] Peri B, Borah B, Dash RK. Effect of RF power and gas flow ratio on the growth and morphology of the PECVD SiC thin film s for MEMS applications. Bulletin of Materials Science. 2015;38(4):1105–12.

[37] Dams R, Vangeneugden D, Vanderzande D. Plasma deposition of thiophene derivatives under atmospheric pressure. Chemical Vapor Deposition. 2006;12(12):719–27.

[38] Schutze A, Jeong JY, Babayan SE, Park J, Selwyn GS, Hicks RF. The atmospheric-pressure plasma jet: a review and comparison to other plasma sources. IEEE Transactions on Plasma Science. 1998;26(6):1685–94.

[39] Groenewoud LMH, Engbers GHM, White R, Feijen J. On the iodine doping process of plasma polymerised thiophene layers. Synthetic Metals. 2001;125(3):429–40.

[40] Heyse P, Van Hoeck A, Roeffaers MB, Raffin JP, Steinbüchel A, Stöveken T, et al. Exploration of Atmospheric pressure plasma nanofilm technology for straightforward bio-active coating deposition: enzymes, plasmas and polymers, an elegant synergy. Plasma Processes and Polymers. 2011;8(10):965–74.

[41] Da Ponte G, Sardella E, Fanelli F, Paulussen S, Favia P. Atmospheric pressure plasma deposition of poly lactic acid-like coatings with embedded elastin. Plasma Processes and Polymers. 2014;11(4):345–52.

[42] Nozaki T, Ohnishi K, Okazaki K, Kortshagen U. Fabrication of vertically aligned single-walled carbon nanotubes in atmospheric pressure non-thermal plasma CVD. Carbon. 2007;45(2):364–74.

[43] Barankin MD, Gonzalez Ii E, Ladwig AM, Hicks RF. Plasma-enhanced chemical vapor deposition of zinc oxide at atmospheric pressure and low temperature. Solar Energy Materials and Solar Cells. 2007;91(10):924–30.

[44] Babayan SE, Jeong JY, Tu VJ, Park J, Selwyn GS, Hicks RF. Deposition of silicon dioxide films with an atmospheric-pressure plasma jet. Plasma Sources Science and Technology. 1998;7(3):286.

[45] Hiramatsu M, Hori M. Aligned growth of single-walled and double-walled carbon nanotube films by control of catalyst preparation. Japanese Journal of Applied Physics. 2011;46(2):12–16.

[46] Shiji K, Hiramatsu M, Enomoto A, Nakamura M, Amano H, Hori M. Vertical growth of carbon nanowalls using rf plasma-enhanced chemical vapor deposition. Diamond and Related Materials. 2005;14(3–7):831–4.

[47] Kogelschatz U. Dielectric-barrier discharges: their history, discharge physics, and industrial applications. Plasma Chemistry and Plasma Processing. 2003;23(1):1–46.

[48] Jiang H, Johnson WE, Grant JT, Eyink K, Johnson EM, Tomlin DW, et al. Plasma polymerized multi-layered photonic films. Chemistry of Materials. 2003;15(1):340–7.

[49] Tibbitt JM, Bell AT, Shen M. Dielectric relaxations in plasma-polymerized hydrocarbons and fluorocarbons. Journal of Macromolecular Science: Chemistry. 1976;A10:519–33.

[50] Beshkov G, Lei S, Lazarova V, Nedev N, Georgiev SS. IR and Raman absorption spectroscopic studies of APCVD, LPCVD and PECVD thin SiN films. Vacuum. 2002;69(1–3):301–5.

[51] Sawa G, Morita S, Ieda M. Dielectric properties of polystyrene formed in a glow discharge. Journal of Polymer Science Polymer Physics Edition. 1974;12:1231.

[52] Zhang X, Chen K-S, Mark Spearing S. Thermo-mechanical behavior of thick PECVD oxide films for power MEMS applications. Sensors and Actuators A: Physical. 2003;103(1–2):263–70.

[53] Domínguez C, Rodríguez JA, Muñoz FJ, Zine N. The effect of rapid thermal annealing on properties of plasma enhanced CVD silicon oxide films. Thin Solid Films. 1999;346(1–2):202–6.

[54] Olbrechts B, Raskin J-P. PECVD oxide as intermediate film for wafer bonding: Impact of residual stress. Microelectronic Engineering. 2010;87(11):2178–86.

[55] Richter F, Schaarschmidt G, Franke DW, Wallendorf T, Rau B. Analysis of the growth process of a-C:H layers. Diamond and Related Materials. 1993;2(10):1344–9.

[56] Perrin J. Plasma and surface reactions during a-Si:H film growth. Journal of Non-Crystalline Solids. 1991;137–138(Part 2):639–44.

[57] Denaro AR, Owens PA, Crawshaw A. Glow discharge polymerization—styrene. European Polymer Journal. 1968;4(1):93–106.

[58] Michelmore A, Whittle J, Short R. The importance of ions in low pressure PECVD plasmas. Frontiers in Physics. 2015;3:3.

[59] Gordillo-Vazquez FJ, Herrero VJ, Tanarro I. From carbon nanostructures to new photoluminescence sources: an overview of new perspectives and emerging applications of low-pressure PECVD. Chemical Vapor Deposition. 2007;13(6–7):267–79.

[60] Jérôme P, Jacques S, Christoph H, Alan H, Laurent S. The physics of plasma-enhanced chemical vapour deposition for large-area coating: industrial application to flat panel displays and solar cells. Plasma Physics and Controlled Fusion. 2000;42(12B):B353.

[61] Shah A, Torres P, Tscharner R, Wyrsch N, Keppner H. Photovoltaic technology: the case for thin-film solar cells. Science. 1999;285(5428):692–8.

[62] Late DJ, Liu B, Matte HR, Dravid VP, Rao C. Hysteresis in single-layer MoS2 field effect transistors. Acs Nano. 2012;6(6):5635–41.

[63] Binetti S, Acciarri M, Bollani M, Fumagalli L, Von Känel H, Pizzini S. Nanocrystalline silicon films grown by low energy plasma enhanced chemical vapor deposition for optoelectronic applications. Thin Solid Films. 2005;487(1):19–25.

[64] O'Mara WC. Liquid Crystal Flat Panel Displays: Manufacturing Science & Technology. New York: Chapman & Hall; 1993.

[65] Förch R, Chifen AN, Bousquet A, Khor HL, Jungblut M, Chu LQ, et al. Recent and expected roles of plasma-polymerized films for biomedical applications. Chemical Vapor Deposition. 2007;13(6–7):280–94.

[66] Pryce Lewis HG, Edell DJ, Gleason KK. Pulsed-PECVD films from hexamethylcyclotrisiloxane for use as insulating biomaterials. Chemistry of Materials. 2000;12(11):3488–94.

[67] Tang L, Wu Y, Timmons RB. Fibrinogen adsorption and host tissue responses to plasma functionalized surfaces. Journal of Biomedical Materials Research. 1998;42(1):156–63.

[68] Sherman S, Wagner S, Mucha J, Gottscho RA. Substrate effect on plasma-enhanced chemical vapor deposited silicon nitride. Journal of the Electrochemical Society. 1997;144(9):3198–204.

[69] Martins R, Aguas H, Cabrita A, Tonello P, Silva V, Ferreira I, et al. New nanostructured silicon films grown by PECVD technique under controlled powder formation conditions. Solar Energy. 2001;69:263–9.

[70] Sriraman S, Agarwal S, Aydil ES, Maroudas D. Mechanism of hydrogen-induced crystallization of amorphous silicon. Nature. 2002;418(6893):62–5.

[71] Marra DC, Edelberg EA, Naone RL, Aydil ES. Silicon hydride composition of plasma-deposited hydrogenated amorphous and nanocrystalline silicon films and surfaces. Journal of Vacuum Science & Technology. 1998:3199–210.

[72] Kessels W, Smets A, Marra D, Aydil E, Schram D, Van De Sanden M. On the growth mechanism of a-Si: H. Thin Solid Films. 2001;383(1):154–60.

[73] Martin-Palma RJ, Gago R, Torres-Costa V, Fernandez-Hidalgo P, Kreissig U, Martinez Duart JM. Optical and compositional analysis of functional SiOxCy:H coatings on polymers. Thin Solid Films. 2006:2493–6.

[74] Service RF. Patterning electronics on the cheap. Science (Washington, DC). 1997;278(5337):383–4.

[75] Kim KN, Lee SM, Mishra A, Yeom GY. Atmospheric pressure plasmas for surface modification of flexible and printed electronic devices: a review. Thin Solid Films. 2016;598:315–34.

[76] Fanelli F. Thin film deposition and surface modification with atmospheric pressure dielectric barrier discharges. Surface and Coatings Technology. 2010;205(5):1536–43.

[77] Seo HJ, Nam S-H, Kim S, Boo J-H. Organic and organic–inorganic hybrid polymer thin films deposited by PECVD using TEOS and cyclohexene for ULSI interlayer-dielectric application. Applied Surface Science. 2015;354:134–8.

[78] Chu S-M, Louh T, Yang L-W, Yang T, Chen K-C, editors. Capacitance density and breakdown voltage improvement by optimizing the PECVD dielectric film characteristics in metal insulator metal capacitors-Chin-Tsan Yeh. In Joint e-Manufacturing and Design Collaboration Symposium (eMDC) & 2015 International Symposium on Semiconductor Manufacturing (ISSM), 2015. IEEE; 2015.

[79] Bae I-S, Cho S-H, Lee S-B, Kim Y, Boo J-H. Growth of plasma-polymerized thin films by PECVD method and study on their surface and optical characteristics. Surface and Coatings Technology. 2005;193(1):142–6.

[80] Zhu W, Neumayer D, Perebeinos V, Avouris P. Silicon nitride gate dielectrics and band gap engineering in graphene layers. Nano Letters. 2010;10(9):3572–6.

[81] Young DJ, Du J, Zorman CA, Ko WH. High-temperature single-crystal 3C-SiC capacitive pressure sensor. Sensors Journal, IEEE. 2004;4(4):464–70.

[82] LaBarbera MA, Zorman CA, Scardelletti MC, editors. Long-term thermal mechanical stability of PECVD amorphous silicon carbide films for harsh environment microelectromechanical systems. In Sensors. Cleveland, OH: Case Western Reserve University; 2014.

[83] Mandal S, Dhar S, Das G, Mukhopadhyay S, Barua A. Development of optimized n-μc-Si: H/na-Si: H bilayer and its application for improving the performance of single junction a-Si solar cells. Solar Energy. 2016;124:278–86.

[84] O'Donnell B, Yu L, Foldyna M, Cabarrocas PRI. Silicon nanowire solar cells grown by PECVD. Journal of Non-Crystalline Solids. 2012;358(17):2299–302.

[85] Yamamoto K, Yoshimi M, Tawada Y, Okamoto Y, Nakajima A. Cost effective and high-performance thin film Si solar cell towards the 21st century. Solar Energy Materials and Solar Cells. 2001;66(1):117–25.

[86] Maycock PD. PV review: world solar PV market continues explosive growth. Refocus. 2005;6(5):18–22.

[87] Parsons R, Vossen JL, Kern W. Sputter deposition processes. Thin Film Processes II. 1991;2:177.

[88] Peng S, Wang D, Yang F, Wang Z, Ma F. Grown low-temperature microcrystalline silicon thin film by VHF PECVD for thin films solar cell. Journal of Nanomaterials. 2015;2015:88.

[89] Kondo M, Fukawa M, Guo L, Matsuda A. High rate growth of microcrystalline silicon at low temperatures. Journal of Non-Crystalline Solids. 2000;266:84–9.

[90] Juneja S, Sudhakar S, Gope J, Lodhi K, Sharma M. Highly conductive boron doped micro/nanocrystalline silicon thin films deposited by VHF-PECVD for solar cell applications. Journal of Alloys and Compounds. 2015;643:94–9.

[91] Engelhardt J, Frey A, Mahlstaedt L, Gloger S, Hahn G, Terheiden B. Boron emitters from doped PECVD layers for n-type crystalline silicon solar cells with LCO. Energy Procedia. 2014;55:235–40.

[92] Matsuda A. Formation kinetics and control of microcrystallite in μc-Si: H from glow discharge plasma. Journal of Non-Crystalline Solids. 1983;59:767–74.

[93] Norris KJ, Garrett MP, Zhang J, Coleman E, Tompa GS, Kobayashi NP. Silicon nanowire networks for multi-stage thermoelectric modules. Energy Conversion and Management. 2015;96:100–4.

[94] Rath J, Brinza M, Liu Y, Borreman A, Schropp R. Fabrication of thin film silicon solar cells on plastic substrate by very high frequency PECVD. Solar Energy Materials and Solar Cells. 2010;94(9):1534–41.

[95] Street R. Large area electronics, applications and requirements. Physica Status Solidi (a). 1998;166(2):695–705.

[96] Yang P, Wirnsberger G, Huang HC, Cordero SR, McGehee MD, Scott B, et al. Mirrorless lasing from mesostructured waveguides patterned by soft lithography. Science. 2000;287(5452):465–7.

[97] Yang C-C, Lin Y-H, Lin G-R, editors. Nonstoichiometric Si1-xGex based tunable saturable absorber for mode-locked erbium-doped fiber laser. In CLEO. OSA Technical Digest [online]. Optical Society of America; 2015. Paper JW2A.70.

[98] Loerke J, Marlow F. Laser emission from dye-doped mesoporous silica fibers. 2002;14(23):1745–9.

[99] Burt MC, Dave BC. An optical temperature sensing system based on encapsulation of a dye molecule in organosilica sol–gels. Sensors and Actuators B: Chemical. 2005;107(2): 552–6.

[100] Barranco A, Aparicio F, Yanguas-Gil A, Groening P, Cotrino J, González-Elipe AR. Optically active thin films deposited by plasma polymerization of dye molecules. Chemical Vapor Deposition. 2007;13(6–7):319–25.

[101] Limb SJ, Lau KK, Edell DJ, Gleason EF, Gleason KK. Molecular design of fluorocarbon film architecture by pulsed plasma enhanced and pyrolytic chemical vapor deposition. Plasmas and Polymers. 1999;4(1):21–32.

[102] Chen K-S, Yang M-R, Hsu ST. Fabrication and characterization of fluorine-containing films using plasma polymerization of octafluorotoluene. Materials Chemistry and Physics. 1999;61(3):214–8.

[103] Kenyon A, Trwoga P, Federighi M, Pitt C. Optical properties of PECVD erbium-doped silicon-rich silica: evidence for energy transfer between silicon microclusters and erbium ions. Journal of Physics: Condensed Matter. 1994;6(21):L319.

[104] Shimizu-Iwayama T, Terao Y, Kamiya A, Takeda M, Nakao S, Saitoh K. Correlation of microstructure and photoluminescence for nanometer-sized Si crystals formed in an amorphous SiO_2 matrix by ion implantation. Nanostructured Materials. 1995;5(3):307–18.

[105] Vinciguerra V, Franzo G, Priolo F, Iacona F, Spinella C. Quantum confinement and recombination dynamics in silicon nanocrystals embedded in Si/SiO2 superlattices. Journal of Applied Physics. 2000;87(11):8165–73.

[106] Rezgui B, Sibai A, Nychyporuk T, Lemiti M, Brémond G. Photoluminescence and optical absorption properties of silicon quantum dots embedded in Si-rich silicon nitride matrices. Journal of Luminescence. 2009;129(12):1744–6.

[107] Merkulov VI, Guillorn MA, Lowndes DH, Simpson ML, Voelkl E. Shaping carbon nanostructures by controlling the synthesis process. Applied Physics Letters. 2001;79(8):1178–80.

[108] Meyyappan M, Delzeit L, Cassell A, Hash D. Carbon nanotube growth by PECVD: a review. Plasma Sources Science and Technology. 2003;12(2):205.

[109] Raveh A, Klemberg-Sapieha JE, Martinu L, Wertheimer M. Deposition and properties of diamondlike carbon films produced in microwave and radio-frequency plasma. Journal of Vacuum Science & Technology A. 1992;10(4):1723–7.

[110] Cui L, Chen J, Yang B, Jiao T. High current emission from patterned aligned carbon nanotubes fabricated by plasma-enhanced chemical vapor deposition. Nanoscale Research Letters. 2015;10(1):1–6.

[111] Dekker C. Carbon nanotubes as molecular quantum wires. Physics Today. 1999;52:22–30.

[112] Hiramatsu M, Shiji K, Amano H, Hori M. Fabrication of vertically aligned carbon nanowalls using capacitively coupled plasma-enhanced chemical vapor deposition assisted by hydrogen radical injection. Applied Physics Letters. 2004;84(23):4708–10.

[113] AuBuchon JF, Daraio C, Chen L-H, Gapin AI, Jin S. Iron silicide root formation in carbon nanotubes grown by microwave PECVD. The Journal of Physical Chemistry B. 2005;109(51):24215–9.

[114] Dresselhaus MS, Dresselhaus G, Eklund PC. Science of Fullerenes and Carbon Nanotubes: Their Properties and Applications. Cambridge, MA: Academic Press; 1996.

[115] Kato T, Hatakeyama R. Formation of freestanding single-walled carbon nanotubes by plasma-enhanced CVD. Chemical Vapor Deposition. 2006;12(6):345–52.

[116] Bo Z, Mao S, Han ZJ, Cen K, Chen J, Ostrikov KK. Emerging energy and environmental applications of vertically-oriented graphenes. Chemical Society Reviews. 2015;44(8): 2108–21.

[117] Jacob MV, Rawat RS, Ouyang B, Bazaka K, Kumar DS, Taguchi D, et al. Catalyst-free plasma enhanced growth of graphene from sustainable sources. Nano Letters. 2015;15(9):5702–8.

[118] Szymanowski H, Sobczyk A, Gazicki-Lipman M, Jakubowski W, Klimek L. Plasma enhanced CVD deposition of titanium oxide for biomedical applications. Surface and Coatings Technology. 2005;200(1):1036–40.

[119] Hájková P, Špatenka P, Krumeich J, Exnar P, Kolouch A, Matoušek J, et al. Antibacterial effect of silver modified TiO_2/PECVD films. The European Physical Journal D. 2009;54(2):189–93.

[120] Ayllon J, Figueras A, Garelik S, Spirkova L, Durand J, Cot L. Preparation of TiO_2 powder using titanium tetraisopropoxide decomposition in a plasma enhanced chemical vapor deposition (PECVD) reactor. Journal of Materials Science Letters. 1999;18(16):1319–21.

[121] Williams L, Hess D. Structural properties of titanium dioxide films deposited in an rf glow discharge. Journal of Vacuum Science & Technology A. 1983;1(4):1810–9.

[122] Maeda M, Watanabe T. Visible light photocatalysis of nitrogen-doped titanium oxide films prepared by plasma-enhanced chemical vapor deposition. Journal of the Electrochemical Society. 2006;153(3):C186–C9.

[123] Rosler RS, Engle GM. Plasma-enhanced CVD of titanium silicide. Journal of Vacuum Science & Technology B. 1984;2(4):733–7.

[124] Candan S, Beck AJ, O'Toole L, Short RD. Effects of "processing parameters" in plasma deposition: acrylic acid revisited. Journal of Vacuum Science & Technology A. 1998;16(3):1702–9.

[125] Beyer D, Knoll W, Ringsdorf H, Wang JH, Timmons RB, Sluka P. Reduced protein adsorption on plastics via direct plasma deposition of triethylene glycol monoallyl ether. Journal of Biomedical Materials Research. 1997;36(2):181–9.

[126] Tarducci C, Schofield W, Badyal J, Brewer S, Willis C. Synthesis of cross-linked ethylene glycol dimethacrylate and cyclic methacrylic anhydride polymer structures by pulsed plasma deposition. Macromolecules. 2002;35(23):8724–7.

[127] Choukourov A, Biederman H, Slavinska D, Hanley L, Grinevich A, Boldyryeva H, et al. Mechanistic studies of plasma polymerization of allylamine. The Journal of Physical Chemistry B. 2005;109(48):23086–95.

[128] van Os M, Menges B, Timmons R, Knoll W, Foerch R, editors. Variations in the film chemistry of pulsed rf plasma deposited allylamine. In 13th International Symposium on Plasma Chemistry. Beijing, China: Peking University Press; 1997.

[129] Schiller S, Hu J, Jenkins A, Timmons R, Sanchez-Estrada F, Knoll W, et al. Chemical structure and properties of plasma-polymerized maleic anhydride films. Chemistry of Materials. 2002;14(1):235–42.

[130] Wu YJ, Timmons RB, Jen JS, Molock FE. Non-fouling surfaces produced by gas phase pulsed plasma polymerization of an ultra low molecular weight ethylene oxide containing monomer. Colloids and Surfaces B: Biointerfaces. 2000;18(3):235–48.

[131] Beier O, Pfuch A, Horn K, Weisser J, Schnabelrauch M, Schimanski A. Low temperature deposition of antibacterially active silicon oxide layers containing silver nanoparticles, prepared by atmospheric pressure plasma chemical vapor deposition. Plasma Processes and Polymers. 2013;10(1):77–87.

[132] Gilabert-Porres J, Martí S, Calatayud L, Ramos V, Rosell A, Borrós S. Design of a nanostructured active surface against Gram-positive and Gram-negative bacteria through plasma activation and in situ silver reduction. ACS Applied Materials & Interfaces. 2015;8(1):64–73.

[133] Taglietti A, Arciola CR, D'Agostino A, Dacarro G, Montanaro L, Campoccia D, et al. Antibiofilm activity of a monolayer of silver nanoparticles anchored to an amino-silanized glass surface. Biomaterials. 2014;35(6):1779–88.

[134] Saulou C, Despax B, Raynaud P, Zanna S, Marcus P, Mercier-Bonin M. Plasma deposition of organosilicon polymer thin films with embedded nanosilver for prevention of microbial adhesion. Applied Surface Science. 2009;256(3):S35–S9.

[135] Vogler EA. Structure and reactivity of water at biomaterial surfaces. Advances in Colloid and Interface Science. 1998;74(1):69–117.

[136] Ninham BW, Kurihara K, Vinogradova OI. Hydrophobicity, specific ion adsorption and reactivity. Colloids and Surfaces A: Physicochemical and Engineering Aspects. 1997;123:7–12.

[137] Jin SB, Lee JS, Choi YS, Choi IS, Han JG. High-rate deposition and mechanical properties of SiOx film at low temperature by plasma enhanced chemical vapor deposition with the dual frequencies ultra high frequency and high frequency. Thin Solid Films. 2011;519(19):6334–8.

[138] Satriano C, Marletta G, Kasemo B. Oxygen plasma-induced conversion of polysiloxane into hydrophilic and smooth SiOx surfaces. Surface and Interface Analysis. 2008;40(3–4):649–56.

[139] Sarapirom S, Lee J, Jin S, Song D, Yu L, Han J, et al. Wettability effect of PECVD-SiOx films on poly (lactic acid) induced by oxygen plasma on protein adsorption and cell attachment. Journal of Physics: Conference Series. 2013;423(1):012042.

[140] Gandhiraman R, Gubala V, Volcke C, Doyle C, James B, Daniels S, et al. Deposition of chemically reactive and repellent sites on biosensor chips for reduced non-specific binding. Colloids and Surfaces B: Biointerfaces. 2010;79(1):270–5.

[141] Gandhiraman RP, Gubala V, O'Mahony CC, Cummins T, Raj J, Eltayeb A, et al. PECVD coatings for functionalization of point-of-care biosensor surfaces. Vacuum. 2012;86(5): 547–55.

[142] De Groot K, Geesink R, Klein C, Serekian P. Plasma sprayed coatings of hydroxylapatite. Journal of Biomedical Materials Research. 1987;21(12):1375–81.

[143] Friedman RJ, Black J, Galante JO, Jacobs JJ, Skinner HB. Current concepts in orthopaedic biomaterials and implant fixation. Journal of Bone and Joint Surgery: American. 1993;75(7):1086–109.

[144] Long J, Sim L, Xu S, Ostrikov K. Reactive plasma-aided RF sputtering deposition of hydroxyapatite bio-implant coatings. Chemical Vapor Deposition. 2007;13(6–7):299–306.

[145] Intranuovo F, Favia P, Sardella E, Ingrosso C, Nardulli M, d'Agostino R, et al. Osteoblast-like cell behavior on plasma deposited micro/nanopatterned coatings. Biomacromolecules. 2010;12(2):380–7.

[146] Quintieri L, Pistillo B, Caputo L, Favia P, Baruzzi F. Bovine lactoferrin and lactoferricin on plasma-deposited coating against spoilage Pseudomonas spp. Innovative Food Science & Emerging Technologies. 2013;20:215–22.

[147] Yu D, Yang Y-Q, Chen Z, Tao Y, Liu Y-F. Recent progress on thin-film encapsulation technologies for organic electronic devices. Optics Communications. 2016;362:43–9.

[148] Feng L, Anguita JV, Tang W, Zhao J, Guo X, Silva SRP. Room temperature grown high quality polymer-like carbon gate dielectric for organic thin-film transistors. Advanced Electronic Materials. 2016;2(3):1–6.

[149] Jeong D-C, Javid A, Wen L, Choi EJ, Park SY, Kim YH, et al. Low-temperature plasma polymerization of dicyclopentadiene for anti-corrosion properties. Polymer. 2016;92:133–9.

[150] Agnès T-M, Satoshi I, Bertrand-David S, Grant AC, Takuya T, Hiroyuki F, et al. Review on thin-film transistor technology, its applications, and possible new applications to biological cells. Japanese Journal of Applied Physics. 2016;55(4S):04EA8.